A WOMAN'S RIGHT TO KNOW

Inside Technology
Wiebe Bijker and Rebecca Slayton, series editors

A list of books in the series appears at the back of the book.

A WOMAN'S RIGHT TO KNOW

PREGNANCY TESTING IN TWENTIETH-CENTURY BRITAIN

JESSE OLSZYNKO-GRYN

THE MIT PRESS CAMBRIDGE, MASSACHUSETTS LONDON, ENGLAND

The MIT Press would like to thank the anonymous peer reviewers who provided comments on drafts of this book. The generous work of academic experts is essential for establishing the authority and quality of our publications. We acknowledge with gratitude the contributions of these otherwise uncredited readers.

This book was set in Stone Serif and Avenir by Westchester Publishing Services. Printed and bound in the United States of America.

Library of Congress Cataloging-in-Publication Data

Names: Olszynko-Gryn, Jesse, author.
Title: A woman's right to know : pregnancy testing in twentieth-century Britain / Jesse Olszynko-Gryn.
Description: Cambridge, Massachusetts : The MIT Press, [2023] | Series: Inside technology | Includes bibliographical references and index.
Identifiers: LCCN 2023002870 (print) | LCCN 2023002871 (ebook) | ISBN 9780262544399 (paperback) | ISBN 9780262371384 (epub) | ISBN 9780262371377 (pdf)
Subjects: LCSH: Pregnancy—Signs and diagnosis—History. | Pregnancy—Great Britain—History—20th century. | Women's rights—Great Britain—History—20th century. | Reproductive rights—Great Britain—History—20th century. | Pregnancy—Signs and diagnosis—Technological innovations.
Classification: LCC RG563 .O47 2023 (print) | LCC RG563 (ebook) | DDC 362.198200941—dc23/eng/20230607
LC record available at https://lccn.loc.gov/2023002870
LC ebook record available at https://lccn.loc.gov/2023002871

10 9 8 7 6 5 4 3 2 1

For my parents, Andy and Goldie

CONTENTS

1

FEELING PREGNANT

Pregnancy testing has never been easier. Waiting on one side or the other of the bathroom door for pink or blue lines to appear has become a modern ritual and rite of passage. If you haven't been there yourself, odds are you've seen it done on television or in the movies (figure 1.1).[1] The ubiquitous plastic stick is implicated in personal choices and public negotiations about all aspects of reproduction, from miscarriage and abortion to infertility and assisted conception. As artist Tracey Emin puts it in the installation *Feeling Pregnant* (2005), "I go to the bathroom, knowing that within three minutes my life might never be the same again."[2] Technologies of fetal imaging and testing have become embroiled in controversy, attracting much attention.[3] So too have technologies of contraception and abortion.[4] Pregnancy tests have transformed experiences of reproduction as much as ultrasound or the pill, but little is known of their history. To better understand the history and provide resources for the present and future, this book reconstructs the surprisingly controversial past of what today may be the least contentious and most commonplace of reproductive technologies.

A Woman's Right to Know tells the story of pregnancy testing—one of the most significant and least studied technologies of reproduction—for the first time.[5] Focusing on Britain, a key player on the global scene, I show how demand shifted from doctors to possibly pregnant women. I explain the remarkable transformation of an esoteric laboratory service

1.1 Self-testing has become a ubiquitous trope on television and in the movies, especially where young unmarried or otherwise transgressive pregnancy is involved. For example, first-generation Hashida (Sarita Khajuria) is in a taboo relationship with her West Indian boyfriend (Mo Sesay) when she tests positive on a daytrip from Birmingham to Blackpool with a group of British Asian women in *Bhaji on the Beach* (Gurinder Chadha, 1993): produced by Channel 4 Films and Umbi films; distributed by Channel 4; see Olszynko-Gryn 2017, 510–512.

into a ubiquitous consumer product. And I revise timelines of innovation that foreground only the most obvious turning points. Like all "revolutionary" technologies, the first home pregnancy tests neither fell from the sky nor remade mainstream experiences overnight. They resulted, in the early 1970s, from a longer commercialization and were not universally embraced by consumers. Crucially, the social innovations of domestic privacy and demedicalization came before, with commercial labs that in the mid-1960s began serving women not as patients but as clients. This was the key development until, thirty years later, sleeker products of Britain's biotech boom and ticking biological clocks (among other new sources of demand) normalized self-testing for a younger generation of consumers.

But first, how did past generations of women determine whether they were pregnant? Starting with Barbara Duden's groundbreaking history of the female body, historians of early modern Europe have recovered a past full of uncertainty and ambiguity, for women and physicians alike.[6] Corporeal signs, including amenorrhea (cessation of menstruation), nausea,

1.2 Watercolor by I.T. (1826) of a physician visually inspecting a urine specimen by the light of the window for signs of pregnancy. A homuncular figure in the flask indicates that conception has occurred. A woman, presumably the unmarried patient, is dabbing her eye as an older woman (probably her mother) admonishes her while pointing at the flask. A man who could be the father skulks behind the half-open door. Then as now, the implications and social relations of pregnancy diagnosis could be fraught. Wellcome Collection 21828i.

abdominal swelling, and even quickening (the feeling of fetal movements usually in the fourth month), suggested but could not guarantee a living entity within the womb.[7] Experiences varied enormously. Some women were keenly aware of their own cycles and even used their knowledge to date conception and predict delivery.[8] For others, birth (or miscarriage) came as a shock.[9] Vernacular texts and domestic practices included diverse techniques for confirming or ruling out pregnancy.[10] Women frequently consulted the uroscopist, or "piss prophet," to have their urine visually inspected, including for signs of pregnancy (figure 1.2).[11]

For women and physicians alike, determining the existence or not of "fruit" in the womb was a tricky business.[12] A medieval woman sentenced to death under English common law could "plead her belly" to forestall execution until after the birth of her child. When this happened, a special "jury of matrons" would be empaneled to establish whether the woman was truly "quick with child."[13] Victorian ladies recorded the feeling of quickening in diaries and letters as well as the hopes and fears engendered by a missed menstrual period.[14] Even as many modern women attempted to evade maternity by taking widely advertised "female pills" and other abortifacients, they also recognized the absence of menstruation and onset of nausea as early, though uncertain, indications of pregnancy.[15] When it comes to reproductive self-awareness, generations of women have employed a range of more or less mediated resources. So where does self-testing come in?

The normalization of self-testing is typically traced back to the invention, in Berlin in the late 1920s, of the Aschheim–Zondek test.[16] The first modern laboratory method for diagnosing pregnancy, it functioned by detecting the presence or absence in a woman's urine of the pregnancy-supporting hormone today known as human chorionic gonadotropin (hCG). In this sense, the Aschheim–Zondek test was the forerunner to today's home tests, which are all based on the same principle. Unlike home tests, however, the Aschheim–Zondek test (and modifications thereof) involved injecting a woman's urine into sexually immature female mice or rabbits and then dissecting the animals to visually inspect their ovaries for signs of precocious sexual maturation induced by the human pregnancy hormone (figure 1.3). Mice and rabbits, the story continues, were eventually supplanted by more efficient frogs and toads, which in turn were replaced in the 1960s by less cumbersome laboratory test kits and finally, in the 1970s, by the first home tests.[17]

Especially since the invention of the Aschheim–Zondek test and spurred by subsequent advances, reproductive scientists have celebrated technical progress in the detection of pregnancy. Heroizing narratives typically construct demand as a transhistorical constant, as if women (or men) were always dreaming of an earlier, faster, cheaper, and more convenient method. Timelines of innovation are long and frequently conflate woman's will to self-knowledge with "man's" scientific curiosity. Consider the

1.3 The "rabbit test," an early modification of the Aschheim-Zondek reaction, was especially popular in the United States and gave rise to the American euphemism for pregnancy, "the rabbit died." It was also adopted more globally. Pictured here is Lida Tabatznik injecting the ear vein of a test rabbit at the laboratory of the Swedish Association for Sexuality Education (RFSU) in Stockholm, *c.*1940s. Note the caged rabbits along the wall and urine specimens on the table. Tabatznik was born in Russia and fled Berlin to Sweden in the early 1940s. She was not permitted to work as a doctor, so found employment as a lab technician instead. See Lennerhed 2002, 72; Ramsey 2021, 73; chapter 3, this book. © Photo: Anna Riwkin/Moderna Museet-Stockholm.

following statement: "There has been a constant demand in the minds of the medical profession, and in the lay mind, also, for signs and tests that would diagnose early pregnancy."[18] Or, this one: "Man's natural curiosity concerning proof of early pregnancy probably extends to the beginning of time; evidence of this interest can be found in the Egyptian medical papyri dating back nearly 4,000 years."[19] Anchored in just a few lines on a thin strip of papyrus held by the Egyptian Museum of Berlin (figure 1.4) is the claim that modern pregnancy tests "still answer a very personal and private question—Am I pregnant?"[20]

A big problem with giving modern tests such long pedigrees is that the question "Am I pregnant?" does not mean today what it did four thousand, four hundred, or even forty years ago. Societies and cultures have changed

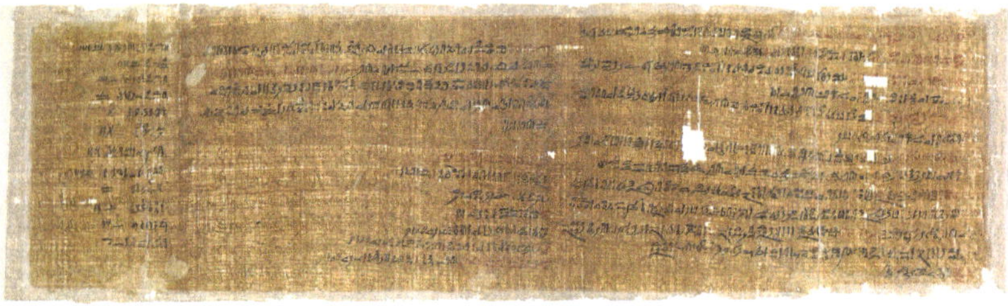

1.4 Medical papyri written in hieratic include tests for fertility, pregnancy, and fetal sex. The most famous of these, "Berlin 199," describes a test to "see [if] a woman will bear a child or [if] she will not bear a child. Emmer (*bedet*) and barley (*it*), the lady should moisten with her urine every day, like dates and like sand in two bags. If they all grow, she will bear a child. If the barley grows, it means a male. If the emmer grows, it means a female. If they do not grow, she will not bear a child": Nunn 2002, 191–192; after Wreszinski 1909, 110. Pap. Berlin P 3030 Vso © SMB Ägyptisches Museum und Papyrus-sammlung Berlin, Photo: Sandra Steiß.

too much, and with them notions of the "personal" and the "private." Narratives of linear progress leave questions of social relations and power dynamics unanswered. They conceal a palimpsest of diagnostic resources that coexisted with uncertainties and ambivalent feelings. And they are silent about the historical specificities of demand and the full range of meanings invested in a (positive or negative) test result. Without a social history, it is not clear what difference, if any, technological improvements made to women's lives or whether women even knew about them.

So how and with what effects did self-testing become thinkable, then a reality, and, finally, a commonplace of everyday life? To answer these and other questions, I construct a social history of technology that takes seriously the gendered experiences and perceptions of women as "lay end users."[21] I do so by combining approaches from social history with those from science and technology studies and applying them to a wide range of sources.[22] Central to my analysis is the progressive creation of public visibility through mass media, especially magazines and newspapers, without which most people would have remained in the dark about laboratory tests.[23] The images contribute visual evidence of change and continuity across a succession of fascinatingly different practices and relations.[24]

To better grasp what changed and what stayed the same, I pay close attention to conditions of possibility, on the one hand, and the infrastructures that enabled new routines, on the other.[25] I demonstrate that old practices, meanings, and experiences were rarely swept away and thematize the distributed nature of "innovation away from the cutting edge."[26] This offers a fresh perspective on the power dynamics of laboratory, clinic, pharmacy, and home as contested spaces of reproductive choice. Not least, I follow historian Andrea Tone in locating agency especially with small-time entrepreneurs and with women, as patients and consumers.[27]

Pregnancy testing in twentieth-century Britain shows steadily increasing demand and supply through the Great Depression, World War II, the creation of the National Health Service, and, with neoliberalism, the rise of consumer culture.[28] Successive generations of women in this period successfully campaigned for improved access to contraception and abortion while gaining financial independence and consumer clout.[29] Men became progressively more involved in childbirth as they moved from "hiding in the pub to cutting the chord."[30] Pregnancy, labor, and delivery became medicalized, especially with the interwar rise of professional prenatal care and hospital birth—a demand-driven trend that, by the 1960s, was provoking a consumer backlash against what sociologist Ann Oakley termed "the captured womb."[31]

The 1960s and 1970s were eventful decades for women's reproductive health in Britain. Between the arrival of the pill (1961) and the birth of the first "test-tube baby" (1978), major pieces of legislation removed obstacles to abortion and contraception.[32] The timing of the first home pregnancy tests (1971) is no coincidence; they were as much a part of broader permissive trends as the pill, and centering them while taking a longer view will shed new light on the history of reproductive technologies (figure 1.5). Neither a critique of medicalization nor a celebration of market capitalism, *A Woman's Right to Know* tells a subtler story about the noninevitable democratization of a reproductive technology more through commerce than as the result of professional surveillance or grassroots activism. I argue that, for better or worse, the private sector met (and later created more of) the demand that doctors and the welfare state rejected.

Although generally suspicious of women's motives for seeking confirmation of a suspected pregnancy, British clinicians were early and enthusiastic adopters of the Aschheim–Zondek test for what they regarded as legitimate medical purposes. The number of tests performed by the Edinburgh

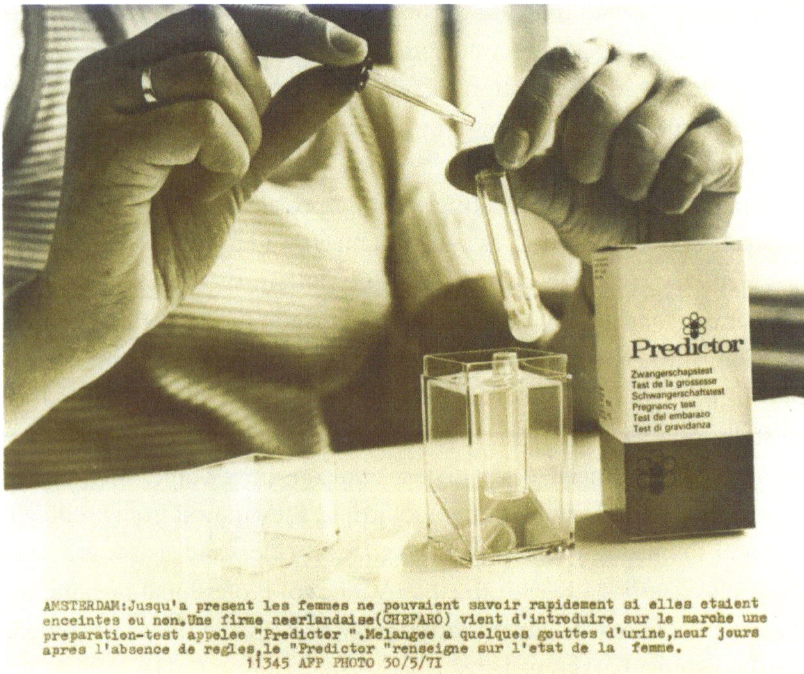

1.5 French press photograph of Predictor, the pioneering home pregnancy test, from the Amsterdam launch on May 30, 1971. Note the prominently displayed wedding band (worn on the right hand by many Dutch women) and the six languages on the packaging, anticipating sales in several European countries. The text reads: "Until now, women could not tell quickly whether or not they were pregnant. A Dutch firm (Chafaro) has just introduced onto the market a test kit called 'Predictor.' Mixed with a few drops of urine, nine days after the absence of menstruation, the 'Predictor' provides information on the condition of the woman." See Olszynko-Gryn 2020a; chapter 10, this book.

pregnancy diagnosis station, a key institution, increased from 840 in 1929, the year it was set up, to over 20,000 in 1964, the year it stopped using animals. The meticulously kept records of a rural general practitioner (GP) interviewed by Oakley show that he ordered pregnancy tests for just over 1 percent of his female patients in the late 1940s and nearly 40 percent in the late 1970s, a thirtyfold increase in three decades (figure 1.6).[33] A laissez-faire government facilitated the arrival, with minimal regulatory oversight, of the first home pregnancy tests in 1971—a full seven years before they were allowed on the American market. In the same year, British women were responsible for around 1.5 million laboratory pregnancy tests.[34] By the

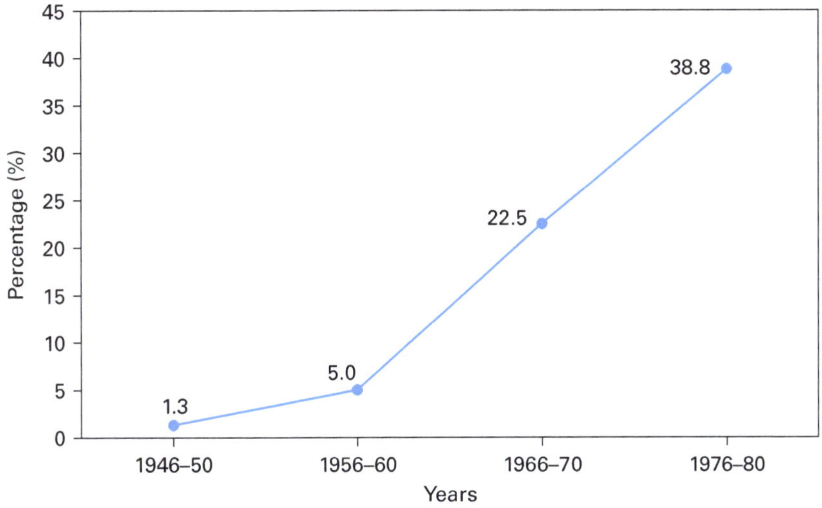

1.6 Percentage of female patients for whom pregnancy tests were ordered by Dr. Edgar Hope-Simpson, Gloucestershire, 1946–1980, based on samples extracted by Ann Oakley from his case notes: Oakley 1984, 230.

early 1990s, the total number of tests had increased to nearly five million, or around seven for every baby born in Britain.[35] At the end of that decade, women were more likely to buy a home test and wait anxiously for the lines to appear than to schedule an appointment with their GP.[36] Clearblue, a British innovation, came to dominate the global market.

Pregnancy tests in some ways followed a similar trajectory to other domesticated technologies of self-management, including thermometers, bathroom scales, and menstrual products.[37] Yet, the story of their ascent to unremarkable ubiquity is less straightforward than it may seem at first blush. For one thing, as chapter 7 shows, the commercial success in the 1960s of diagnostic hormone tablets that ruled out pregnancy by inducing uterine bleeding in the patient herself significantly complicates linear accounts of technological progress.[38] For another, as historian Lara Freidenfelds recently observed, "The more sensitive [pregnancy tests] get the less accurately they predict the birth of a baby in nine months, and the more likely they are to create the experience of miscarriage."[39] Chapter 11 takes up this irony of improvement, a direct result of the increased sensitivity of home tests that in the 1980s raised profound questions about the (unstable) ontological status of pregnancy, on the one hand, and the social pressures endured by women of childbearing age, on the other.

A Woman's Right to Know covers three major and, to some extent, overlapping regimes of pregnancy testing: animal assays (late 1920s to mid-1960s), laboratory test kits (since the early 1960s), and home tests (since the early 1970s). Four chapters (3–6) chart the rise and fall of animal assays between 1929 and 1964. Chapter 7 investigates the controversial use of hormone tablets in pregnancy diagnosis, a troubling and unresolved episode in the history of women's health. The next two chapters (8 and 9) reconstruct the adoption of immunological test kits by commercial laboratories and pharmacies. Chapters 10 and 11 recover the creation and growth of the retail market for home pregnancy tests. The final chapter reprises the main themes and arguments of the book to consider the broader legacy of pregnancy testing in the twenty-first century. Chapter 2, which establishes a baseline for lay and medical knowledge before the Aschheim–Zondek test, begins with Mrs. B., a Londoner whose period didn't come on Tuesday, June 16, 1925, and the story she told of how she knew she was pregnant.

2

BEDSIDE AND BENCH

Mrs. B., a middle-aged mother of three, "should have been unwell" on June 16, 1925, the day she fell and broke her ankle, but "never saw any colours or anything." Suspecting pregnancy, she told her doctor after two weeks had gone by, but he dismissed her concern, possibly on account of her age (Mrs. B. was past forty), as did her attendants at the infirmary, where she was convalescing from the fall. When, to her "horror," she "felt a movement in the body," she first sent her sister's friend and then her husband to take a sample of her urine to a "water doctor" (uroscopist), who claimed on both occasions that there was "no sign of pregnancy" but that her kidneys "were in a poor condition." Next, she had her family doctor examine her "properly" and, although "he could not tell for a long time," he eventually "felt a tiny movement" and confirmed her suspicion. Finally, Mrs. B. "had another examination at the infirmary by a specialist" who proclaimed she "was 28 weeks pregnant" and when she "got home" she "felt the child turn" and "ever since then" could "get no peace" for it seemed always "on the move." Mrs. B. reckoned she was "about a month now to being confined."[1]

Mrs. B.'s narrative showcases the range of diagnostic resources that were available to a possibly pregnant woman in the early twentieth century—from bodily signs, including amenorrhea (she "should have been unwell" but "never saw any colours or anything") and quickening (when she first "felt a movement in the body"), to medical and other consultants (infirmary attendants, her family doctor, a specialist, the "water doctor").[2] Significantly, Mrs. B. had had at least three pregnancies under her belt, possibly more if she had ever miscarried, so she knew what to expect, hence her persistence despite multiple misdiagnoses. Her pregnancy realization narrative, as related in a letter to birth control pioneer Marie Stopes, is gradual, ambivalent, and irreducible to a single moment of clarity.

In this chapter, I complicate and add nuance to the view that "pregnancy did not really begin for a nineteenth- or early twentieth-century woman until she felt the sensation of the fetus moving, sometime in the fourth or fifth month."[3] I do this first by examining medical textbooks and advice manuals to recover the various diagnostic resources available to women, midwives, and doctors in the early twentieth century. I next turn to the laboratory and the clinic to examine the two most prominent "scientific" pregnancy tests before Aschheim and Zondek's, then conclude with a brief summation of the rise of reproductive endocrinology and "biological" tests for pregnancy, up to and including the famous Aschheim–Zondek reaction, which dominated in the 1930s.

SIGNS AND SYMPTOMS

An eighteenth-century physician might have taken the pulse of his patient, but otherwise, there would have been little physical contact between the two. Following René Laennec's invention of the stethoscope in 1816, manuals of physical diagnosis published in the midcentury canonized the four main procedures of inspection, palpation, percussion, and auscultation. Not only Paris but also German and Austrian universities became favorite destinations of British medical teachers, who imported continental practices such as histology.[4] GPs in Britain adopted the stethoscope and specialists made additional use of newly invented instruments, including the ophthalmoscope, otoscope, and laryngoscope. In obstetrics and gynecology, the socially awkward and physically unpleasant vaginal speculum, along with special examination tables and stirrups, became iconic of hospital practice.[5] Gynecological diagnosis was particularly fraught. James Young Simpson, famous for having introduced chloroform as pain relief in childbirth in 1847, also used it to avoid the embarrassment of a pelvic examination.[6]

District nurses and (female) midwives were not generally called on to confirm early pregnancy. For example, Augustus Calder's *Questions and Answers on Midwifery for Midwives* (1906), a study guide for the London Obstetrical Society's examination, posed questions on pregnancy diagnosis in "about the fifth month," the "latter months," and "at full term" but did not ask about the difficult first trimester.[7] Physicians preferred to rely on trust when they could. "When a woman engages you to attend her," explained Glasgow professor of midwifery Robert Jardine, "you naturally believe her statement that she is pregnant, and you do not examine her before labour, unless there is some reason for doing so." But when "dealing with unmarried women," it was important to be "exceedingly careful."[8] Jardine described one dramatic instance of pregnancy denial and deception up to childbirth:

> In one case, a girl, who had not menstruated for seven months, consulted me as to her condition. She had all the signs and symptoms of pregnancy, and as I distinctly felt foetal movements, and heard the foetal heart, I told her she was seven months pregnant. I was indignantly told this was quite impossible. Next day her mother called on me, furious that I had dared to say such a thing about her daughter. As I was absolutely sure of my diagnosis, I advised her to wait a couple of months, and then to come and discuss the matter with me. About two months later I was called to see the girl one night by the indignant mother, who had been diligently poulticing the daughter's abdomen for cramp until a child had been expelled, rupturing the perineum in its exit. The girl had kept up the farce to the very end, and completely deceived her mother, until the arrival of the infant made further deception impossible.[9]

Although childbirth, stillbirth, miscarriage, or abortion would ultimately, although retrospectively, confirm pregnancy, abdominal growth on its own could not guarantee the existence of a fetus within. Tumors, cysts, and moles could mimic pregnancy, and the pregnant belly did not bulge visibly until the fourth month. Determining pregnancy in the first trimester was medically challenging and risked social embarrassment and professional disaster. Victorian practitioners relied on women's self-reporting and were reluctant to perform a vaginal examination to diagnose pregnancy or for any other reason.[10] Edinburgh-trained physician Thomas Watts Eden warned in the *American Journal of the Medical Sciences* that "except in the case of old women and little girls," the gynecologist and general practitioner "must keep the fear of pregnancy ever before him." In some circumstances, it was better to avoid the matter altogether as nothing tempted "disgrace" more than an error in pregnancy diagnosis. Eden's work at the outpatient department of the Chelsea Hospital for Women provided him with "abundant" material to master the art; of his last 1,000 cases, fifty had involved the early diagnosis of pregnancy.[11]

Textbooks of obstetrics and (male) midwifery typically dedicated an entire chapter to pregnancy diagnosis.[12] In addition to diagnostic uncertainty, authors emphasized the social difficulties of early diagnosis. William Playfair's *The Science and Practice of Midwifery* warned that pregnancy determination, which was "often beset with great difficulties," could jeopardize the patient's "moral character" and the practitioner's reputation.[13] Not only was a "correct opinion" of "extreme importance" to patients, but according to Alfred Lewis Galabin's *A Manual of Midwifery*, time would "inevitably" reveal the "medical man's skill, or want of skill." Overlooking or mistaking an advanced pregnancy would "incur ridicule," and a practitioner might find the result "still more unpleasant" if he erroneously accused a "virtuous unmarried woman."[14] "Never venture an opinion without making a thorough examination," cautioned Jardine's *Clinical Obstetrics*, "and do not say a woman is pregnant unless you are absolutely sure of your diagnosis."[15]

The canonical signs and symptoms of pregnancy were typically classified into "presumptive," "probable," and "positive."[16] These ranged in degree of certainty from a missed period, which could result from just about any constitutional disturbance, to the fetal heartbeat, generally regarded as the surest sign of all. Many textbooks, including Robert William Johnstone's *A Text-Book of Midwifery*, tabulated the principal signs and symptoms in order of occurrence (figure 2.1). Presumptive signs included a missed period, morning sickness, breast changes, and quickening. The absence of menstruation, or amenorrhea, was considered indispensable for estimating the date of delivery. But menstrual irregularities could also result from anemia, menopause, malnutrition, tuberculosis, and various other conditions, so a missed period could only be considered suggestive, unless corroborated by other signs.

For Galabin, an obstetric physician at Guy's Hospital in London, amenorrhea was usually the sign that precipitated suspicion of pregnancy in

DIAGNOSIS OF PREGNANCY 95

TABLE OF THE PRINCIPAL SIGNS AND SYMPTOMS OF PREGNANCY IN THE ORDER OF THEIR OCCURRENCE.

	Months.								
	1	2	3	4	5	6	7	8	9
Suppression of menstruation .	×	×	×	×	×	×	×	×	×
Irritability of bladder . .	×	×	?						×
Morning sickness . . .	?	×	×	×	?	?	?	?	?
Enlargement of breasts . .		×	×	×	×	×	×	×	×
Changes in size, shape, and consistency of uterus . . .	? ×	×	×	×	×	×	×	×	×
Vaginal pulsation . . .		×	×	×	×	×	×	×	×
Mammary areola . . .			×	×	×	×	×	×	×
Softening of cervix . . .			×	×	×	×	×	×	×
Intermittent contractions .			×	×	×	×	×	×	×
Apparent shortening of cervix .			×	×	×	×	×	×	
Discoloration of vagina . .			?	×	×	×	×	×	×
Progressive enlargement of abdomen				×	×	×	×	×	×
Uterine souffle				×	×	×	×	×	×
Ballottement				×	×	×	×		
Perception of active movements				?	×	×	×	×	×
Fœtal heart				?	×	×	×	×	×

2.1 A typical table of the canonical signs and symptoms of pregnancy. Johnstone 1913, 95, Cambridge University Library Ant.c.39.137.

his patients. But a deceptive (unmarried) woman wishing to "conceal" her condition could easily "deny the suppression of the menses" or even "artificially stain [her] linen to simulate menstruation."[17] Conversely, many women continued to bleed lightly even after conception, and others managed to conceive during amenorrhea. Jardine documented the case of one pregnant patient who had not menstruated in twelve months:

A well-nourished young woman consulted me on account of amenorrhoea of twelve months' duration. A year previously she had been very anaemic, and had taken a course of Blaud's [iron] pills. When I saw her there was no evidence of anaemia, and I was struck with her plump appearance. She stated that she had got very much stouter lately, not only in the abdomen, but all over. Her breasts were very large, and there was a distinct areola. Palpation of her abdomen revealed foetal parts and distinct movements, and auscultation gave foetal heart-sounds. She was about seven months pregnant. Conception had occurred during the amenorrhoea from anaemia. She was delivered of a full-time child some two months later.[18]

Nausea, often referred to as "morning sickness," although it could strike at any time of day or night, was also generally regarded as symptomatic of early pregnancy, especially if combined with other signs.[19] One of Jardine's patients invariably continued to bleed for months after conception but could self-diagnose based on the "severe sickness which attacks her from the very first."[20] Breast changes (increased size, firmness, tenderness) were especially useful in the unmarried patient because a practitioner could inspect the breasts "in passing," to help decide whether further diagnostic procedures were warranted.[21] Other potentially idiosyncratic and thereby less significant signs included the violet or "port wine" color of the vulva and cervix, uterine contractions, peevishness, despondency, frequent urination, toothache, pigmented patches on a pregnant woman's face, the linea nigra (a thin dark line that vertically bisects the abdomen in some pregnancies), lactation, headache, heartburn, skin eruptions, insomnia, stretchmarks, the cervical plug, and food cravings.

Probable signs included palpable or audible changes in the uterus and cervix detected by vaginal examination or auscultation with a stethoscope. The most important of these was "Hegar's sign," a soft, compressible area between the cervix and the uterus.[22] After the gravid uterus lost its distinctive pear shape, bimanual examination could reveal a "globular" form about the size of a Jaffa orange.[23] Eliciting Hegar's sign depended on "the *tactus eruditus* gained by practice," and Galabin, for one, encouraged students to take every opportunity of "becoming familiar with the feel of the uterus in the early stage of pregnancy."[24] Although admitting that it required "skill" and "experience" to detect, Eden's *Manual of Midwifery* rated Hegar's sign as of "very great" value (figure 2.2).[25]

Previously known as "placental souffle," authors often described uterine or funic souffle as a "musical" murmur probably caused by blood supplying the uterine arteries.[26] At times, it might be "composed of several notes," forming a "sort of chord."[27] Although it could be detected at an earlier stage than the fetal heartbeat, it was not as diagnostically certain because uterine fibroids (common benign growths of muscle and fibrous tissue sometimes known as "myomas") and other tumors could cause a

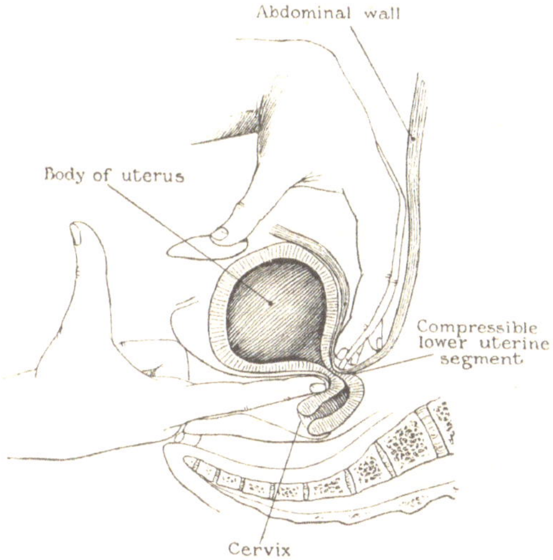

FIG. 33.—SCHEMATIC REPRESENTATION OF HEGAR'S SIGN.
(EDGAR.)

2.2 Nearly every major textbook of midwifery and obstetrics included a diagram of Hegar's sign. This line drawing in Eden's *Manual of Midwifery*, taken from the American gynecologist James Clifton Edgar's lavishly illustrated *The Practice of Obstetrics*, first published in 1903, is typical in depicting a doctor's disembodied hands bimanually examining a patient's cross-sectional pelvis. Equivalent diagrams in other books sometimes added flourishes such as the patient's pubic hair or the doctor's gloved hands or sleeved arms. Eden 1906, 57, Cambridge University Library Ant.c.39.107.

similar sound. External ballottement involved resting one hand on one side of the abdomen and giving a "single sharp pat" with the other, on the other side.[28] Internal ballottement, which was generally considered more reliable, involved sending the fetus floating up in the amniotic fluid with a "smart push upwards" and then waiting a few seconds to feel the distinctive sensation of "something lightly falling on the finger."[29]

Positive signs occurred only later in pregnancy and so were of little use in early diagnosis. They were, however, generally considered decisive and so highly rated. Based on the direct detection of a living fetus in the womb by a doctor or midwife, positive signs included feeling the movements of the fetus by palpation and hearing the fetal heart sounds by auscultation.[30] Although most authors agreed that the faint beating of the fetal heart was the only truly reliable sign of pregnancy, it was not easy to detect. Textbooks often compared it to the muffled ticking of a watch heard through a pillow. It was only audible over a small area, so careful exploration in perfect stillness and silence (with all ticking clocks stopped) was often required to confirm its presence or absence. Some authors preferred the intensifying sound of the binaural stethoscope, but others argued that intensified muscular sounds were more likely to confuse and so recommended instead the ordinary cedar stethoscope.[31] Some textbooks encouraged students to practice by listening to the heartbeat of newborn infants.[32]

Textbooks typically described quickening not as a sharp kick but rather as a feeble fluttering like that of a small bird in the hand. Later in pregnancy, the fetal movements became more distinct and even visible. Quickening was diagnostically valued, but authors also warned that women could be intentionally misleading or even unintentionally deceived by flatulence or wishful thinking. Some authors distinguished between the patient's sensation of quickening and the doctor's detection of fetal movements.[33] If verified by a midwife or physician, quickening or "stirrage," as it was sometimes called, was highly rated as certain evidence of a living fetus in the womb.

HINTS TO MOTHERS

First published in 1684, *Aristotle's Masterpiece,* the most widely circulated source of sexual and reproductive knowledge in Britain, could still be found, little altered, in sleazy London sex shops as late as the 1920s.[34] Edith Hinson, a Stockport mill girl born in 1910, first learned about the symptoms of pregnancy in a copy found under her mother's mattress.[35] The chapter on "how a woman may know whether she hath conceived or not" noted visible or painful changes in and around the eyes, breasts, and face, as well as a method of keeping urine in a glass for three days and then inspecting it for the presence of "small living creatures." Green nettle could also be added to the urine overnight, and "if the woman be with child, it will be full of red spots on the morrow; if not, it will be blackish."[36]

In contrast to the notoriously illustrated "masterpiece," a genre of respectably unillustrated domestic health and marriage manuals promising a scientific explanation of the "facts of life" to middle-class women was flourishing by the mid-nineteenth century.[37] As with midwifery textbooks, they typically devoted an entire chapter to the canonical signs and symptoms, thereby setting the stage for subsequent chapters on the progress of gestation, lying-in, childbirth, and infant care. Thomas Bull's *Hints to Mothers*, the leading Victorian manual, claimed that many possibly pregnant women experienced "much difficulty in attaining certainly" and "suffered months of anxiety and doubt."[38] Henry Allbutt's *The Wife's Handbook*, better known for advertising contraceptive devices, lamented that newlyweds were generally "ignorant," warned that no married woman under forty-five was "*safe*," and promised that knowledge of the "*subjective*" and "*objective*" signs could "save her from much bad health."[39] Allbutt atypically recommended using a looking glass to verify the change in vaginal hue from rosy to violet as a "very early" and "extremely valuable" sign of pregnancy, but most manuals stuck to the canonical signs found in medical textbooks: amenorrhea, nausea, breast changes, quickening, and the fetal heartbeat.[40]

Most manuals emphasized the significance of quickening even as they explained that the sensation of fetal movements did not mean the child had come to life. For instance, Charles Glasson's *Motherhood*, praised in the *Lancet* as a "useful little book" for the "young married woman," referred to the "very great importance" of "quickening" even as its author, a London physician, clarified that the "child is alive from the *very first*."[41] *Dr. Chavasse's Advice to a Wife on the Management of Her Own Health* identified quickening as one of the most valuable signs because there was "less likelihood of a miscarriage *after*, than *before it*."[42] Taking a hard line on abortion, it also rectified the "old-fashioned" and "mistaken" notion that the "child" was not alive before quickening: life began from the "very commencement of his formation," and the "heinous sin" of early abortion was "as much murder as though the child were at his full term, or as though he were butchered when he was actually born."[43]

"The first point of importance," according to Ada Sarah Ballin's *The Expectant Mother*, was determining that a woman was "actually in what the Germans call 'The blessed condition' ('Selige Zustand')."[44] As with medical textbooks, Ballin explained that the causes of amenorrhea included not only pregnancy but also anemia and other conditions and that a "discharge," indistinguishable from menstruation, could persist in pregnancy.[45] She also advised the reader to engage a nurse as soon as she knew she was pregnant because the "best" ones were "always engaged long in advance."[46]

The Edinburgh obstetrician, teratologist, and "apostle" of prenatal care, John William Ballantyne, presented his hefty manual, *Expectant Motherhood: Its Supervision and Hygiene*, as combating ignorance that could endanger both mother and child.[47] Lack of knowledge and, worse still, "dangerous" misinformation could "easily make havoc" with a woman's "happiness" and "her hopes as a mother." If she failed to recognize the earliest signs of pregnancy, she might persist in risky activities "such as taking long bicycle rides or undertaking big pieces of social or philanthropic work, with the result that abortion is threatened or actually brought about." Or she might take purgatives to wash away the "obstruction" if her period had "not come on within six weeks after marriage." Ballantyne did not distinguish between menstrual regulation and miscarriage, which always meant the "death of an unborn child."[48] Although Ballantyne argued that the symptoms experienced by the mother and the signs detected by her physician were sufficient, in practice, "to be acted upon," he admitted that it was "as yet impossible to be so certain of the existence of early pregnancy as to swear to its presence in . . . a court of law."[49]

THE PROMISE OF SEROLOGY

Diagnostic tests of all kinds proliferated through the slow rise of laboratory medicine that in some ways culminated in the early twentieth century with the routinization of the famous Wassermann reaction for syphilis.[50] Around the same time, serologists working in Germany on therapeutic sera for eclampsia, a dangerous and still poorly understood pathology of late pregnancy, also announced new serodiagnostic methods of pregnancy diagnosis.[51] Researchers published preliminary results with various new and experimental pregnancy tests, including a modification of Wassermann's and a cobra venom reaction, but these were marginalized in 1913 by the great interest in Emil Abderhalden's methods.[52] A Swiss biochemist based at the University of Halle in the Prussian province of Saxony, Abderhalden based his test on two principles: first, that the human body reacts to an injection of albumen (protein) by producing a defensive "ferment" (enzyme) to digest the foreign substance and, second, that during pregnancy, the chorionic epithelium circulates in the woman's blood. He argued for the existence of a specific enzyme, found only in the blood of pregnant women, that divided the placental albumen into peptones and amino acids.[53]

Abderhalden proposed not one but two diagnostic methods. The first, called the "optical" method, depended on a change in rotation of the plane of polarized light before and after incubating a pregnant woman's serum together with placental peptone. Very few used Abderhalden's "difficult" optical method, which involved the time-consuming production of peptone (figure 2.3). The basis of the second "dialyzation" method was the impermeability of animal membrane to albumen, on the one hand, and its permeability to products of "proteolytic digestion," on the other. The end point of this method was a visible color-change reaction: ninhydrin, the chemical today used in fingerprinting, was supposed to turn the incubated solution blue or violet in a positive result while a control solution remained colorless. In practice, however, laboratory workers found themselves comparing between shades of violet.[54]

Even the "simpler" dialyzation method was technically demanding and required "extreme care" and "scrupulous exactness." Preparing the reagents was labor intensive. Fresh human placenta, "washed absolutely free from blood," needed repeated boiling in water with two drops of acetic acid until the water became negative to the biuret reaction.[55] Laboratory workers struggled to streamline these elaborate procedures into a simple, practical, and reliable blood test for pregnancy.

Although Ballantyne admitted that the method remained "essentially a laboratory test and not one to be done by the general practitioner," he hoped that it might be used not only to detect pregnancy, especially in unmarried women, but also for differential diagnosis. Serology held promise in distinguishing between normal gestation, on the one hand, and a range of pathologies, on the other. Of special concern were fibroids, amenorrhea (caused by lactation, tuberculosis, or diabetes), eclampsia, and chorioepithelioma (a rare placental cancer).[56] Moreover, research on laboratory diagnosis had already opened up new vistas in the physiology and pathology of pregnancy and would lead to a better understanding of the "complex and wonderful relationship between mother and unborn infant, which some have called a harmonious symbiosis, others a prejudicial parasitism, and others an immunity reaction."[57]

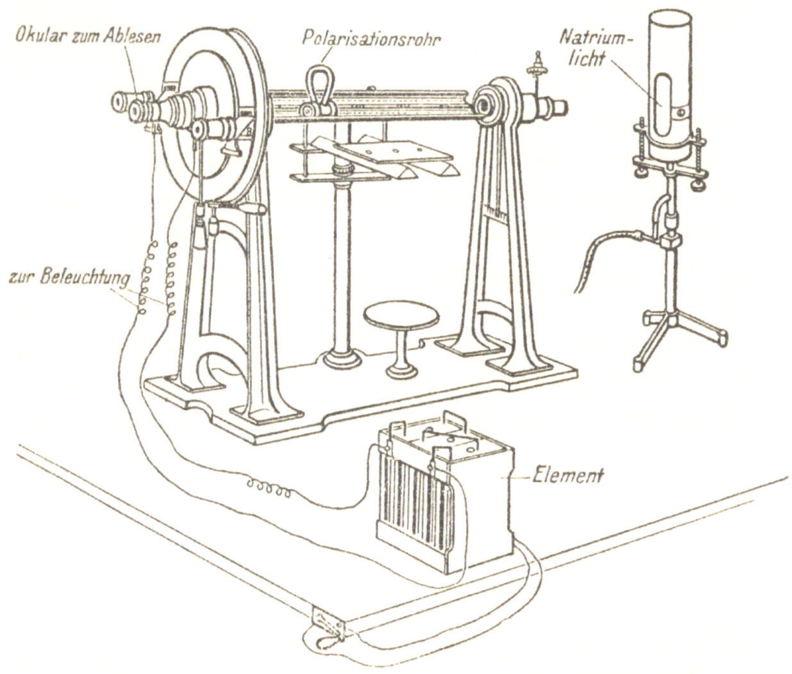

2.3 Diagram of Abderhalden's polarization apparatus showing the ocular for taking readings, polarization tube, sodium flame, and a battery connected to wires for illumination. Abderhalden 1914a, 335, Cambridge University Library 385:1.c.90.14.

Herbert Williamson, an obstetrician at St. Bartholomew's Hospital ("Barts"), argued that if a pregnant woman's blood contained a specific, constantly present, and easily detectable ferment, and if diagnostic error could be avoided, then Abderhalden's test would be of "great value" to clinicians. Together with the chemical pathologist Robert Lauder Mackenzie Wallis, Williamson experimentally tested fifty pregnant patients, recently delivered patients and patients with suspected ectopic pregnancy, pelvic and abdominal tumor, chorioepithelioma, chorea, puerperal sepsis, heart disease, and nephritis.[58] In their presentations to the Royal Society of Medicine, Williamson and Mackenzie Wallis concluded that Abderhalden's test was especially useful for early pregnancy diagnosis, differential diagnosis (between fibroids and pregnancy), and the diagnosis of chorioepithelioma.[59]

A house surgeon at Barts further proclaimed in the *Lancet* the "practical advantage" of the test as a diagnostic aid in cases of stomach and bowel carcinomas to enable early surgical intervention.[60] But others were more cautious. For instance, a clinical pathologist at the Sheffield Royal Infirmary argued that although a significant trial period had corrected technical errors, it was still too soon to ascribe "clinical value" to the test.[61] Nevertheless, in his review of over eighty (mostly German) articles in the *Journal of Obstetrics and Gynaecology of the British Empire*, Mackenzie Wallis concluded that researchers had "clearly" and "sufficiently" demonstrated the value of Abderhalden's tests.[62] These were not yet perfected, and it was only by cooperation "between the clinician and the chemical pathologist" that further progress would be made.[63]

"Scientific medicine," as Ballantyne put it to the York Medical Society in early 1914, had been for several years "trembling on the brink" of the discovery of a reliable biochemical test for early pregnancy. Abderhalden's reactions had made such a test seem more likely than ever. Nevertheless, the careful preparation of placental albumin was laborious, and Ballantyne worried about diagnostic error. So far, the blood of some patients with diseases, including those with cancer, had tested positive, and the blood of some definitely pregnant women was negative. Laboratory workers were, however, improving the technique; there were fewer false results, and the reliability of the reaction was becoming established.[64]

Meanwhile, Abderhalden's fame continued to grow. Springer published two German editions of Abderhalden's *Abwehrfermente* in less than one year. The first, praised in the *Lancet* in January 1913 as "interesting" and "suggestive," was recommended to gynecologists and biologists on account of the "biological diagnosis of pregnancy."[65] In November, the *Lancet* wished the second expanded and costlier edition "as rapid a success as its predecessor."[66] In 1914, John Bale published the first English edition of *Defensive Ferments of the Animal Organism*, translated by Jacob Gavronsky of the Hale Clinical Laboratory, London Hospital, from the third German edition (the second had been exhausted in less than three months). A review in the *Lancet* emphasized the practical applications of Abderhalden's "discoveries" to the serodiagnosis of pregnancy, cancer, and other diseases.[67]

The circulation of these new "weapons of research," as Gavronsky called them in the preface to *Defensive Ferments*, had been made possible by

Abderhalden's willingness to accommodate visitors at his institute in Halle, promptly answer written inquiries, and freely supply reagents (albumen and peptone) prepared in his laboratory.[68] The pharmaceutical company Höchst marketed a placental peptone for testing pregnancy. Less reputable "carcinoma extracts" and other dubious reagents were marketed "by some people who [were] apparently in a great hurry to make the best out of Abderhalden's promising scientific work."[69]

In early 1914, the Berlin correspondent to the *British Medical Journal* (*BMJ*) warned that the pregnancy test was heading for controversy. In a "surprising" and "dramatic" turn, Leonor Michaelis, the Jewish director of a small bacteriological laboratory of a municipal hospital in Berlin, reported on a "very large" study in the prestigious *Deutsche medizinische Wochenschrift.*[70] Michaelis's assistant had learned the method directly from Abderhalden in Halle, and together they argued that pregnant women's blood did not react differently from the blood of nonpregnant women "or even men." They denied the existence of a specific ferment of pregnancy. The editorial concluded that Berlin physicians were awaiting Abderhalden's reply with the "liveliest interest."[71]

Abderhalden responded in the *Wochenschrift* that Michaelis's results diverged from those of "numerous investigators" at university clinics, where the reliability attributed to Abderhalden's tests varied but was never less than 90 percent.[72] In the following weeks, however, the *BMJ* reported on two confirmations in the *Münchener medizinische Wochenschrift* of Michaelis's "negative opinion."[73] Faith began to collapse in Britain. Archibald Leitch, a pathologist at the Cancer Hospital in London, reported "adverse results" that contrasted strikingly with those of Abderhalden and his European and American "disciples," whose claims he now considered "amazingly mistaken."[74] William Bullock of the Imperial Cancer Research Fund contrasted Abderhalden's "numerous" supporters with the "few" who rejected the existence of specific ferments. In his view, even with technical adjustments, Abderhalden's method failed to distinguish normal from pregnant or cancerous sera.[75]

In 1915, the number of publications about Abderhalden's defensive ferments exceeded 300 (of which only fourteen were in English), and Gavronsky predicted that opinion over the specific ferments and their clinical applications would remain divided for years.[76] Gavronsky had twice visited

Abderhalden's institute in Halle to learn the technique before attempting it on blood samples obtained from the London Hospital and Bethnal Green Asylum. "Paradoxical at it may sound," lamented Gavronsky, "the more one follows Abderhalden's directions . . . the less one is likely to meet with specific reactions." The problem of specificity became acute when working with randomly selected ward patients. Even so, Gavronsky remained "full of admiration" for Abderhalden's theory and hoped that in time it would be vindicated.[77]

Even as they were primarily driven by an interest in diagnosing cancer and other diseases, pathologists and physicians preferred to experiment with pregnancy. This is because it was easier to obtain the necessary "materials," and the result would be conveniently confirmed or disproved in nine months or fewer. By using pregnant women, rather than patients with cancer, laboratory staff acquired a working knowledge of the method. Mackenzie Wallis's own positive experiences convinced him that there was a "placental splitting ferment in the blood of pregnant women." After eighteen months of tinkering, he was able to obtain "fairly reliable" results and considered the method to be "really quite simple" and within reach of "any trained laboratory worker."[78] But even as Mackenzie Wallis praised the usefulness of Abderhalden's reactions in scientific research, he admitted that, from a clinical perspective, they merely added to a "history of failures." In 1916, he hoped that modified forms of the tests would someday be of "greater value," not only in pathological research but also in clinical diagnosis.[79]

By the end of World War I, many of those who had previously supported Abderhalden, including his English translator, had changed their tune. "Were it actually possible to demonstrate the presence of specific ferments in the blood serum," wrote Gavronsky in the *Lancet* in 1918, Abderhalden's test would be the "greatest and the most useful discovery ever made in the domain of medical science." But clinical laboratories had not adopted Abderhalden's methods. This was not because the techniques were too "complicated"; they required skill but could be mastered by an "average laboratory worker." Rather, it was because the methods were "*of no use for clinical purposes*." Abderhalden's test was "not a clinical test at all." The "whole story," Gavronsky claimed, was that of a "great scientist" who prematurely announced a discovery, dug in, and used "personal influence" to induce others to "repeat his assertions." Hundreds of published results had confirmed those of Abderhalden, but in this case, the "few" and "not the majority" were right.[80]

An anonymous critic of the use of Abderhalden's reaction to diagnose psychiatric disorders remarked in the *Lancet* in 1921 that its history would make an "interesting study" in the "dominance of German opinion in Europe and America until quite recent times." First announced as a pregnancy test, it quickly generalized to the diagnosis of disease and was crafted into an "instrument" capable of testing "hypotheses" and "theories." But it now seemed clear that Abderhalden's theoretical framework had been accepted too hastily by a credulous "medical scientific world." Perhaps someday, the critic concluded, the error would "leak through into Germany."[81]

In his contribution on bloodwork to *A New System of Gynaecology*, London bacteriologist William Topley advised clinicians to "await further developments" before turning to Abderhalden's reaction. No diagnostic test, he argued, could ever become "generally useful" if it was reliable in the hands of only an "elect few." It was "absurd," he went on, to blame failure on the "inexperience" or "incompetence" of so many laboratory workers.[82] In his *Manual of Midwifery*, Eden portrayed Abderhalden's "discovery" as the "greatest" recent advance in the "biology of pregnancy" but then added that "some observers" had "failed" to corroborate its clinical value. He recommended the reaction as a screening test: because cancer and other pathologies could produce false positives, a negative result could be depended on to "exclude pregnancy," but a positive result ought to be greeted with caution.[83]

The second edition of *A Guide to Gynaecology in General Practice* explained that Abderhalden's test was of "very questionable" value and unlikely to be accepted in a court of law.[84] The fourth edition of William Robertson's *Manual of Medical Jurisprudence and Toxicology* dismissed the methods as "too elaborate for description" and untrustworthy.[85] The *Combined Textbook of Obstetrics and Gynaecology*, by four Scottish teachers, lamented that Abderhalden's test was positive in too many conditions other than pregnancy to be of use in clinical practice.[86] The fourth edition of Johnstone's popular *Text-Book of Midwifery* briefly mentioned the test as "impracticable" and of "theoretical interest only."[87] Samuel Cameron's *Glasgow Manual of Obstetrics* similarly dismissed the technique as too "complicated" to be clinically useful.[88] And Sydney Smith's *Forensic Medicine* explained how the "doubtful" test worked in some detail, only to dismiss it as "worthless during early pregnancy, when a diagnosis is most difficult."[89]

TOO MUCH CERTAINTY

Advice manuals published after World War I did not mention Abderhalden's test and continued to rely on the canonical signs and symptoms. For example, the authorized English edition of Chicago obstetrician and sex radical Alice Bunker Stockham's *Tokology: A Book for Every Woman* emphasized amenorrhea, abdominal growth, quickening, and the fetal heartbeat.[90] Birth control pioneer Marie Stopes's *Radiant Motherhood*, the follow-up to her bestselling *Married Love*, claimed that although some women could sense the "actual moment of conception," the majority were "less completely cognisant of the voices of their own organism" and for the first two or three months "almost unaware that anything different from the usual course of their life is taking place."[91] Stopes included a brief appendix on the "physical signs of coming motherhood" for the woman who suspected pregnancy but for whom medical confirmation was unavailable.[92] Alice Lady Lovat's *Marriage and Motherhood* advised the reader not to wait until quickening to engage a doctor or nurse.[93] And *For Women Only*, attributed to the anonymous "physician" author of *How to Be Healthy*, described the fetal heartbeat as the only "absolute proof."[94]

In the 1920s, the most promising alternative to the awkward intimacies of physical examination was obstetric radiography. A pioneering American handbook praised X-rays as a "very valuable aid in the diagnosis of pregnancy," especially for differential diagnosis, but also to "dissipate" the "scandalous" stories told by "venomous gossip-mongers" about unmarried women or widows, as well as in court for settling lawsuits, libel cases, and divorce proceedings.[95] Fetal bones, however, did not cast shadows until about the sixteenth week, and the demand for X-rays in pregnancy diagnosis significantly declined following the introduction of pregnancy testing.[96] No laboratory method was "absolutely and infallibly diagnostic," and so tests were used mainly to bolster the "already present suspicion of pregnancy, or the probability of its absence."[97]

As interest in Abderhalden's reaction was fading, German and American researchers began experimenting with a new kind of test, which supposedly exploited the fact that women in the early months were prone to glycosuria, the excretion of sugar in the urine. In 1923, John Cooke Hirst and Charles-Francis Long at the William Pepper Laboratory of the University of Pennsylvania in Philadelphia, published a preliminary report of thirty-nine cases using a sugar tolerance test in the *New York Medical Journal*.[98] Of particular interest was a recently proposed test based on phlorizin, a glucoside derived from domestic apple tree bark available as a popular drug for lowering the kidney sugar threshold in lab animals. The protocol was to fast a patient for twelve hours and then inject her with 2 milligrams of phlorizin; the appearance of sugar in urine within two hours indicated pregnancy. The "simplicity" of the technique, as Hirst and Long explained,

2.4 Half-page ad for Maturin in *Münchener medizinische Wochenschrift* 42, 1922, 4. The hexagram-like logo is a combination of the alchemical symbols for fire, water, and salt: Thore Grimm to author, Sept. 21, 2022. Kamnitzer and Joseph practiced at Krankenhaus Moabit, a "reform" hospital and a center of Jewish doctors, many of whom were prominent in newer and less prestigious fields including reproductive endocrinology and neurology (the famous Kurt Goldstein practiced there): Pross and Winau 1984; Stürzbecher 1997; Joseph: Press 2000, 117. Schering Archives, Bayer AG, B2 1677.

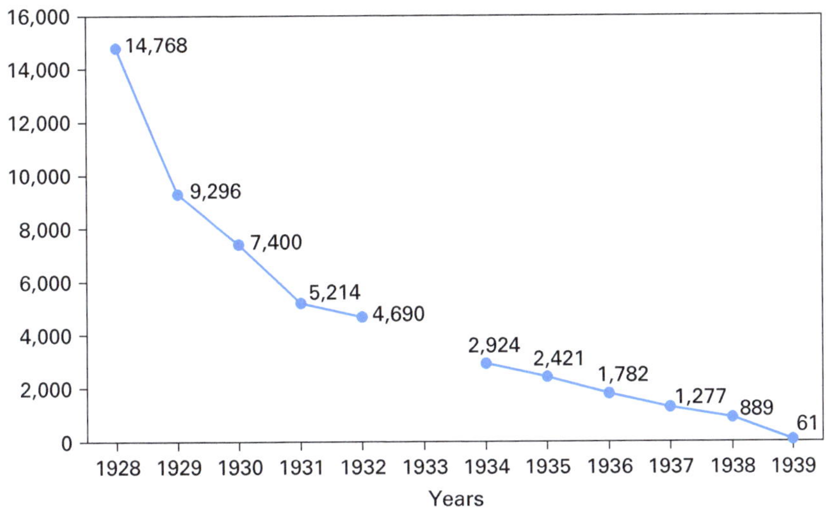

2.5 Maturin sales "units," 1928–1939. Schering produced over 35,000 ampoules of Maturin in 1926. Although from 1928 the only figures recorded are sales units ("Wert"), not ampoules, it is clear that production declined precipitously to virtually nothing by the start of World War II. This was not a general trend: the manufacture of many other more successful Schering products increased in the same period. As we shall see in the next chapter, this period of decline correlates with the rise of the Aschheim-Zondek test and related "bioassays," which appear to have displaced Maturin and other glycosuria tests. Based on data from Schering Archives, Bayer AG, S1 63–64.

was further enhanced by "Maturin," a proprietary solution of phlorizin in ampoules launched in 1921 by the Berlin pharmaceutical company Schering (figures 2.4 and 2.5).[99]

Upon surveying published reports of the glycosuria test, however, Hirst and Long concluded that it resulted in too many false positives. They preferred to administer their patients a dose of table sugar dissolved in lemon-flavored water and then collect the urine one or two hours later. This was less invasive than injections but also had drawbacks. Several patients poured the "lemonade" down the sink or "out the window," and others became nauseous and vomited, ruining the test. The test, when it was trustworthy, presented Hirst and Long with a different kind of dilemma. When one patient, an unmarried woman whose period was a few days late, tested positive, she "induced an abortion" a few weeks later. From the Philadelphia doctors' perspective, the glycosuria test provided "unscrupulous characters" with "too much certainty at an early date in pregnancy."[100]

Despite the paradoxical hazard of providing the wrong women with "too much certainty," Hirst and Long promoted their "extremely simple" test as a useful aid in the diagnosis of early pregnancy. In a series of 150 patients, they reported an error of 6 percent in pregnant women (false negatives) and 8 percent in nonpregnant women (false positives). This beat all other methods, they claimed, and compared favorably with the famous Wassermann test for syphilis.[101] But some experts were not convinced of the need. Catholic gynecologist and infertility specialist John Rock dismissed sugar tolerance tests as unreliable and argued in the prestigious *New England Journal of Medicine* that pregnancy diagnosis was "not always of immediate importance. Time is probably still the surest aid: indeed, for all practical purposes it may be considered certain."[102]

UNMISTAKABLE BLOOD POINTS

By the late 1920s, neither serology nor glycosuria tests seemed likely to provide clinicians with a practical alternative to physical examination. Although some doctors, like Rock, were content to simply wait and see, others adopted new laboratory tools and techniques made available by the increasingly prominent science of endocrinology. The most promising of these was the Allen–Doisy test, a vaginal smear for estrogenic activity announced in 1923 by anatomist Edgar Allen and biochemist Edward Adelbert Doisy at the Washington University School of Medicine in St. Louis, Missouri.[103] Prominent New York gynecologist Robert T. Frank attempted to adapt the technique for pregnancy diagnosis but concluded that the serum of pregnant women was insufficiently rich in "female sex hormone" to be useful in this regard.[104]

In February 1928, the less famous Cleveland physician, Alcines Clair Siddall, announced a promising new method in *JAMA*, the prestigious journal of the American Medical Association.[105] Siddall reasoned that if some unknown hormone caused the changes in a pregnant woman's body, then similar changes ought to occur in the uterus and breasts of a female mammal injected with her blood. He expected the blood from a nonpregnant woman to yield a negative result. Siddall first allowed his patients' blood to clot in a sterile tube before injecting the serum subcutaneously into a sexually immature virgin white mouse once a day for four or five days. He then performed the Allen–Doisy test to establish the phase of the estrous cycle of the mouse and then killed the test animal and weighed it on a chemical balance. Next, he dissected out the uterus and ovaries and weighed those. Finally, he divided the weight of the mouse by the weight of its reproductive organs: a ratio below 400 was positive for pregnancy while a ratio above 400 was negative. After a preliminary report on forty-five patients, Siddall concluded that his test seemed "reliable."[106] A *BMJ* editorial welcomed Siddall's study, commenting that a "simple" and "satisfactory" pregnancy test would be "most valuable, not only to the obstetrician, but also to the general practitioner," and hoping for further confirmation on a "larger series" "with controls."[107]

The ninety-seven mouse tests reported a few months later by Siddall generally confirmed his preliminary results and further promoted his technique as a method of determining the potency of proprietary extracts of ovary, placenta, and pituitary.[108] Using his test, Siddall had determined that only one of the seven commercial preparations he assayed was hormonally active. Finally, he highlighted the usefulness of the test in the monitoring and management of patients receiving infertility treatment. Following a round of artificial insemination, "Case 89," a twenty-four-year-old woman, had missed her next anticipated menstrual period. But her pregnancy test was negative, and a few days later, she experienced a "perfectly normal menstruation." Siddall modestly concluded that his "hormone test" was "not a specific test for pregnancy" but rather detected the circulation of hormones that "probably increased" during the "gravid state."[109]

Despite initial enthusiasm, interest in Siddall's test was short-lived. In February 1929, an editorial in the *BMJ* reported on the "appearance" of yet another "reputed" method: "We referred last summer (June 2nd, 1928, p. 952) to A. C. Siddall's report of the discovery in the blood of a pregnant woman of a hormone which caused enlargement of the uterus and breasts. Now, in the *Zentralblatt für Gynäkologie* for January 5th (p. 15), S. Aschheim describes a technique and results of the test which he has devised with B. Zondek."[110]

For some years, the gynecologist Selmar Aschheim and the physiologist Bernhard Zondek, working together at Berlin's famous Charité hospital,

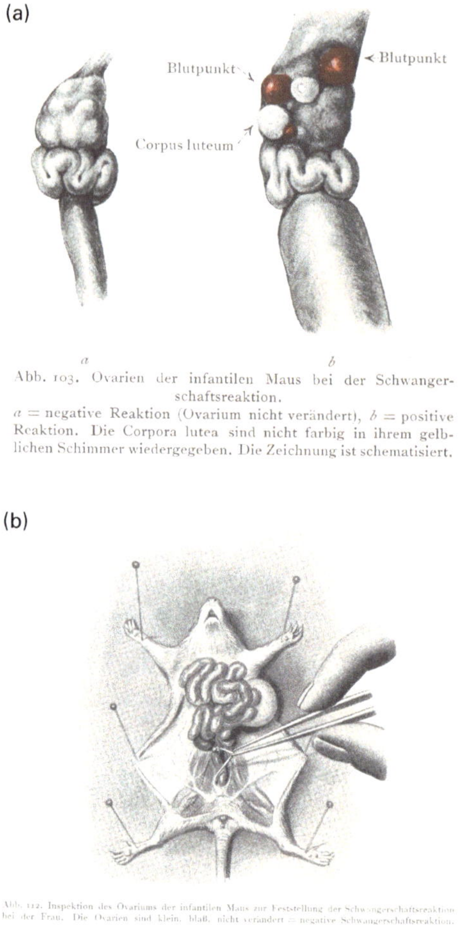

2.6 (a) Artist's comparison of "negative" (left) and "positive" (right) Aschheim-Zondek reactions; note the conspicuous blood points colored in red. (b) Dissected Aschheim-Zondek test mouse, showing a negative reaction. Zondek 1931, 302, 306, Cambridge University Library 379.c.93.47, with permission of SNCSC.

had used the Allen–Doisy reaction to test the hormonal activity of commercial ovarian products.[111] In 1927, they presented their discovery that the pituitary gland contained an ovary-stimulating hormone at a meeting of the German Society of Gynecologists in Bonn.[112] The following year, Aschheim announced the new pregnancy test based not directly on the "female sex hormone" but on the presence of an ovary-stimulating hormone in pregnant women's urine.

In contrast to Siddall's procedure of weighing mice and calculating ratios, Aschheim and Zondek based their test on the visual detection of "blood points" in the hormonally ripened ovaries of immature mice. A single test involved injecting a batch of five mice with urine extract twice a day for three days in a row (a total of thirty injections). After that, the mice were dissected and their ovaries visually inspected. Aschheim and Zondek interpreted the presence of blood points (a sign of sexual maturity) in at least one mouse as a positive reaction. Immature organs meant a negative result (figure 2.6).

With cautious optimism, the *BMJ* editorial claimed that the "reliability" of Aschheim and Zondek's test appeared, at least "in the hands of Aschheim," to be "considerably greater" than that of the other tests.[113] Frank corresponded with Aschheim, sourced immature mice from a dealer, and performed the test with "considerable success." The results impressed him: "The blood points in the ovaries are unmistakable, and a positive reaction is recognized. This is by far the best test for pregnancy as yet discovered."[114] As an American commentator later observed, Siddall's "important observation" was "completely overshadowed" by the Aschheim–Zondek test, as it came to be called.[115]

3

THE BUSINESS OF DIAGNOSIS

Although not the first "scientific" test for pregnancy, the Aschheim–Zondek reaction was the first to become prevalent on a large scale. As others have noted, by the mid-1930s, a diagnostic service in Edinburgh was performing thousands of tests every year for clinicians and hospitals around Britain.[1] In her classic history of prenatal care, sociologist Ann Oakley credited the Aschheim–Zondek test with launching the "modern era in which obstetricians would eventually be able to claim a knowledge superior to that possessed by the owners of wombs themselves, as to the presence of a guest, invited or uninvited, within."[2] Yet beyond the fact that the test was invented in Berlin and implemented on a large scale in Edinburgh, surprisingly little is known about how it worked in practice or the purposes for which it was used.

Above all, there is the problem of demand. Many women were aware of their menstrual cycles and familiar with the early signs of pregnancy, especially if they had already borne children.[3] In the early twentieth century, they rarely called on doctors or attended prenatal clinics before the second or third trimester, so it was unusual for medical practitioners to be involved in the early stages of pregnancy.[4] A woman who did seek out medical advice to confirm or allay her suspicions was typically told to return in a month's time, unless an Aschheim–Zondek test was medically indicated.[5] Women who were contemplating abortion probably did not let on to their doctor.[6]

And many took steps to regularly bring on menstruation, a practice they did not necessarily equate with aborting a fetus.[7] So if neither women nor doctors relied on the laboratory to help detect pregnancy, what was the Aschheim–Zondek test used for?

In this chapter, I explain the adoption and institutionalization of the Aschheim–Zondek test in terms not of the medicalization of ordinary pregnancy but of clinicians' increasing reliance on laboratory services for differential diagnosis. Crucially, the test did not in fact detect a viable fetus but rather placental tissue and so was "strongly positive" for pathological growths such as hydatidiform mole or chorioepithelioma.[8] Conversely, a weakly positive reaction could predict "spontaneous abortion," as miscarriage was then called.[9] I show how the Aschheim–Zondek test was made, less into a binary test for normal pregnancy and more into a versatile tool for differential diagnosis, calibrated to monitor placental tumors and hormonal deficiencies.[10] Pregnancy tests were, as sociologist Adele Clarke put it, "early and important technoscientific products of the reproductive sciences."[11] But innovation is not the whole story, and my account will focus more on the establishment and maintenance of everyday routines.[12]

I also situate the Aschheim–Zondek test within the business of diagnosis.[13] Historians have explained the rise of modern laboratory testing in terms of statist efforts to rationalize health care.[14] This kind of analysis explains well the role of the laboratory in public health campaigns such as mass screening programs for syphilis or cervical cancer.[15] But it neglects the commercial market for clinical diagnostic testing, which was established by the 1920s.[16] This chapter looks outside the laboratory and beyond the managerial state for the crucial determinants of financial and other forms of viability.[17] As we shall see, the success or failure of the Edinburgh service hinged on whether it could generate sufficient demand from doctors as clients.

TESTING THE TEST

An expectant mother who visited the prenatal clinic or was seen at home by a midwife in the early twentieth century might have had her blood pressure taken, her urine examined for albumin or sugar, or her blood tested for syphilis, but it was not routine to test the urine of an apparently healthy woman to confirm pregnancy.[18] By 1914, nearly half the adult population of Britain was covered by the 1911 National Health Insurance Act. Most women, all children, the elderly, and self-employed were, however, excluded, and benefits to women workers were cut in 1915 and again in 1932.[19] Because they were unlikely to be covered by health insurance, working-class women did not usually visit a doctor except in an emergency.[20] The 1911 act made no provision for laboratory services, so patients who could afford them paid out of pocket for diagnostic tests. Basic urinalysis was a side-room practice performed by a GP, nurse, or midwife, but bacteriological and biochemical tests were left to clinical pathologists.[21] The wartime campaign against syphilis created state demand for mass Wassermann testing, and the introduction of insulin and liver treatments in the 1920s increased interest in biochemical and hematological testing.[22] Routine analysis became increasingly structured around new divisions of labor and new specialties, such as X-ray and laboratory technicians, who provided diagnostic services not directly to patients but to doctors.[23]

The Aschheim–Zondek reaction was first established in Britain at Francis Crew's Department of Animal Breeding Research (later the Institute of Animal Genetics) at the University of Edinburgh.[24] Of the three animal breeding research institutes in 1920s Britain (at Cambridge, Edinburgh, and Reading universities), this was the only one to branch into medical research.[25] Best known for his work on sex reversal and intersexuality in the domestic fowl, Crew aspired to make a name for himself as an expert in human heredity, eugenics, and social biology.[26] But first he needed to medicalize his department, which was beholden to the Ministry of Agriculture. With help from Edinburgh professor of physiology Sir Edward Sharpey-Schafer, Crew attracted public and private donors for medical research, including controversial work on chemical spermicides.[27] When Thomas B. Macaulay, a wealthy Canadian financier with Scottish ties, paid for a lectureship in endocrinology, Crew hired Bertold P. Wiesner, a young Austrian physiologist and "rejuvenationist" he had met in 1926 at the Berlin Congress for Sex Research (figure 3.1).[28]

A product of Eugen Steinach's controversial Institute of Experimental Biology in Vienna (the "Vivarium"), Wiesner modeled the Macaulay Laboratory on that institution.[29] His work contributed to the "air of impropriety" about Crew's institute, which was reputed to be "slightly immoral and absorbed with sex."[30] When the Medical Research Council (MRC) refused

3.1 Portrait photograph by Shackleton, Piccadilly, of Bertold Wiesner as a visionary scientist, c.1930s; courtesy of Jonathan Wiesner.

Crew's request for funding on the grounds that his institute was too agricultural, Crew turned to Robert W. Johnstone, the influential chair of the midwifery department, for support.[31] Swayed by Johnstone, the MRC agreed to finance Wiesner's work for one year.[32] Wiesner and Crew began to collaborate with Johnstone, exchanging valuable research material (pregnant women's urine and placentas) and access to patients for experimental therapeutic products (made from the urine and placentas) and access to laboratory animals.[33]

During the endocrine "gold rush" of the 1920s and 1930s, drug companies isolated and mass-produced the internal secretions of the ovaries, testicles, pituitary, and placenta.[34] The Aschheim–Zondek test was a by-product of this "heroic age" of "sex physiology," as reproductive endocrinology was then called, and first Wiesner used the reaction not as a test for pregnancy but to verify the potency of potentially therapeutic substances.[35] Impressed by its efficacy in drug standardization, he then proposed to offer diagnostic testing as a routine service for doctors, beginning with Johnstone. He had three main reasons. First, the station would evaluate the test on a large number of clinically unselected patients, thereby demonstrating the value of the agricultural institute to medical practitioners and researchers. Second, any surplus (hormonally rich) pregnancy urine sent to the station could be redirected toward research (injected into rats). Third, the station would charge a fee and so was expected to be self-financing or even to turn a profit that could be ploughed back into research, an economic strategy that other university and hospital laboratories were then adopting.[36]

Collaborating with Wiesner offered Johnstone several clear advantages too. First, with sex hormones a novelty in gynecology, Wiesner supplied Johnstone with new and experimental therapeutic substances. The chance to test the expensive extracts on his private patients placed Johnstone at the forefront of clinical research. He also gained access to a new and potentially powerful diagnostic tool that could be tested on his hospital (and private) patients. A controversial specialist in infertility treatment, Johnstone used the Aschheim–Zondek test not simply for pregnancy diagnosis but to calibrate hormone injections in cases of endocrine deficiency believed to cause miscarriage.[37] Last but not least, Johnstone needed Wiesner for animal injections, which were forbidden on infirmary property.[38]

Animal experiments, including routine injections, were permitted only in labs registered by the Home Office under the 1876 Cruelty to Animals Act and regularly spot-checked by medical inspectors. Every year, hundreds of thousands of animal injections were performed by the MRC, public health authorities, and private companies (under the Therapeutic Substances Act of 1925) in the routine production, testing, and standardization of millions of doses of drugs, sera, and vaccines.[39] These accounted for 95 percent of all licensed animal experiments in Britain and required "Certificate A" (in addition to the license) to forego the use of anesthetics in mice and other species. As antivivisectionists gained public support in the late 1920s, hospital administrators became increasingly wary of losing the voluntary contributions of wealthy patrons and tended to keep animals away from hospital property.[40] For instance, the Middlesex Hospital in London used the animals kept at the Courtauld Institute of Biochemistry next door, and the Royal Infirmary of Edinburgh fostered a cooperative attitude toward offsite laboratories.[41]

The Aschheim–Zondek test, Johnstone later quipped, raised mice to the "rank of obstetrical consultants."[42] The increasing demand for laboratory mice was met in Britain chiefly by the specialist commercial breeder and distributor, A. Tuck & Son's "Mousery" in Rayleigh, Essex.[43] The agricultural correspondent of the *News Chronicle* called Mr. Tuck the "uncrowned king of mice fanciers," and the *Daily Mirror* reported that his "farm" housed 200,000 mice and dispatched up to 3,000 of "all sizes, shapes and colours" daily.[44] Tuck supplied young, female mice for use in Edinburgh, where Crew's staff initially followed Aschheim and Zondek's original technique to the letter (figure 3.2).

Aschheim and Zondek intended the use of multiple test animals to mitigate the variability of individual mice and so increase the sensitivity of their test, which required several days to perform because infant mice would not tolerate an injection of the required amount of extract all at once. Preparing the urine was also time-consuming, but failing to do so often resulted in dead mice before a conclusive result could be obtained. Crew's staff initially sectioned the ovaries and inspected them under a microscope. To further simplify, streamline, and speed up the procedure, they soon abandoned microscopy in favor of naked-eye inspection, which was usually adequate. In borderline cases, an intact ovary could be pressed between coverslips and examined under a hand lens or held up to the light, where small and deeply embedded blood points could usually be distinguished from even the densest yellow bodies without going to the trouble of slicing.[45] For the first three months, Crew and Wiesner tested urine specimens provided by Johnstone and then, satisfied with their results, decided to go postal.

REDESCRIBING ERRORS

A *Lancet* editorial had first mentioned the Aschheim–Zondek reaction in October 1928 as a "specific" new test for the "presence or absence" of early pregnancy. The editorial anticipated the "very great value" of the test, assuming the promising results obtained in Berlin would be confirmed by others.[46] A few months later, the *Lancet* and *BMJ* carried a letter from

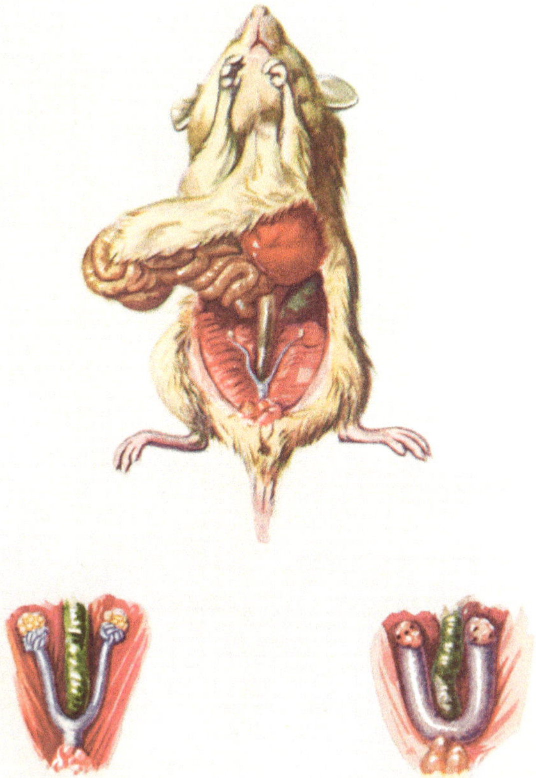

3.2 Colorful illustration of the Aschheim-Zondek reaction from the seventh edition of Johnstone's popular textbook. Johnstone 1934, University of Cambridge, Ant.c.39.137.

Johnstone explaining that by indiscriminately testing any specimen sent by a doctor, Crew and Wiesner would investigate the sensitivity and specificity of the Aschheim–Zondek test. This was said to be trustworthy from two weeks after a missed period, and the only requirements were a few ounces of urine, a cover letter with clinical data, and a postal order for the fee. Results would be returned in about a week.[47] A supportive *BMJ* editorial amplified Johnstone's hope that many doctors would take advantage of the station and endorsed the fees as "very moderate." Laboratories in Germany and other countries were beginning to evaluate the test and to publish their reports in research journals (table 3.1). However, the editorial argued

Table 3.1 Pregnancy tests with location of laboratory in descending order of number of tests; based on Zondek (1931, 315).

Place	Number of tests
Berlin (Aschheim and Zondek and others)	1,200, 200, 109, 12
Frankfurt am Main	1,080
Dresden	413
Edinburgh (Crew)	400
Düsseldorf	249
Göttingen	243
London (Dickens)	207
Cologne	139
New York	132, 100
Marburg	129
Breslau	127
Munich	110
Moscow	100
Italy	91, 36, 30
Prague	79, 30
Kiel	51
Münster	49
Greifswald	46
Würzburg	36
Utrecht	33
Vienna	30
Paris	30
Buenos Aires	24
Total	5,515

that a large-scale trial on unselected material was still needed to confirm the "clinical value" of the test in Britain.[48]

In the six weeks following the publication of Johnstone's letters, the station received around ninety specimens. This was a fair start, but there were some logistical problems, so Crew provided additional guidelines in another letter. Mice had to be purchased and looked after, and some doctors failed to pay up, so he reminded them that the service was not free. Private cases were charged a "modest fee" of five shillings, intended to permit a reduced hospital fee of one shilling and sixpence. The station required 2 ounces of fresh morning urine in a clean bottle enclosed in a sturdy package, accompanied by case notes, especially the date of the patient's last menstrual period, but doctors frequently posted "too much, too little, or too stale urine," often in packages that broke in transit.[49]

The General Post Office, Britain's largest employer in the 1920s, allowed urine and other normally prohibited substances to be sent to any recognized medical institute or qualified practitioner.[50] Diagnostic laboratories typically appointed a medical superintendent to oversee operations, a position filled by Edwin Robertson in Edinburgh. Every year, tens of thousands of packets containing pathological specimens (mostly urine) circulated in the post. Many reached the Clinical Research Association (CRA), a large London-based commercial laboratory that supplied doctors with regulation containers and ready-addressed envelopes or boxes for return.[51] The frequency of broken and spilled packages induced the Postmaster General repeatedly to specify regulations in the *BMJ*.[52] Specimens needed to be securely packed in a strong wooden, leather, or metal case to prevent shifting about and with sufficient absorbent sawdust or cotton wool to prevent leakage. The container had to be conspicuously marked "Pathological specimen—Fragile with care," and any packet found to contravene regulations would be destroyed and the sender liable to prosecution.[53]

Nurtured by the requirements of life insurance companies for urinalysis, the CRA and other commercial labs scaled up diagnostic services to meet an increasing demand from doctors.[54] A pregnancy test cost about as much as a hemoglobin estimation or Wassermann reaction, which ranged from two shillings a test for panel patients and their dependents to ten and six for the well heeled.[55] Specimens that survived the trip to Edinburgh were filtered on arrival by laboratory workers into numbered bottles.[56] Crew's staff then entered the particulars in a special logbook with perforated pages to produce numbered labels for the urine container and mouse cage, record cards for injection and filing, and "result" and "follow-up" letters. No later than six days after receipt of the specimen, a secretary would post the "result" letter to the sender. Two months after that, she would post a reminder letter to find out if the doctor had corroborated or contradicted the laboratory diagnosis by clinical evidence of pregnancy or its absence (figure 3.3).[57]

Other labs had reported a disturbingly large error of up to 5 percent, which provoked debate over the specificity and clinical value of the Aschheim–Zondek reaction. Delegates from the Edinburgh station defended the test in January 1930 at a London meeting of the prestigious Royal Society of Medicine. John Hannan, a registrar at the Soho Hospital for Women, had used rats instead of mice and reported a 7 percent error. He doubted the usefulness of any method that was not "absolutely reliable" and preferred the "old method of seeing the patient in a month's time." Wiesner insisted that the Aschheim–Zondek reaction could only be evaluated fairly if the original unmodified method was tested with "sufficient material collected under clinical conditions." This had been done, he claimed, not in London but in Edinburgh, where the error was a satisfactory 2 percent. But he emphasized that a positive reaction was a "sign of placental activity" only and looked forward to the day when a "chemical test" would be able to detect the "presence of a living foetus." Meanwhile, Wiesner was the first to admit that the Aschheim–Zondek reaction was simply "not a pregnancy test, *sensu strict[o]*."[58]

Influential obstetric surgeon Louis C. Rivett claimed that clinical diagnosis was "easy" in 99 percent of cases and that an expert could usually handle the doubtful 1 percent without recourse to the lab. He had provided biochemist Frank Dickens at the Courtauld Institute with over 200 specimens collected from Queen Mary's Hospital, where East End women competed for limited beds by applying for accommodation at the first sign of pregnancy.[59] Dickens was reasonably satisfied with the reliability of the test, but like Hannan, he discontinued routine testing to free up laboratory animals for more prestigious pituitary research.[60] Arthur Giles, a well-known gynecologist at the Chelsea Hospital for Women, amplified Rivett's criticism about lack of specificity. The test gave positive results for nonpregnant women in a "considerable variety of conditions," and most gynecologists,

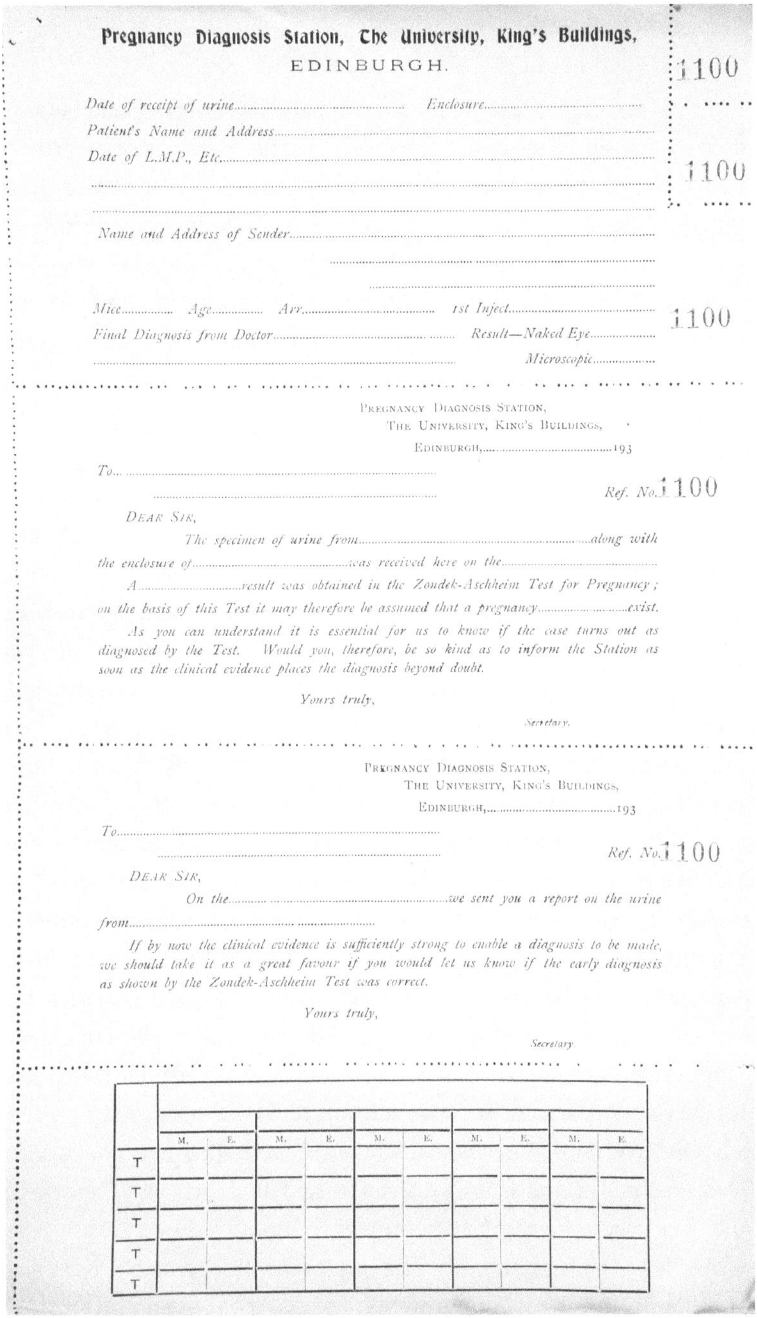

Pregnancy Diagnosis Station, The University, King's Buildings,
EDINBURGH.

1100

Date of receipt of urine.. *Enclosure*..

Patient's Name and Address..

Date of L.M.P., Etc...

1100

Name and Address of Sender..

Mice............... *Age*............... *Arr.*............................ *1st Inject*..

1100

Final Diagnosis from Doctor.. *Result—Naked Eye*....................

Microscopic....................

PREGNANCY DIAGNOSIS STATION,
THE UNIVERSITY, KING'S BUILDINGS,
EDINBURGH,..193

To..

Ref. No. 1100

DEAR SIR,

The specimen of urine from..*along with the enclosure of*..*was received here on the*..

A..............................*result was obtained in the Zondek-Aschheim Test for Pregnancy; on the basis of this Test it may therefore be assumed that a pregnancy*..............................*exist.*

As you can understand it is essential for us to know if the case turns out as diagnosed by the Test. Would you, therefore, be so kind as to inform the Station as soon as the clinical evidence places the diagnosis beyond doubt.

Yours truly,

Secretary.

PREGNANCY DIAGNOSIS STATION,
THE UNIVERSITY, KING'S BUILDINGS,
EDINBURGH,..193

To..

Ref. No. 1100

DEAR SIR,

On the..*we sent you a report on the urine from*..

If by now the clinical evidence is sufficiently strong to enable a diagnosis to be made, we should take it as a great favour if you would let us know if the early diagnosis as shown by the Zondek-Aschheim Test was correct.

Yours truly,

Secretary.

	M.	E.	M.	E.	M.	E.	M.	E.	M.	E.
T										
T										
T										
T										
T										

3.3 The Edinburgh pregnancy diagnosis station printed result, follow-up, and reminder letters. "L.M.P." stands for "last menstrual period" and, though it would become conventional to refer to the Aschheim-Zondek test, in these forms it was referred to as the "Zondek-Aschheim" test. The National Archives FD 1/2816.

he claimed, would probably agree that "for the present they had better trust to their fingers and their senses generally for the diagnosis of pregnancy."[61] He did, however, praise the ability of the test to detect placental cancer.

Rarely, in the early stages of pregnancy, the fingerlike protrusions of the placental membrane (chorionic villi) transform into bunches of grape-like cysts (figure 3.4). As the "hydatidiform mole" grows, the embryo usually dies and is reabsorbed. At first, a "molar pregnancy" looks and feels normal, but then the uterus begins to grow abnormally fast and becomes soft and boggy to touch, with no fetal parts to feel or heartbeat to hear.[62] Before the Aschheim–Zondek test, the only foolproof diagnostic criterion was a discharge containing tiny cysts, resembling "white currants in red currant juice."[63] Once diagnosed, a mole could be manually squeezed out, but any retained bits were liable to develop into a highly malignant trophoblastic cancer known as chorioepithelioma or choriocarcinoma, which could rapidly and fatally spread to the lungs. So, following surgical removal or spontaneous delivery, a patient would be instructed to check in regularly for up to a year or at once if there were any irregular bleeding.

Aschheim had been one of the first to report a positive reaction in a case of chorioepithelioma following the expulsion of a hydatidiform mole.[64] Although chorioepithelioma was rare, cancer specialists nevertheless embraced his test as a significant breakthrough in diagnostics.[65] Early detection and treatment (with some combination of surgery, radium, and chemotherapy) was a cornerstone of the "crusade" against cancer in early twentieth-century Britain.[66] Yet few GPs saw many patients with malignancy, which made early diagnosis a real challenge.[67] Hopeful researchers announced new serological tests for cancer on a regular basis, and by 1930, over twenty serodiagnostic methods had been proposed.[68] "Unfortunately," as Liverpool gynecologist William Blair-Bell lamented, "none had proved specific for malignancy." Even as he "doubted" whether "science" would ever produce a test "so delicate as to indicate the existence of a few cancer cells in the human body," he implored "biochemical investigators" to "not lose sight of the immense importance" that would attach to such a discovery.[69]

Robertson, who had also been at the London meeting, echoed Rivett's hopes for cancer monitoring and control in an address to the Edinburgh Obstetrical Society. One local patient with chest symptoms caused by a

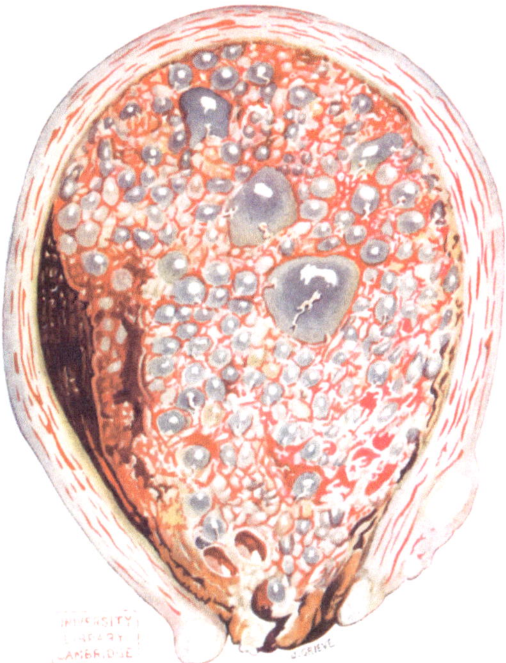

FIG. 1.—Hydatid Mole *in utero*. Shows expanded villi lying in fluid blood. To the left the uterine cavity is seen filled with blood-clot. Small fibroids are seen in the lower part of the muscular wall.

3.4 Cross section of uterus with hydatidiform mole, illustrated by J. Grieve. The uterus was removed from Mrs. C., a previously healthy woman of forty-nine years with a history of miscarriage, by Dr. Haig Ferguson at the Nairn Cottage Hospital in the North of Scotland in December 1912. Ferguson 1913, 261, Cambridge University Library T323.c.12.38.

metastatic mass had tested positive, demonstrating how repeated testing at regular intervals could be used to monitor the results of surgery or other treatment. Leading Edinburgh gynecologists were easily persuaded of its value: Theodore Haultain was having one of his patients tested on a weekly basis after she had delivered a hydatidiform mole, and James Young proposed that interval testing should be made routine in all such cases. The president of the society congratulated Robertson, who "had only to ask" if he needed specimens, "for those who had listened to him and to his facts would be only too glad to help to further the uses of such a test."[70]

Despite this locally warm reception, however, the Edinburgh station incurred a deficit of £135 in its first year and was threatened with closure.

Some doctors had failed to pay up, and dozens of tests had been repeated when batches of mice were killed by toxic urine or the visible changes in their bodies were ambiguous. Retesting with second and third specimens was costly and usually fruitless. An increasing demand suggested that the station was "appreciated by hospitals and practitioners," but this did not necessarily justify its continued existence. Wiesner informed the MRC that the station had met its stated research goal of evaluating the test and that he would need to propose new research aims to justify any continued funding. On the other hand, standards of animal stock had been established and the necessary infrastructure built to support a routine service independent of any research agenda. This relatively well-equipped and smoothly running laboratory was now "ready for use by anybody" willing to uphold the necessary standards.[71]

At this critical juncture, Wiesner was the first to declare that the station could simply be shut down. But he stood by the value of the service and advised its relocation to some other adequately equipped institution, such as the Laboratory of the Royal College of Physicians of Edinburgh (RCPE). Alternatively, he estimated that doubling the fees would cover expenses in a second year of operation. He also expressed an interest in continuing to work with the test and with the surplus urine it brought him. Crew's weak position within the British medical establishment, in an agricultural department far from the great London teaching hospitals, enhanced for him the value of Wiesner's initiative, and in the end, the station remained in Crew's institute, which moved into a new building in March 1930 (figure 3.5). Wiesner promised to tighten his bookkeeping, and the MRC agreed to cover the station for a loss of up to £50 for one year only.

Crew's first annual report announced that fees would be increasing to ten shillings for private cases and three for hospitals (still well within the range of a Wassermann test). This was a winning strategy, and in one year, the station had become financially "self-supporting," even generating a "small balance" to be "carried forward as reserve" (figure 3.6).[72] Crew's report further clarified the potentially misleading use of the word "pregnancy" in communications by the station. A few doctors had complained that a negative result was followed by miscarriage, proving that the patient had been pregnant (with a dead fetus) at the time of the test.[73] Rather than

DEPARTMENT OF ANIMAL GENETICS

3.5 Official photograph of Crew's institute at the King's Buildings Site, viewed from the northeast. "The Department of Animal Genetics," *University of Edinburgh Journal*, Autumn 1930, 35–40, Cambridge University Library L985.b.42.

STATEMENT OF ACCOUNTS OF THE PREGNANCY DIAGNOSIS STATION,

From 1st January, 1930 to 18th February, 1931.

INCOME	£ s. d.	EXPENDITURE	£ s. d.
(1) Balance carried forward from 1929	5 : : 3	(1) Wages and Insurance	174 : 12: 1
(2) Specimen fees	264 : 5: 9	(2) Purchase of Food, Hay, etc.	12 : 5: -
(3) Sale of Mice, etc.	53 : 18:11	(3) Purchase of Mice	91 : 16:11
(4) Repayment by Section E	20 : 13: 4	(4) Telegrams, Postage Telephone	16 : 3: 4
		(5) Bank Charges	1 : 13: 3
		(6) Stationery (including printing of forms)	8 : 16:10
		(7) Equipment & Chemicals	5 : 7: 1
		(8) Sundries	3 : 3:10
		(9) Money lent to Section E	20 : 13: 4
		(10) Repayment of advanced sums	4 : -: -
		(11) Balance in Hand: Cash 4:10:5, Bank 16:2	5 : 6: 7
	£343 : 18: 3		£343 : 18: 3

3.6 Annual income and expenditures of the pregnancy diagnosis station in 1930–1931. The National Archives FD 1/2816.

admit error, Crew creatively reinterpreted "false" negatives as positive indications of a hormonally deficient pregnancy that would probably not go to term.[74] Far from discouraging, such "errors" opened a window of opportunity for Crew and Wiesner, who began to calibrate the test so that laboratory results would match clinical expectations.[75] In addition to the asset of "false positives" in cancer diagnosis, they redescribed "false negatives" as positive predictors of "fetal death" and began to remake the Aschheim–Zondek test into a detector of women who were likely to miscarry.[76]

POSTAL PATHOLOGY

In the late 1920s, the well-connected physician Sir Thomas Horder had lamented the "existence of laboratories in which the personal element as between doctor and pathologist is quite eliminated," even as he admitted that they were "necessary" and had "come to stay."[77] A decade later, the *Practitioner* generally recommended working with a local pathologist, rather than relying on a remote laboratory, a practice later derided as "mail order" or "postal" pathology.[78] Despite the distance, its many southern clients generally welcomed the Aschheim–Zondek reaction and the Edinburgh station. This was a significant achievement at a time when some diagnostic tests were renowned for their "great reliability" and others "definitely black-listed."[79] The procedure for collecting a specimen was lauded as the "simplest imaginable" (it did not require a catheter as with urine for bacteriological tests), and the manageable error was "easily guarded against by ordinary clinical observation."[80] One article in the *Clinical Journal* recommended London hospitals for pregnancy testing, but Crew's service was usually singled out.[81] Although Liverpool gynecologist Arthur Gemmell cautioned that the station was not "always accurate" (he had received two incorrect results), he did not reject the test but instead recalled that it was "not a test for pregnancy, but for the presence of living chorion, and that its reported result must be carefully considered in connexion with the clinical findings."[82]

As we have seen, a few elite gynecologists trusted their own senses more than a test that gave the wrong answer in one of every fifty or even twenty cases. But there was no consensus on the error, which varied by laboratory, and Crew and Wiesner were creatively redefining mistakes to convert the liability of nonspecificity into the advantage of versatility. Furthermore, family doctors had their own reasons for preferring a postal service to the delicacies of pelvic examination. A note in the *Lancet* in 1930 recommended the Aschheim–Zondek test as "sufficiently reliable for all clinical purposes" and for the "further advantage that in delicate circumstances it can be done without the knowledge of the patient or her friends." The note predicted that, although the "technique needs practice," it was "likely to be acquired by clinical pathologists" now that its "value" had been "confirmed." "The family doctor," it concluded, will be "grateful for the simplicity of his share, which consists only in collecting morning urine from the patient and possibly adding a drop of tricresol as a preservative."[83]

For the ordinary family doctor, pelvic examination was complicated by the ever-present possibility of normal pregnancy, which generally needed to be confirmed or excluded. Light bleeding, however, could complicate a diagnosis, and the presence of fibroids challenged even the "most erudite."[84] As discussed in the previous chapter, the most important clinical method of early pregnancy diagnosis involved the bimanual palpation of the uterus, but as a somewhat later commentator made explicit, "attempts to elicit Hegar's sign [could] be as effective in terminating a pregnancy as the abortionist's curette."[85] Perhaps even more important, a mutual feeling of "delicacy and sensitiveness" between a patient and her doctor strongly discouraged the practice of pelvic examination unless absolutely necessary.[86]

Textbooks began providing practical instructions on how to collect and post a urine specimen for pregnancy diagnosis. The second edition of Haultain and Fahmy's *Ante-natal Care* claimed that the Aschheim–Zondek test could be performed only "in a laboratory, by expert observers," and specifically mentioned Edinburgh.[87] The sixth edition of Johnstone's textbook instructed doctors to post specimens, a brief history, and ten shillings to the "Pregnancy Diagnosis Station, University—King's Buildings, Edinburgh."[88] The fourth edition of Blair-Bell's *Principles of Gynaecology* enthusiastically proclaimed that the Aschheim–Zondek test had "revolutionized" pregnancy diagnosis.[89] And Aleck Bourne's *Midwifery for Nurses*, recommended as a study guide for the Central Midwives Board examination, suggested posting urine to Edinburgh "with the name and age of the woman, the date of dispatch, date of her last menstruation, and a postal order for 10*s*."[90]

As with X-rays and the Wassermann test in mass screening, the cost of performing an Aschheim–Zondek test decreased as demand increased.[91] But some critics objected to the organization of pregnancy testing in Britain. In his public speech at the opening of Crew's institute in 1930, Sir Edward Sharpey-Schafer complained that the resources of a research institute "should not be diverted to a routine method of diagnosis which might as well be done anywhere else," an objection that was repeated in the *Scotsman* under the subheading "Certificate for a mouse."[92] Crew's institute was licensed for vivisection, but pregnancy testing as such was not specifically addressed by the Home Office until 1932, when an inspector advised a doctor to obtain a license and Certificate A, setting a precedent for subsequent would-be pregnancy testers.[93]

Even as the *BMJ* complained that doctors were forced to rely on "special centres" that concentrated and maintained "large stocks of mice" and "skilled service," it doubted that pregnancy testing would ever become practical as a side-room technique. So the search continued for the "ideal test," one that was not "unpleasant to patient or physician, but simple, capable of being used by the geographically isolated general practitioner, cheap in time and money, and, of course, reliable."[94] Researchers at London hospitals and Crew's student Cecil Voge in Edinburgh investigated cheap, quick, and simple biochemical reactions, but after hundreds of tests on surplus pregnancy urine, they were forced to admit that infant mice beat their in vitro tests.[95] Others experimented with adult mice and (male and female) rats, but the next major breakthrough came in 1931, when researchers in Philadelphia announced a new rabbit test.[96]

The "Friedman test" used one or two large, female adult rabbits instead of a batch of five tiny, immature mice. Because rabbits only ovulate immediately after mating (or when one doe "jumps" another), an isolated animal with a known history could be used at any time without fear of a false positive from spontaneous ovulation. Rabbits, like mice, had to be sacrificed but were comparatively easy to handle and inject in the ear vein, an already standard procedure in bacteriological testing and vaccine production. They could also tolerate larger doses of urine and soon became the pregnancy test animal of choice in American laboratories.[97] Compared to mice, housing rabbits individually in cages (to prevent ovulation) was expensive and required more space, but Friedman's test dramatically reduced the waiting time for a result from five days to twenty-four hours, offering doctors a more flexible service in urgent cases.

The Edinburgh station soon experimented with the Friedman test, charging one pound, ten shillings to private doctors and one pound to hospitals to cover the higher cost of rabbits and telegraphic communication of the results.[98] Contrary to Crew's expectations, demand for Friedman testing in Edinburgh remained low, mainly because it was expensive and because large teaching hospitals in London and other cities managed to establish facilities of their own.[99] Crucially, the use of rabbits facilitated the establishment of local alternatives to Crew's remote (for clients outside Edinburgh) service.

By 1935, most London teaching hospitals were equipped for the Friedman test. Ronald Kelson Ford's *Short Ante-natal and Post-natal Handbook* called it the "more generally used" pregnancy test in Britain, and the *BMJ* claimed it was "well established in clinical midwifery practices."[100] A pathologist at St. Thomas's Hospital praised the "much simpler" Friedman test, reporting over seven hundred reactions in 1936.[101] Unlike "delicate to handle" and "difficult to obtain" mice, rabbits were "much more satisfactory" to work with at St. John's Hospital, Lewisham. There, a specially constructed box was used to bunch up the rabbit's back and prevent it from kicking at one end while holding its neck between two boards ("after the manner of an old-fashioned pillory") at the other (figure 3.7).[102]

Peter Bishop, a clinical endocrinologist at Guy's, modified the Friedman test by introducing a delicate surgical procedure to identify spontaneous ovarian blood points that otherwise might have led to a misdiagnosis.[103] This involved operating on each rabbit before and after every test. Bishop's modified technique was considered impractical in Edinburgh, where Friedman's test was combined with a confirmatory Aschheim–Zondek, a control that required "much less surgical skill."[104] The Edinburgh station had been made for mice, which were more convenient to house on a large scale. Rabbits, in contrast, were locally expensive, "difficult to breed, to procure, and to accumulate in large numbers."[105] In Crew's words, different tests

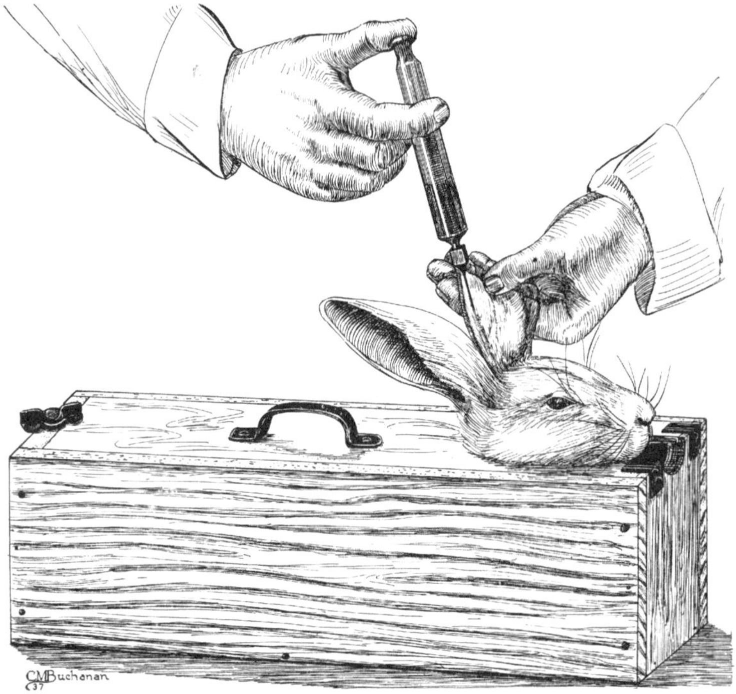

FIG. 198. Injecting the marginal ear vein of a rabbit with urine in performing a Friedman test for pregnancy. Note the style of wooden box which restrains the animal in a satisfactory way.

3.7 Line drawing of a rabbit injection with restraining box from an American textbook of clinical pathology. Kracke 1938, 513, Cambridge University Library 309:2.c.90.4.

were "equally satisfactory in the hands of different people."[106] When it came to pregnancy testing (and diagnostic tests more generally), each lab implemented its own protocols, locally adapted to suit particular needs and constraints.

CALIBRATING MICE

Even as Johnstone claimed that the station was "not a commercial undertaking" and that it served the "interest of the [medical] profession and of science," Wiesner's research program had become marginalized within Crew's institute and was finally shut down in 1934.[107] Crew had come under increasing government pressure to use national funds for work with farm animals only, and the economic depression dried up Macaulay's money.[108] The new financial situation strained Crew's relationship with Wiesner, whose work on sex hormones had embarrassingly led to the development of a placenta-based drug by their chief competitor, the Montreal biochemist James B. Collip.[109] Crew later recalled that Wiesner's research on the maternal behavior of rats, which had little relevance to "either animal genetics or animal breeding," was "getting out of hand," and so Crew was not "unhappy to see it come to an end."[110]

Wiesner moved to London to set up an infertility clinic with his second wife Mary Barton.[111] Artificial insemination by donor was becoming more widely used in British clinics as a medical fix for male infertility in married couples, and Wiesner integrated the Aschheim–Zondek reaction (as an early pregnancy test) into infertility diagnosis and treatment regimens.[112] He also circularized clients of the Edinburgh station to inform them that he was taking it with him to London. Crew responded in the *BMJ* that testing would not stop just because Wiesner was leaving. The station was larger than the "personal activities of one man" and would continue under the supervision of Wiesner's assistant, John M. Robson.

Crew put on a show of confidence, but scaling up had made the station vulnerable to competition. Thousands of tests had to be made annually to cover the running costs. Solvency thus depended on custom from London and southeast England. To keep serving Scotland, Crew would have to serve England as well, and he was unwilling to give up that lucrative share of his market without a fight. Crew admitted that if endocrinology were a more advanced science, "there would of course be room for more diagnostic laboratories." But for now, he claimed, a centralized, noncommercial service was needed to produce knowledge about the "unusual" and "exceptional" cases that would someday lead to breakthroughs in hormone therapy.[113]

By 1936, the Aschheim–Zondek test was "becoming one of the everyday tools of the practitioner."[114] The third edition of *Recent Advances in Endocrinology* called it "probably the most accurate biological test known."[115] A handbook for GPs on the early diagnosis of cancer claimed it was "so reliable that a positive result must be accepted as proof of the presence of chorion epithelioma."[116] Even the previously skeptical Hannan had begun to recommend fortnightly testing in "all cases where the histological picture is suggestive of chorion carcinoma."[117] Crew declared that the "widespread demand" for pregnancy diagnosis had been "successfully met" and predicted that "as their usefulness [became] more generally known," the number of tests performed every year would continue to increase.[118] Competition had intensified, but so too had demand.

Meanwhile, several laboratory workers and clinicians in Britain, dissatisfied with the impracticalities of mice and rabbits, tried out the new methods that continued to be reported in American and German journals (table 3.2). In 1936, Gladys Dodds, a physician at University College Hospital and Clapham Maternity Hospital, investigated the Visscher–Bowman test of Cleveland, Ohio, but concluded that too many false positives rendered it worthless.[119] Jocelyn Patterson, a biochemist at Charing Cross Hospital, compared the Schmulovitz–Wylie (estriol) test of Baltimore, Maryland, to rabbits. Although reliable, it was tedious and labor intensive, and Patterson did not expect it to become routine.[120] In 1937, a *BMJ* editorial cautioned that the perfect record claimed for a new spectroscopic test by its German inventor was too good to be true.[121] Drs. Alan Morton Gill and John Howkins of Middlesex Hospital condemned an intradermal sensitivity test for pregnancy test using "Antuitrin S," a commercial gonadotropic product marketed by Parke, Davis & Co., as "valueless" (figure 3.8).[122]

Albert Sharman, a gynecologist and fertility specialist at the Royal Samaritan Hospital for Women in Glasgow, and Nora Keevil of the Royal Free Hospital in Hampstead independently tested the Antuitrin S test before abandoning it as unreliable.[123] Scientists in Europe and Japan, where the carp-like bitterling was plentiful, had for some years induced color change in the male and ovipositor extension in the female to assay hormone preparations when three Chicago doctors proposed the fish as a pregnancy test animal (figure 3.9).[124] In Britain, Stanley Way of the General Lying-in Hospital in Portsmouth initially had "great success" with the bitterling (fish) test until a batch he purchased from a dealer was already "in oestrus" and "giving positive results" even before they could be put to use. Although the bitterling test caused some excitement in the United States, Way found the fish to be a "great nuisance to look after." He also found the Antuitrin S test to be "hopelessly inaccurate" and abandoned it after about just twenty-five

Table 3.2 Number of tests by kind of test, hospital, and published source, 1930–1938

Number	Test	Hospital	Source
700+	Friedman	Guy's	Bishop (1934)
700+	Friedman	St. Thomas's	Bamforth (1936)
395	Friedman	Guy's	Bishop (1933)
380	Friedman	University College	Dodds (1930)
265	Biochemical	St. Bartholomew's	Hannan (1930)
237	Biochemical	Queen Mary's	Allan and Dickens (1930)
234	A-Z	Royal Free	Keevil (1937)
180	Biochemical	University College	Dodds (1936)
147	Intradermal	Middlesex	Gill and Howkins (1937)
98	Friedman	Soho	Hannan (1930)
65	Biochemical	Charing Cross	Patterson (1937)
53	Friedman	University College	Dodds (1931)
50	A-Z	St. Bartholomew's	Brewer (1934)
25	Intradermal	Portsmouth	Way (1937)
?	Friedman	St. John's	Ralph (1934)

attempts.[125] Whether they liked it or not, pregnancy testers continued to rely on mice and rabbits.

The unique selling point of Crew's station over competitors was the degree to which laboratory workers calibrated test mice to produce a graded series of reactions from strongly positive to unequivocally negative via standard, weak, and extremely weak positive. Graded results produced information beyond the "existence or non-existence of normal pregnancy," for instance, by revealing the hormonal deficiency in early pregnancy thought to be a predictor of miscarriage. They could also "distinguish between true pregnancy and the endocrine repercussions of abnormal emotional states, and between pregnancy and menopausal conditions," as well as track the "stages of recrudescence of chorion epithelioma and hydatidiform mole."[126]

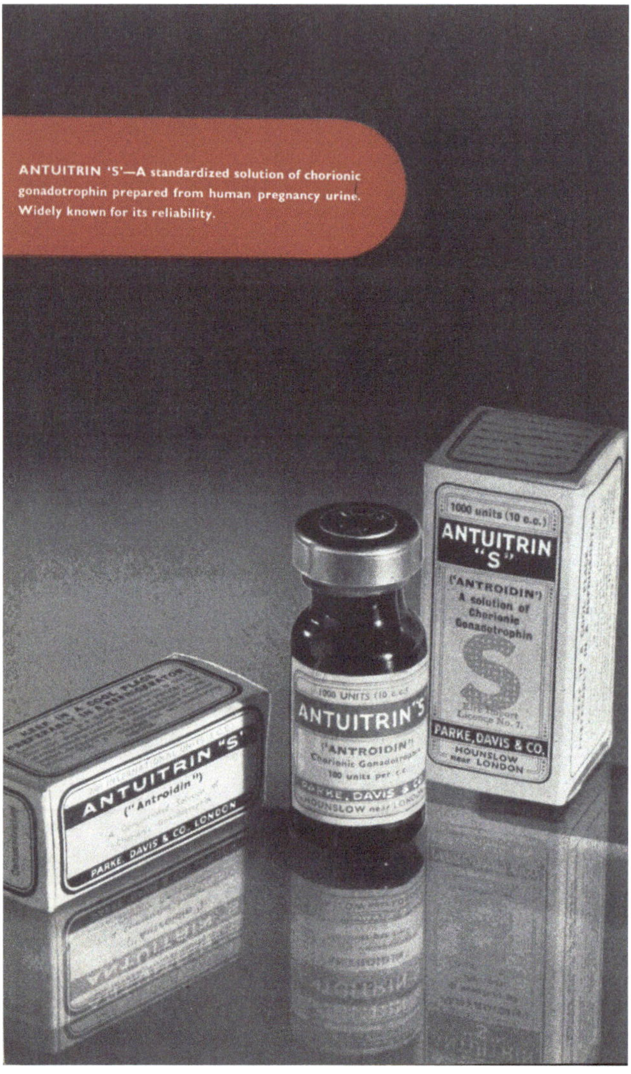

3.8 Advertisement for Antuitrin S, the standardized hCG product made from pregnancy urine and marketed in Britain by the Hounslow-branch of the American company Parke, Davis & Co. On the company: Deeson 1995. *Descriptive Catalogue and Therapeutic Index* (Hounslow: Parke, Davis & Co. c.1949), Wellcome Collection, London, M4774.

3.9 Specimen jar containing three female Japanese bitterlings for pregnancy testing, c.1930s, demonstrating "positive" (top), "negative" (middle), and partially dissected "normal" (bottom) ovipositors. See McEnroe and Hurley 2019, 50–51. In the US, an imaginative *Time* magazine reader predicted that "every standard American home" would soon be "equipped with an aquarium containing a female bitterling that would be as handy as the radio, the vacuum cleaner, the bottle of antiseptic, etc., in maintaining the even tenor of existence, especially in times when the budget of most households does not permit haphazard payments to obstetricians": "Bitterling possibilities," *Letters*, Feb. 18, 1935, 1–2. Science Museum, London: Object 1994–1009 (85 × 85 × 35 mm).

In the case of a suspected placental mole or malignancy, the station also offered special dilution tests. For example, an Edinburgh lab report sent to Alan Brews, a leading gynecologist at the London Hospital, stated, "We have examined the specimen of urine and have found that the concentration of gonadotrophic hormone is very high, dilutions of 1 in 200, giving positive reactions when the normal doses are employed. The result supports your diagnosis of chorion-epithelioma."[127] Others complained that in their hands, the test was "capricious," but for Brews, it was a valued diagnostic aid in cases of hydatidiform mole and "as a means of excluding the subsequent growth of a chorion-carcinoma."[128] By 1939, he had used the Aschheim–Zondek test in six cases, "where no part of the mole had escaped from the uterus; in 5 a positive reaction was obtained in a dilution of 1/200 (in 1 case up to 1/800) and in the remaining case a negative reaction was obtained in undiluted urine."[129]

The number of urine specimens sent to Edinburgh for pregnancy testing increased from around 840 in 1929 to over 10,000 in 1939 (figure 3.10). About half the demand came from private cases, the other from hospitals. About half were for nonpregnant women (negative results), many of whom were near menopause. The other half tested positive. Although I have found no records that further break down this demand quantitatively, it is possible to put together a qualitative picture from published reports. Doctors called on the station when patients were unmarried, when obesity or vaginismus impeded ordinary physical examination, in cases of unusual amenorrhea or vomiting, if fetal death was suspected, and when differential diagnosis was difficult—for instance, between ordinary pregnancy and an abdominal tumor, ectopic pregnancy, pseudocyesis (phantom pregnancy), or fibroids. They also requested tests when therapeutic abortion was indicated in expectant mothers with tuberculosis or toxemia (preeclampsia) and, occasionally, in medicolegal circumstances—to establish or exclude pregnancy in cases of criminal abortion, rape, or divorce. Sometimes a doctor requested a test for allegedly domestic reasons as when a woman was planning to "accompany her husband" to the tropics but would stay home instead if she happened to be pregnant.[130] For those who could afford it, testing was used to calibrate the expensive hormone treatment of infertility.[131]

PREGNANT WITH PROMISE

The Edinburgh station "quite commonly" received brilliant green urine specimens posted by doctors that were lethally toxic to mice, which Crew attributed to "single women" trying to "avoid pregnancy" by chemical means.[132] By the end of the decade, the station received and refused to test five or six urine specimens every week from women "who send it in themselves, or chemists, or men."[133] These two or three hundred rogue

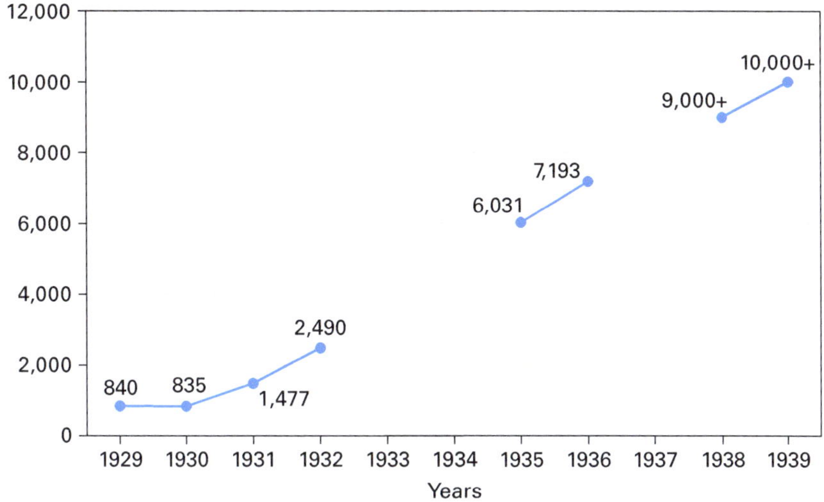

3.10 Annual number of tests performed by the pregnancy diagnosis station, Edinburgh, 1929–1939. Based on published reports: Crew 1930, 1936a, 1937a, 1937b; Wiesner 1931, 1932, 1933; for comparison, the total number of diagnostic tests of all kinds performed by the older Laboratory of the RCPE was 14,798 in 1929 and 16,714 in 1939: Ritchie 1953, 154.

specimens per year suggest that at least a minority of women had learned of the station, despite the evident lack of publicity. Crew rejected this demand and continued to deal exclusively with the medical profession in order to maintain the respectability of his diagnostic service.[134] In practice, however, women who knew about the service and could afford to reimburse a sympathetic doctor could order a test for any reason whatsoever: the Edinburgh service was "unrestricted" in this sense and "never made a distinction between the medical and social reasons for doing a test."[135]

When the British Congress of Obstetrics and Gynaecology convened in Edinburgh in April 1939, Crew boasted that the large volume of urine handled by his laboratory was a "measure of the quality of the service that pregnancy diagnosis offers to the clinician, great numbers of whom regarded it as an essential item of their diagnostic equipment."[136] With a view toward further expansion, Regina Kapeller-Adler, a refugee biochemist from Vienna who had recently joined Crew's team and one of the few female research scientists in this history, was working on a promising new histidine reaction (figure 3.11).[137] Demand had increased to the point that

3.11 Photograph of Regina Kapeller-Adler by Eric Karling, a Swedish pharmacist and visiting associate studying her pregnancy test, probably taken in the clinical laboratory of the Sanatorium Hera in Vienna, c.1937; courtesy of Liselotte Adler-Kastner.

Crew confidently recommended the creation of new facilities in London, Leeds, Manchester, Glasgow, Dublin, and Belfast. In addition to providing routine diagnostic services, these laboratories could also actively research new tests for sex hormones. The future was, Crew punned, "pregnant with the promise of great discoveries."[138]

4

KNOWLEDGE AND IGNORANCE

Pregnancy testing, as Edinburgh geneticist Hugh Donald later recalled, was a "thoroughly unmentionable subject" in the 1930s.[1] In her March 1931 address on "Birth Control and the Right to Abortion" to the Women's Ethical Union, socialist feminist Stella Browne questioned why knowledge of a "modern biochemical technique [that] could establish the fact of impregnation at an extremely early date [was] kept from women who needed it."[2] Browne had particular reasons for promoting the pregnancy test. A founding member of the Abortion Law Reform Association (ALRA), she insisted that abortion was a woman's "absolute right" and should be available on request, "ideally in the first three months of pregnancy."[3] Her address, more broadly publicized in the Malthusian League's monthly journal, *New Generation*, argued that the Aschheim–Zondek test could potentially avert the "weeks of anguished uncertainty so many women endured."[4] Yet, statements by historians about the absence or lack of pregnancy tests in the 1930s, although technically incorrect, attest to their general invisibility and inaccessibility in this decade.[5]

Explicit discussion of sex, pregnancy, and contraception, which had been taboo before the first publication of Marie Stopes's *Married Love* in 1918, became increasingly accessible in the interwar years.[6] Yet, as historian Kate Fisher has persuasively argued, ignorance remained an "important identity for many; women in particular sought to preserve and maintain a state of naivety in defiance of the spread of information. Ignorance implied moral purity, innocence and respectability."[7] This is why the women she and historian Simon Szreter interviewed were able to insist "they were ignorant, while at the same time presenting details of the information they did obtain."[8]

This chapter asks what women knew and didn't know about pregnancy testing in the 1930s. It investigates the available information within a public culture of sexual ignorance and silence, on the one hand, and an expanding market for frank and biologically detailed knowledge about pregnancy, on the other.[9] Picking up on the discussion of advice manuals in chapter 2, I examine women's diagnostic resources (books and magazines) up to the end of World War II. As we shall see, practical information about self-diagnosis became more widely available in more formal and commercial forms after the invention of the Aschheim–Zondek test. I first argue that maternity experts communicated pregnancy testing to married women not to encourage reliance on the laboratory but to discourage it as expensive and unnecessary. I then trace and explain the increasing visibility and acceptability of early pregnancy diagnosis in terms of the interwar debate over abortion law reform and the emphasis of wartime propaganda on patriotic motherhood.

DREADFULLY WELL

Visibly pregnant women were hidden from public view in the 1930s, and the word "pregnancy" was not mentioned in polite conversation. Popular euphemisms included "making bread," "a bun in the oven," "a kick in the back," "fallen," "carrying," "in the family way," "clicked," "caught," "missed," "like that," "done up," "up the stick," and "up the spout."[10] Pregnancy was a "sackable offence," and some factories routinely sent married employees "up to the surgery to be examined."[11] A Lincoln woman recalled, "When you were late, you went to the doctor. They examined you inside; it wasn't nice. You always felt embarrassed."[12] Girls sometimes complained to their mothers of "morning sickness, not realizing it was a symptom of pregnancy."[13]

Fisher asked the women she interviewed about the "process of working out that one was pregnant" but received "very little 'narrative' as a result."

She speculates that, before World War II, many women initially "suspected they might have been pregnant and then such suspicions grew, gradually becoming more of a certainty by the time they were 2 or 3 months into their pregnancy," perhaps earlier for those with serious morning sickness.[14] Memoirs don't provide much narrative either. Vera Brittain coyly acknowledges the "malaises of pregnancy" and Elizabeth Longford's jokes: "I had morning and evening sickness, and the smell of our new distemper at Stairways made me feel sick in the middle of the day too."[15]

Novels unsurprisingly provide more narrative. A particularly rich source is Rosamond Lehmann's controversial novel, *The Weather in the Streets* (1936). Olivia Curtis, the déclassé protagonist who will have a backstreet abortion, begins to worry as soon as she is a few days late:

> I was happy . . . till I got worried. Even after that of course; because, of course, there's no need to worry. Six, seven days late . . . I'm worried. But it's happened once before, the first year Ivor and I were married; over a week then, I was beginning to be sure—but it was a false alarm. . . . That was in August too—so I expect it's the time of year, I'm sure I've heard it does happen sometimes; or all that long cold bathing, lake water's very cold, that might easily account for it . . . I'm worried. Falling for one, Mrs. Banks calls it. "When I fell for our Doris . . ." I feel a bit sick. Train-sick, I expect. I've never been train-sick in my life. This morning when I got up, suddenly retching as I began to wash. . . . Nerves. Lying down like this I feel fine. Be all right tomorrow. Sleep. Thank God for lying down, a sleeper to myself. Supposing I'm sick when I get up to-morrow. . . . That would clinch it. No, it wouldn't. A long journey like this often upsets people.[16]

Olivia's pregnancy wasn't planned, and her reaction to a missed period and nausea is profoundly ambivalent. She recognizes the telltale signs for what they are and rationalizes them away. A *Times* review that mentions the abortion described the novel not as provocative but as "completely typical of the day."[17]

Most fiction published between the wars was produced and consumed by women. Middle-class novelists wrote for similarly leisured women, for whom novel reading was a major pastime.[18] The commercialization of publishing during the economic depression transformed novels from luxury goods into cheap commodities, and mass unemployment encouraged escapist reading, which was cheaper than theater or cinema.[19] Located in most large towns, Boots and W. H. Smith provided a cheaper alternative to exclusive London libraries and the *Times* Book Club. By the mid-1930s,

Boots had become the largest circulating library in Britain with over 400 branches and 500,000 subscribers; 35 million books were exchanged among Boots branches in 1939.[20]

The middlebrow novels that dominated publishing in and beyond the interwar years often drew on personal experience to transgressively narrate women's hitherto private reproductive lives with a new candor, sometimes provoking the ire of critics.[21] For example, in Rose Macaulay's domestic satire, *Crewe Train* (1926), recently married Denham is motorcycling in Buckinghamshire when she suddenly feels "sick and faint." The next day, she "gloomily" confides to novelist husband Arnold the "horrid suspicion" of a "baby coming on." Unhappy and bored with the idea of motherhood, Denham not only hopes it will "pass off" but also exerts herself (through tennis, rowing, and more motorcycling) in a deliberate and ultimately successful plan to miscarry.[22] But it was Macaulay's "extreme frankness" about the banal "discomfort" routinely experienced by countless women "in the good cause" that literary critic Patrick Braybrooke regarded as ill served by the "publicity of cold print." He incorrectly predicted that her "type" of "modern woman writer" would "die out."[23]

Instead, intimate narratives of morning sickness, the solitary or shared anxiety caused by a late period, and successful or failed attempts to restore menstruation proliferated and became more explicit through the 1930s. Characters were often portrayed as understanding all too well the early signs of pregnancy, even as they wishfully hoped it was the weather, illness, or indigestion—self-knowledge that did not deter them from taking measures to evade maternity. For instance, in F. Tennyson Jesse's *A Pin to See the Peepshow* (1934), Julia takes "every kind of patent medicine that urged married ladies to end irregularities and delays now," but to no avail.[24] Likewise, in *Jew Boy* (1935), Simon Blumenfeld's deliberately provocative East End novel, Olive takes hot plunges in public baths, runs up and down stairs, and doses herself with Epsom salts before turning to an older married friend to help "get rid of it somehow!"[25]

In addition to gritty realism, novelists drew on the keyword "anxiety," an increasingly prevalent signifier of all manner of malaises, to embed reproductive worries in narratives of middle-class domesticity.[26] Naomi Mitchison's controversial novel, *We Have Been Warned* (1935), establishes anxious days of waiting as an expected feature of married life: "Like most of her married friends and contemporaries, Dione would occasionally have two or three days of anxiety, sometimes acute and very oppressive." Significantly, she does not portray this experience as exclusively the domain of women. Dione's husband Tom is "nearly as well aware of her times as she was" and "would usually share her anxiety towards the end, but not so immediately or continuously." This time, however, "it was five days late." A "large dose of castor oil" had produced only "vague pains," and she was "feeling dreadfully well again." Tom suggests a sudden turn in the weather may be to blame, but "more to cheer her up than because he thought so." A mother of four, Dione's wishful denial collapses when she begins to "feel absolutely definite signs of *malaise*, a dislike of certain foods, all the old things." Only then does she admit to herself and to Tom that the "little wretch has taken root in me and it's so tough that it won't move for quinine and stuff. It's got the will to live." Determined not to have another child, she plans an abortion in Paris and meanwhile busies herself with "heavy digging in the garden," hoping to "bring it on."[27]

These and many other fictionalized accounts support Kate Fisher's argument, based on oral testimonies, that despite euphemistic language shrouding the womb, many women "took abortifacients when it was clear that they were actually pregnant and they did not ignore the realities of what they were doing."[28] They also constitute confirmatory evidence that most people, even well-educated and well-connected women, probably didn't know about the Aschheim–Zondek reaction. Literary pregnancies in the 1930s were not confirmed by a positive test result but by some combination of a missed period, morning sickness, and the failure of abortifacients to take effect.

IS THERE A BABY ON THE WAY?

As more women took on traditionally male roles, newspaper articles, many written by female journalists, increasingly discussed the feminization of modern society and public culture. The new image of woman as citizen and mother, wage earner and wife, reflected the increased presence of (unmarried) women in the workforce, suffrage reforms, ascendant consumer culture, and the mainstreaming of psychoanalysis and sexology.[29] Women's magazines traditionally focused on society and fashion, but in the interwar years, they expanded their remit to embrace practical domestic advice.[30] The Hearst Corporation launched Britain's first "service" magazine in March 1922. Priced at one shilling, the British version of *Good Housekeeping* offered practical domestic and consumer advice to middle-class households with an annual budget of at least £1,000. It soon attained a circulation of 150,000, and a single (monthly) issue might contain up to one hundred pages of advertisements. Rival houses targeted families earning less than £500 a year by launching down-market versions for sixpence an issue.[31]

The industry changed significantly in 1932, when Newnes launched *Woman's Own*, the first mass-circulation weekly women's magazine. Priced at twopence, it established a winning formula that was soon emulated by Amalgamated Press's *Woman's Illustrated* (1936) and Odhams's *Woman* (1937).[32] In 1937, Odhams first used photogravure to rapidly produce large runs of *Woman*, a full-color illustrated magazine priced to compete with the black-and-white letterpress weeklies. A game changer, *Woman* made competitors look drab and captured a mass market almost on a par with newspapers. By the end of the decade, there were over fifty titles, several of them printed in runs of hundreds of thousands. The relatively niche *Mother* (1936) sold 115,000 every month by 1939 and *Woman*, the leader of the pack, over one million copies weekly in 1940.[33]

Woman's Own soon replaced its "unscientific" childcare advice page, written by the "Mumsie," with the "more professional-sounding Nurse Vincent" and in 1934 launched Nurse Vincent's "Baby Circle": for a shilling, members were sent two exclusive booklets promoting motherhood and infant welfare centers.[34] Women from diverse backgrounds purchased and read these magazines, especially the problem pages, not only for diversion and entertainment but also as a valuable source of information that was unavailable elsewhere. Although largely silent on sex and contraception, maternity experts writing in the new women's magazines often supplied detailed information on early pregnancy diagnosis in response to letters attributed to readers who inquired directly about the "first signs of motherhood."[35] Whether genuine, edited, or fabricated, these exchanges were well placed to mediate how women experienced and interpreted the uncertain and ambiguous physical changes in their own potentially pregnant bodies (figure 4.1).[36]

(a)

WOMAN'S OWN October 15, 1932

A HELPFUL PAGE CONDUCTED BY "MUMSIE" WHO GIVES YOU HER PERSONAL EXPERIENCE. WRITE TO HER IF YOU WANT ADVICE

A Baby is Coming to Town

"MUMSIE" is the wife of a well-known Children's Doctor—herself the Mother of two bonny Babes. Write to her for advice about the care of your own little ones. You can rely upon her wise and kindly help

(b)

WOMAN'S OWN November 11, 1933

QUESTIONS You Ask Our Doctor

Consult "Woman's Own" Medical Adviser about the family's health. He is anxious to help you all—from baby to grandpa! Send your letters to Doctor, c/o "Woman's Own," 18 Henrietta Street, Covent Garden, London, W.C.2, enclosing a stamped addressed envelope for a quick reply.

Worried About Baby

Afraid of Childbirth

A Difficult Age

A Case of Rickets

NEXT WEEK "Winter Colds in Children"

SHE MARRIED FOR ADVENTURE

4.1 (a) The iconic stork remained symbolic of pregnancy for "Mumsie." *Woman's Own*, Oct. 15, 1932, 30. (b) Anonymous doctor reading letters sent to "Questions You Ask Our Doctor." *Woman's Own*, Nov. 11, 1933, 138. British Library LOU.LON 432 and LOU. LON 525, licensed by Future Publishing Ltd.

Mother's maternity expert Nurse Crawford cautioned that it was "not always possible to be certain that Baby is on his way until there have been two months without a period, and by that time there should be other signs of pregnancy."[37] However, even for a knowledgeable woman attuned to her menstrual cycle, disruptions caused by anemia, breastfeeding, or menopause could confound self-diagnosis. A middle-aged Hull mother whose youngest child was sixteen confided to *Woman's Own* that her own "irregular menstruation" caused her to suspect that she "might again be pregnant." The magazine's medical adviser reassured her that she was not pregnant and instead recommended hormone tablets for menopausal symptoms.[38] One request for information about the "first signs of pregnancy" came from a woman whose legs had been "rather puffy" for a couple of days.[39]

A breastfeeding woman wanted to know if she was pregnant again so that she could start weaning her firstborn but was "not quite sure" how to "tell for certain."[40] Another anxiously breastfeeding woman asked, "When do periods normally return?" Nurse Crawford explained that conception could happen at any time unless reliable contraceptives were used.[41]

Bleeding in pregnancy or while breastfeeding sometimes caused anxiety. One reader who suspected pregnancy despite the persistence of menstruation asked whether "any other signs" could confirm her suspicion. Nurse Crawford explained menstruation in pregnancy as "nature's warning that [a pregnant woman's] muscles are not as strong as they should be" and recommended medical treatment to prevent miscarriage.[42] "Is there any way in which I can tell if I am pregnant?" asked a mother who had menstruated twice since the birth of her baby Pamela but not after that. She had no other signs of pregnancy but intended to wean Pamela if another baby was on the way.[43] This time, Nurse Crawford suggested that her "periods" might have been residual bleeding from childbirth and advised her to see a doctor "if only to ease your mind."[44]

Advice columns frequently reminded women that it was not "fussy" to see a doctor in the "months of waiting," a pattern that suggests many expectant mothers shared this concern. *Woman*'s "Questions Mothers Ask" page advised one reader who was two months overdue to visit her doctor even if she was feeling "quite well." Nurse Agnes Patterson testified to the social awkwardness of pregnancy diagnosis when she observed in *Woman's Illustrated* that many young wives were "probably too shy to discuss the matter with a friend, and will not consult a doctor until more proof is forthcoming." She advised the expectant mother to see a doctor or midwife after two missed periods, not because she needed "special care" but rather to establish a friendly rapport with the professional who would eventually attend to her confinement (figure 4.2).[45] This was not "fussy" but "simply a matter of routine."[46]

Although quickening occurred relatively late in pregnancy and so was not considered much use as an early sign, its absence could be a source of anxiety. *Woman's Illustrated* published a letter from a "very worried" woman whose baby was due in four months but had "not yet felt any movements." Her doctor had reassured her that everything was fine, yet she still anxiously turned to the magazine's "baby expert," Mrs. Ruth Best, who advised

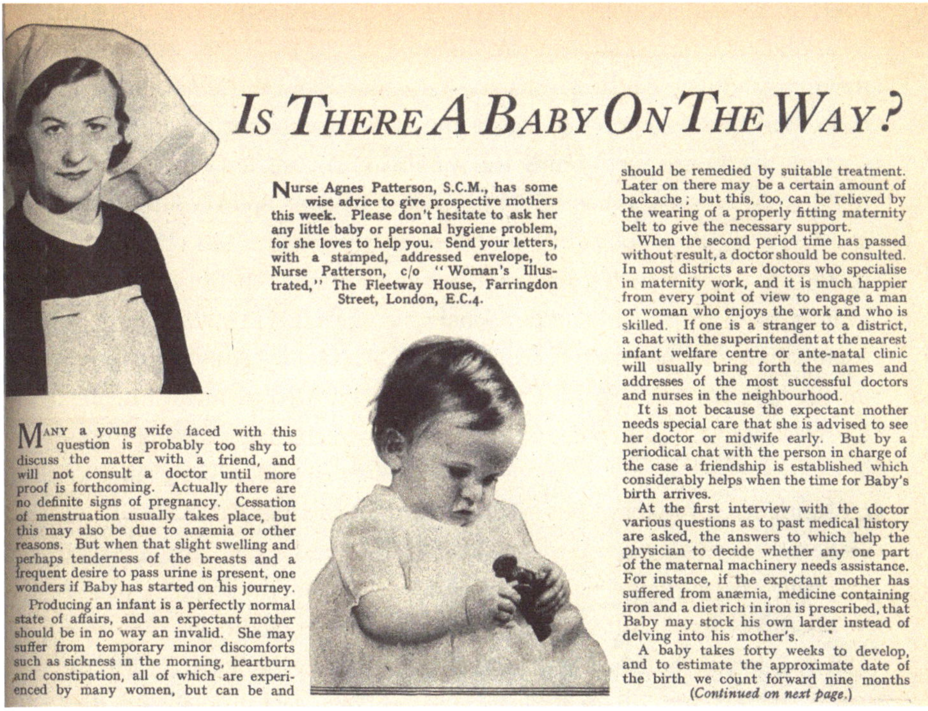

Is There A Baby On The Way?

Nurse Agnes Patterson, S.C.M., has some wise advice to give prospective mothers this week. Please don't hesitate to ask her any little baby or personal hygiene problem, for she loves to help you. Send your letters, with a stamped, addressed envelope, to Nurse Patterson, c/o "Woman's Illustrated," The Fleetway House, Farringdon Street, London, E.C.4.

Many a young wife faced with this question is probably too shy to discuss the matter with a friend, and will not consult a doctor until more proof is forthcoming. Actually there are no definite signs of pregnancy. Cessation of menstruation usually takes place, but this may also be due to anæmia or other reasons. But when that slight swelling and perhaps tenderness of the breasts and a frequent desire to pass urine is present, one wonders if Baby has started on his journey.

Producing an infant is a perfectly normal state of affairs, and an expectant mother should be in no way an invalid. She may suffer from temporary minor discomforts such as sickness in the morning, heartburn and constipation, all of which are experienced by many women, but can be and should be remedied by suitable treatment. Later on there may be a certain amount of backache; but this, too, can be relieved by the wearing of a properly fitting maternity belt to give the necessary support.

When the second period time has passed without result, a doctor should be consulted. In most districts are doctors who specialise in maternity work, and it is much happier from every point of view to engage a man or woman who enjoys the work and who is skilled. If one is a stranger to a district, a chat with the superintendent at the nearest infant welfare centre or ante-natal clinic will usually bring forth the names and addresses of the most successful doctors and nurses in the neighbourhood.

It is not because the expectant mother needs special care that she is advised to see her doctor or midwife early. But by a periodical chat with the person in charge of the case a friendship is established which considerably helps when the time for Baby's birth arrives.

At the first interview with the doctor various questions as to past medical history are asked, the answers to which help the physician to decide whether any one part of the maternal machinery needs assistance. For instance, if the expectant mother has suffered from anæmia, medicine containing iron and a diet rich in iron is prescribed, that Baby may stock his own larder instead of delving into his mother's.

A baby takes forty weeks to develop, and to estimate the approximate date of the birth we count forward nine months

(Continued on next page.)

4.2 This typical headline shows that women's magazines took seriously women's anxieties about the early signs of pregnancy. *Woman's Illustrated*, Feb. 6, 1937, 45, British Library LOU.LON 623, licensed by Future Publishing Ltd.

her to trust her doctor: "movements are not felt at all until well into the fifth month, and often later, so you have no cause for alarm."[47] A similarly concerned woman, "nearly five months on the way," wrote to *Woman* that she had "not yet felt any movements" and was advised that a doctor would be able to ease her mind by "listening to Baby's heart."[48]

Women's pages in newspapers were dominated by housewifery, and articles on motherhood typically discussed child psychology and childhood education, rather than pregnancy management.[49] The national daily press, read by an estimated two-thirds of Britain's adult population, was socially conservative and rarely printed the word "pregnancy"; court reports that dealt with illegitimacy or abortion would refer to a "certain condition."[50] Following the lead of *News of the World*, the mass-circulation dailies began picturing semi-nude young women but rarely discussed potentially educational matters such as contraception and venereal disease.[51] Designed to appeal to a mixed audience of young and old readers of both sexes, newspapers did very little to alleviate ignorance about the female body.

Despite this policy of self-censorship, it was occasionally difficult for journalists to avoid mentioning the Aschheim–Zondek test in connection to other news items. When the test inescapably came up in connection to another story, the opportunity to discuss its significance was consistently passed up. When Francis Crew's department of animal genetics was opened in June 1930, Sir Edward Sharpey-Schafer's address described the "Zondek-Aschheim test for pregnancy" in some detail and blamed antivivisectionists for the fact that one departmental laboratory diverted resources intended for research to a "routine method of diagnosis which might as well be done anywhere else."[52] A lengthy article in the *Scotsman* mentioned the "method of testing for pregnancy" but emphasized the "professor's defence of vivisection."[53] Less extensive coverage in the *Times* and other papers focused exclusively on the vivisection angle, to the exclusion of the pregnancy test.[54]

When Bernhard Zondek was dismissed from his Berlin post in 1933, twenty British scientists, including Crew and Huxley, signed a letter to the *Times* encouraging Britain to welcome Germany's Jewish scientists; they obliquely explained that Zondek's "contributions to sex physiology" had "banished" the "anxieties" of "countless sufferers."[55] In 1934, some months after Manchester's Jewish Hospital had given all three Zondek brothers honorary roles, the *Guardian* described Bernhard as the "celebrated gynaecologist responsible with Asc[h]heim in Germany for the famous 'Asch[h]eim-Zondek' test" but did not say what the test was for.[56]

In 1937, Dr. Ivor Beaumont of the *Daily Mirror* advised worried first-time mothers to put their "whole pregnancy under the supervision of experts and leave any worrying to them" but elided the diagnostic process: "You realised you were going to have a baby. The realisation frightened you."[57] The pregnancy test was infrequently mentioned in newspaper coverage of the famous 1938 trial of gynecologist Aleck Bourne, a landmark in abortion law reform known as the "Horse Guard case."[58] The *Daily Telegraph* and *Hull Daily Mail* quoted Bourne in an interview with two Scotland Yard detectives on the day of the operation as having said, "The girl was brought to me by her mother to my house at Wimpole-street on May 31. She was admitted to St. Mary's Hospital and placed in my ward, under my care, on June 4 or thereabouts. Since then she has been waiting for a pregnancy test, which was positive."[59] The *Times* acknowledged that, in light of the 1929 Infant Life (Preservation) Act, it was crucial that Bourne had operated on a "girl in the earliest stages of pregnancy" but did not link this to his use of the test.[60] Behind the scenes, Sir Bernard Spilsbury used up-to-date obstetrics textbooks to verify the existence of a "test which can be made with the patient's urine, the Aschheim Zonde[k] or the Friedman test, which is claimed to be very reliable and which enables pregnancy to be diagnosed about a fortnight after its commencement," suggesting that even the famous forensic pathologist had been unaware of the test.[61]

NOT INFALLIBLE

A little-studied genre of encyclopedic domestic health manuals aimed at young married women and first-time mothers flourished in the 1930s. Although better known to historians for the sexual and contraceptive knowledge they contained, their foremost stated aim was to prepare women for pregnancy and motherhood. Not merely a respectable camouflage for more risqué topics, although they were that too, advice manuals promised to deliver up-to-date scientific information to hopefully expectant mothers about their own changing bodies.[62]

Many public libraries refrained from stocking "indecent" books, and some London bookstores would sell only to a doctor or a medical student, so publication did not guarantee availability.[63] Cheap mail-order services became an important means of conveying information to wives and mothers in a rapidly expanding network of women's magazines, mothercraft centers, prenatal clinics, and baby clubs.[64] For women who were too embarrassed or otherwise reluctant to visit a doctor, the post office offered an attractively anonymous means of obtaining the up-to-date knowledge required of modern motherhood (figure 4.3). As literacy improved and books became cheaper, women increasingly learned about sex, contraception, pregnancy, childbirth, and mothercraft by reading. Truby King and Grantly Dick-Read became (middle-class) household names.[65] Some mothers bought their daughters a "doctor's book" in preparation for

4.3 Advertisement for a "home doctor" book. *Woman*, Jan. 21, 1939, 38–39, British Library HIU.LON 448, licensed by Future Publishing Ltd.

marriage, while boarding school girls from bookish homes surreptitiously looked up "confinement," "miscarriage," "after-birth," "puberty," "menses," and "life, change of" in health guides and encyclopedias.[66]

Available to *Woman's Own* readers for gift stamps and a postal order, John Dixon Comrie's *Woman's Own Home Doctor* dispelled the "popular" misapprehensions that breastfeeding prevented pregnancy and that quickening meant the child had come to life. Comrie, an Edinburgh physician and medical historian, considered the canonical signs to be "important" but singled out the fetal heartbeat as the "only absolutely certain sign of pregnancy."[67] The Amalgamated Press's *Concise Home Doctor* informed readers that only a "physical internal examination" could definitely confirm or rule out pregnancy.[68] Dr. Gladys M. Cox's *Woman's Book of Health* portrayed amenorrhea as "very suggestive" but inconclusive and cautioned that first-time mothers were apt to mistake quickening for the "colicky movements of the bowel." Cox, a medical officer to the Walworth and East London birth control clinics, advised the reader to "place herself in the care of a doctor or qualified midwife as soon as she realises that she is pregnant."[69] Odham's *Universal Home Doctor Illustrated* listed the conventional signs and warned of "danger signals" including bleeding, fits, and the cessation of fetal movements in the womb, which required immediate medical attention.[70]

In *Why Be Childless?* (1929), Mrs. Cicely Quicke Erskine, a controversial proponent of prenatal sex determination and wife of the independent conservative politician Sir James Monteith Erskine, claimed that some women suspected pregnancy immediately after conception while others remained unaware until after quickening. She dismissed the view that there was "more '*life*' at quickening" than at conception.[71] In *The Ideal Management of Pregnancy* (1930), natural childbirth advocate Cyril Pink portrayed self-diagnosis in terms of the feeling of "malaise" coupled with the "shock" that the routine of menstruation "has been suddenly broken."[72] Neither book mentioned the recently invented Aschheim–Zondek test. From the mid-1930s, however, domestic health manuals began mentioning the test. But they did so less to promote the diagnostic laboratory than to subordinate it to established methods of self-diagnosis.

In *Everywoman's Home Doctor* (1934), the London physician and founder of the New Health Society, Sir William Arbuthnot Lane, mentioned a "special examination of the urine" that would settle "the question" with "great accuracy" but also presented the "conjunction" of amenorrhea with morning sickness as an "almost certain indication" of pregnancy.[73] Herbert Meredith's *The Modern Home Doctor* (1935) revealed that pregnancy could be diagnosed "as early as the third week . . . by means of an examination of the urine" but similarly presented the conventional signs as "quite sufficient to justify a positive diagnosis."[74] Waverley's *The New People's Physician* (1936), edited by the medical writer Douglas Scott and his assistant Dorothy Allmand, mentioned the test but placed greater emphasis on the ability of a "skilled observer" to feel the position of the unborn child's head, perceive its movements within the womb, and eventually detect its heartbeat. Scott instructed the "average laywoman" to see a doctor "as soon as pregnancy is discovered (and the earlier the better). . . . After that, the expectant mother need not worry too much about her signs and symptoms."[75]

British books on the science of sex and reproduction marketed as "popular" were no more informative on pregnancy testing than "home doctor" books. Julian Huxley as well as H. G. Wells and his son, G. P. Wells, did not mention the Aschheim–Zondek test in *Reproduction, Heredity and the Development of Sex* (1935), the fourth volume in their successful "Science of Life" series, and neither did Crew in his own general-interest books on sex and reproduction.[76] Crew gave several BBC radio talks on heredity, eugenics, and the "rights of the unborn" in the 1930s but kept silent about pregnancy testing.[77] Even Wiesner mentioned the test only in passing in *Sex* (1936), his contribution to Thornton Butterworth's Home University Library.[78] British readers were somewhat more likely to discover a detailed explanation of the Aschheim–Zondek test in books by American authors or those translated from German or Dutch.

Stella Browne's translation of Theodoor Hendrik van de Velde's *Fertility and Sterility in Marriage*, published by Heinemann in 1931, described the test in a technically dense appendix to a chapter on reproductive physiology; the Dutch gynecologist maintained that most women would in any case feel pregnant "from the moment that the fruit has taken root—or adhered."[79] The imaginary pupil in *School of Biology* (1935) had "read somewhere that by the use of hormones it is possible to ascertain earlier than in any other way whether a woman is pregnant or not."[80] An anonymous reviewer complained in *The Listener* that the "popular book on biology" had "suffered in translation" and that the "Aschheim-Zondek pregnancy test" was "rightly described" and "wrongly explained" in the same paragraph.[81] Edward Griffith's *The Childless Family: Its Cause and Cure* (1939) unreservedly endorsed the "extremely reliable," "very useful," and "cheap and easy to perform" test in connection with infertility treatment. A Harley Street physician and pioneer of sex education and marriage counseling, Griffith was personally acquainted with Wiesner and recommended the test not for pregnancy diagnosis in healthy women but to help doctors decide whether to administer hormone injections to prevent a likely miscarriage.[82]

The small amount of space devoted to pregnancy testing in these books and, occasionally, in reviews would have been easily missed by all but the most attentive reader. A far more extensive discussion of the Aschheim–Zondek test was to be found in the Left Book Club edition of Drs. Hannah and Abraham Stone's *A Marriage Manual: A Practical Guide to Sex and Marriage* (1936). The Stones structured their marriage manual, first published in New York by Simon & Schuster, as a series of questions and answers between patient and doctor. Allies of Margaret Sanger and pioneers of family planning in their own right, the Stones revealed that the test had been invented by "two German physicians" and involved injecting a patient's urine into a "young female mouse or rabbit." The Stones argued that women were neither "emotionally nor physically" aware of the early stages of pregnancy, and although they considered the test to be 95 percent accurate, they cautioned that it was "not infallible" and should be considered only in relation to other signs.[83]

As late as 1939, it was still possible to dispense pregnancy diagnosis advice without mentioning or endorsing the laboratory test. For instance, pediatrician Lindsey Batten's *The Single-Handed Mother* simply advised readers to see a doctor "early in pregnancy—as soon as two periods have been missed, if not before—to confirm, as far as may be, the fact that a child has been conceived."[84] Pink's *The Foundations of Motherhood* (1941) acknowledged that in "recent years a certain laboratory test has been devised which involves the use and eventual killing of guinea-pigs [*sic*]." But, according to Pink, some doctors preferred an "electrical test," relying on a modified version of Albert Abrams's box or "oscilloclast," and most women were in any case satisfied with their doctor's ability to "diagnose pregnancy at a very early stage by mere examination."[85] Pink was a theosophist, a vegetarian, and an antivivisectionist, so his preference for electrical diagnosis over animal dissection almost certainly reflects his decidedly marginal allegiances.[86] In any case, most authors did not present Aschheim and Zondek's bioassay or Abrams's box as particularly necessary or desirable accessories of early pregnancy diagnosis.

DEMOCRATIZATION

Anxiety over the incidence of criminal abortion increased dramatically in the mid-1930s, when it was considered a significant cause of the perceived rise in maternal mortality, especially among the poor.[87] The National Council of Women demanded a government inquiry into abortion in 1936, and a Ministry of Health report on maternal mortality in 1937 resulted in the establishment of an interdepartmental committee on abortion by the Home Office and Ministry of Health known as the Birkett Committee.[88] Francis Crew was invited to submit a memorandum, and in June 1938, he appeared before the committee to answer questions about pregnancy testing. He had previously argued in a prestigious American journal that the state could mobilize the Aschheim–Zondek test in the struggle against the falling birth rate and "dwindling" population: the Edinburgh station was large enough to charge fees on a sliding scale that would make the test "available for all, irrespective of income category, and, at the same time, accumulate data sufficient for profitable analysis."[89] Now that he had the attention of the state, Crew promoted the expansion of pregnancy diagnosis services in the interwar language of democratization.[90]

According to Crew, pregnancy tests were performed only for a small minority of those who could benefit from them; they had "not been properly democratised." Ideally, the service would be available to every woman at the first suspicion of pregnancy. He recommended the establishment of specialized services in large urban areas that would deal exclusively with pregnancy diagnosis and the "quantitative estimation of the sex hormones." Only by processing at least fifty specimens a day could a technician acquire the "necessary skill" to interpret the various "grades of reactions" and relate these to the various "clinical conditions."[91] Once pregnancy diagnosis had been sufficiently democratized, doctors would be able to distinguish patients who were likely to spontaneously abort (miscarry) from those who were not. "We in Edinburgh started it," Crew boasted to the committee. The famous test had been "elaborated and exploited and democratised" in Edinburgh first. Nearly a decade later, he estimated that 2,000 specimens every year were tested in laboratories around Britain in addition to the 9,000 processed in Edinburgh. For Crew, this was an "entirely desirable democratisation" that justified his own initiative. Expansion demonstrated the existence of an unmet need for tests that were not available to "all those who could profit from them."[92]

Lady Ruth Balfour, who had studied medicine at Newnham College, Cambridge, and worked in biochemistry research at the Lister Institute, asked about the cost of expansion.[93] Crew contrasted the "considerable" running cost of a small-scale laboratory, which would need to charge at least one guinea per test, to a large-scale operation such as the one in Edinburgh, which was able to reduce the cost to around three shillings per test. Crew argued that pregnancy testing should be a "State affair," until bioassays were superseded by a more efficient "chemical test" that did not require signing a Home Office license for 56,000 animals annually.[94] In response to questions by Sir Comyns Berkley, an obstetrician at Middlesex hospital, and Lady Juliet Rhys-Williams, a writer and Liberal politician, Crew estimated that his service had prevented around fifty miscarriages in the past year because it had put doctors in a better position to treat pregnant patients. Weakly positive reactions were "very common" and indicated precarious pregnancies. Crew did not, however, dispense advice on hormone therapy, which was at the doctor's discretion. Hormone testing could be extended until the "danger period" was over, but after the fourth month, there was "no point" in continuing.[95]

The official report of the committee in 1939 noted that "endocrine tests" were "commonly employed" with a "very high degree of accuracy" in the determination of early pregnancy and that Edinburgh demonstrated that scaling up "greatly reduced" the cost. Beyond pregnancy diagnosis, the expanded use of bioassays could reduce the incidence of miscarriage, and the committee officially recommended exploring the possibility of expansion, with a view to facilitating access, "irrespective of income."[96] Although the majority report of the committee rejected legalizing abortion for social or economic reasons, Mrs. Dorothy Thurtle, a social worker, birth control activist, and the only untitled woman member of the committee, prepared a dissenting minority report that proposed allowing abortions for mothers of four or more children.[97] The report contained a memorandum by Joan Malleson, a Harley Street doctor and founding member of the ALRA, which argued that apart from their usefulness in preventing "spontaneous abortion" (miscarriage), a fully democratized service would also be of "great value to worried women who damage their health by taking abortifacient drugs when their periods are overdue."[98] According to Malleson, many of these women were not in fact pregnant and so risked their health and wasted their money unnecessarily on dangerous and illegal substances.

Malleson's memorandum, which was subsequently published in the *Lancet*, proposed that mothers' welfare centers, based on a Danish model, should provide working-class women with access to pregnancy tests and, if necessary, referrals for therapeutic abortion. Many institutions, Malleson claimed, were able to provide an Aschheim–Zondek test for four shillings and sixpence, which was cheaper than purchasing "expensive abortifacients whenever a menstrual period is late!" The "reassurance" of a negative result in "certain circumstances" was "inestimable." This was especially true of menopausal women, who were often the "most desperate" in their "fear of pregnancy."[99] Stella Browne rejoined with her own letter in the *Lancet* that the diagnostic service had to guarantee "anonymity" to prevent positive reports from being used for purposes of notification.[100] In his address to the National Association of Maternity and Child Welfare Centres, Carlos P. Blacker, the general secretary of the Eugenics Society, proposed a "regional system" to alleviate the "mental stress" and "ill-health" of possibly pregnant women: "If the public could be educated as to the very high degree of reliability of these tests, and could be induced to avail themselves fully of such pregnancy-diagnosis services, the sales of abortifacients and the practices of abortionists might be substantially curtailed."[101]

In her address to the Eugenics Society and in a slim book, *Abortion: Right or Wrong?* (1940), Thurtle tactically allied pregnancy diagnosis with nationalist maternalism and the campaign against illegal abortion. She too argued for embedding pregnancy tests in routine prenatal care on the grounds that many women took abortifacients when, in fact, they were not pregnant; menstrual irregularities, she claimed, were the culprit. Unless she was "really desperate," a positive-testing woman could be persuaded to go through with her pregnancy if she was also promised postpartum contraceptive advice. Instead of feeling like a "trapped animal with no one to help her," she would then have the "courage to go through it once more, in the knowledge that in the future she will not be so helpless."[102] A negative result could save a nonpregnant woman's health and "probably much expense" too. Many women thus "saved from abortifacient drugs and violence by means of a pregnancy test may well become mothers later on, with their health in a correspondingly better state than if they had been left to their own devices."[103] Meanwhile, the start of World War II had added new urgency to the argument for democratization. The "future of all social services looks very black," concluded Thurtle, "but if the country is to survive, the health of our mothers must be maintained, and we can only hope that our statesmen, recognizing this, will eventually take the necessary steps to secure healthy, happy motherhood."[104]

DARE YOU HAVE A WAR BABY?

World War II dramatically reconfigured the administration of medical research, public health, and laboratory services in Britain.[105] Crew set up a pregnancy diagnosis service for the Women's Land Army, and the Royal Air Force Medical Services used the Aschheim–Zondek test to rule out early pregnancy in women who ceased menstruating on joining the Women's Auxiliary Air Force.[106] In 1939, the Ministry of Health established the Emergency Medical Services (EMS) in anticipation of air-raid casualties, epidemics, biological warfare, and the need for blood transfusion.[107] In addition to a maternity service for evacuees, the EMS also comprised an extensive network of public health and pathological laboratory services, linking mainly university clinical laboratories around England and Wales (Scotland independently set up its own system in parallel) and placing them under control of various MRC subcommittees.[108]

As Norah Schuster, a clinical pathologist whose career stretched from 1916 to 1960, later recalled, the EMS changed everything. Old laboratories were refurbished, new ones built, and London pathologists, including Schuster, were redeployed in the home counties. Diagnostic laboratory testing became available to many local doctors for the first time, and clinical pathologists found themselves in demand from rural hospitals, nursing homes, and private houses. Although the Ministry of Food promoted rabbit meat for human consumption, making it difficult for pathologists to obtain supplies for pregnancy testing, Schuster "managed with the help of a local resident to collect them from small holdings in the district."[109] She later recalled that pregnancy testing "became fairly frequent" during the war.[110] James Alfred Giles, the chief inspector at the Home Office responsible for enforcing the Cruelty to Animals Act, similarly perceived the scale of pregnancy testing in "all parts of the country" to have increased "out of all measure" in the early years of the war.[111]

Quantitative data are scant, but it seems likely that the upward trend documented for the 1930s continued, possibly intensifying with the emergency services making pregnancy testing more widely available. The reasons for wanting to confirm a suspected pregnancy at an earlier stage also multiplied. Sexual relations changed as men joined the forces, mothers and children were evacuated, and women were put to work in factories or as

4.4 A joint Ministry of Food and Ministry of Health leaflet explains an expectant mother's "duty" to take advantage of the extra nourishment recommended by doctors and provided by the state. "Extras needed by mother and child in wartime," Sept. 1944, Second World War Experience Centre, LEEWW: 2001.906.2.3 Venables, A.

land girls.[112] Illegitimate births increased, and criminal abortions known to police quadrupled.[113] At the same time, wartime propaganda and journalism promoted traditional maternity as a valiant patriotic duty.[114] In 1942, Winston Churchill warned of the "dwindling birth-rate" in a radio broadcast, and the Beveridge Report, a key document in the construction of Britain's welfare state, concluded that "housewives as mothers have vital work to do in ensuring the adequate continuance of the British race and of British ideals in the world."[115] Leaflets distributed to local food offices reminded mothers that a child's life "starts nine months before birth" (figure 4.4).

As Ann Oakley put it, World War II was the "best thing that had happened to pregnant women for a long time."[116] Householders were paid extra for taking in an evacuated pregnant woman, and on production of a medical certificate of pregnancy, expectant and nursing mothers were issued an additional green (child's) ration book to collect coupons from the food office.[117] Between 1940 and 1942, first milk and then orange juice, cod-liver oil, vitamin tablets, meat, eggs, oranges, and bananas were subsidized for pregnant women. From August 1941, they were entitled to fifty coupons to buy materials for a baby's layette.[118] In September 1941, *Mother* magazine announced that, upon production of a certificate from the doctor or midwife booked to attend the birth, an expectant mother could obtain the certificate from a prenatal clinic or Maternity and Child Welfare Centre: "Where a positive diagnosis of twins has been made and confirmed, a double number of coupons will be issued."[119] In practice, ration books and coupons encouraged and formalized the early medical confirmation of a suspected pregnancy but without help from mice or rabbits.

After an initial drop to a record low in the first two years of the war, fertility began to increase precipitously.[120] As the number of babies born, illegitimate childbirths, and illegal abortions increased, so did the public visibility of pregnancy testing. Having previously dispensed advice on the signs of pregnancy and clinical examination, women's magazines first began to comment on the Aschheim–Zondek test. In October 1939, soon after the start of the war, *Mother* magazine launched a new column for expectant mothers. Advice columns, previously the domain of female nurses, had typically aimed at young mothers, not pregnant women. This one, however, featured an anonymous male expert presented as a "distinguished maternity doctor." It marked a greater emphasis on the nine months before birth and on the increasing exposure given to scientific knowledge and medical advice.[121]

In July 1940, *Mother*'s maternity doctor revealed that pregnancy could be determined "within a few days of conception" by means of "the urine test."[122] Far from an unqualified endorsement, however, his description of the test was accompanied by caveats that it was expensive and unnecessary: "Your doctor could send a specimen of your urine to certain laboratories, and a report could be made in about a week's time. The urine is injected into young female mice, and after a few days the mice are killed and their ovaries examined. If the urine came from a pregnant woman, there would be definite changes in the ovaries of the mice. So if it is urgent for you to know at the earliest possible time if you are pregnant, and you can afford to have the urine test, your doctor can arrange the matter for you."[123]

Mother may have introduced the test to tens of thousands of women for the first time, but its debut was lukewarm. The physiological knowledge that underpinned the test was left unexplained, and readers were left in the dark about where the laboratories were located (so they were not enabled to post their own specimens). But its essence as portrayed in the magazine (mouse injections, dissections, ovary inspections) did not differ substantially from equivalent passages in medical textbooks. Animal experimentation was generally perceived as objectionable to women, so it is possible that the technically accurate but otherwise superfluous detail was included to discourage women from inquiring further about the test.[124]

Even as *Mother*'s maternity doctor discussed the test for the first time, he endorsed self-diagnosis as "almost certain," portrayed quickening as the "most dramatic and conclusive sign that Baby is on the way," and advised "anxious" women to wait patiently until a second or third period had been missed before seeing a doctor. So why bother with early diagnosis (self or otherwise) in the first place? As the doctor explained, confirming pregnancy by the "end of the second month" would leave "seven more months to get everything ready for Baby." In practical terms, this meant getting busy with "knitting needles and work basket."[125] For the predominantly aspirational working-class and lower-middle-class readers of *Mother*, pregnancy meant hard work making "little garments" from knitting patterns, the stock-in-trade of women's magazines. An earlier diagnosis was not a gateway to more medical surveillance, but it did leave more time for knitting (figure 4.5).

From 1941, after Winston Churchill had become prime minister and the Battle of Britain had been waged, wartime conditions of motherhood became a dependable fixture of *Mother* magazine. In January 1941, Nurse Crawford offered guidance on "welcoming the war-time baby," the "little one" born in "difficult times."[126] And in February, *Mother*'s maternity doctor addressed the apprehensive mother who despaired at bringing a child into the world "with this dreadful war raging" (figure 4.6). This was "exactly what the Nazis want you to think," he countered: "Hitler and his followers would doubtless rejoice at the prospect of the British race dying out." This was no time to "shirk motherhood." On the contrary, mothers ought to take "special pride" in their "great service" to the nation.[127]

To discourage women from considering abortion and to clarify the medical position, *Mother*'s maternity doctor recounted the story of a distressed

Great New Series for Mothers-in-Waiting

the wonderful promise

Our Maternity Doctor tells you the signs and wonders which will show you whether there is a baby on the way. If you are expecting a little one and need medical advice on any point, write to him at the address on page 55, enclosing a stamped, addressed envelope.

CAN it really be true? Is there a baby on the way? Many anxious wives ask themselves these questions when the period is a few days late, and some rush off to see a doctor at once to confirm their hopes.

If you are in such a position, don't waste your time and the doctor's by consulting him when your period is only a week or two overdue. It is impossible for the doctor to detect a pregnancy before the sixth week, so you should restrain your impatience until you have missed the second period and preferably until the third is nearly due. Then, on examination, your doctor will probably be able to detect an enlargement of the womb and so will be able to diagnose a pregnancy. Baby is so very tiny in the early stages of development that there is no detectable enlargement of the womb until he is at least six weeks old.

There is a means of diagnosing a pregnancy within a few days of conception—that is the urine test.

Your doctor could send a specimen of your urine to certain laboratories, and a report could be made in about a week's time. The urine is injected into young female mice, and after a few days the mice are killed and their ovaries examined. If the urine came from a pregnant woman, there would be definite changes in the ovaries of the mice. So if it is urgent for you to know at the earliest possible time if you are pregnant, and you can afford to have the urine test, your doctor can arrange the matter for you.

GENERALLY, the first sign of pregnancy is cessation of the menstrual periods. To miss a period is strongly suggestive of pregnancy in a woman in normal health who menstruates regularly.

However, there are various conditions of ill-health (glandular disorders, acute anæmia and other diseases) in which menstruation ceases for many months. Absence of the period alone is not conclusive evidence of pregnancy and although, normally, menstruation ceases throughout pregnancy, some women have monthly periods during part or sometimes the whole of their pregnancies.

Your doctor knows the combination of signs and symptoms on which a certain diagnosis can be made.

PERHAPS the next commonest sign of early pregnancy is morning sickness. Here again, there are exceptions; some women are never sick.

But if you have missed a period and at about the time when the next period is due you feel sick or actually vomit as soon as you rise from bed in the morning, pregnancy is almost a certainty.

You must bear in mind that morning sickness may be the result of dyspepsia and liverishness: it is when it occurs together with a missed period that it is so strongly suggestive of pregnancy.

Sometimes as early as the first month of pregnancy, there is a tingling sensation in the breasts, and a feeling of fullness, but many women experience these sensations before each menstrual period. It is not until the second month of pregnancy that the characteristic darkening around the nipple occurs, a brownish tinge which becomes progressively darker as pregnancy advances. In the third month of pregnancy, secretion can be squeezed from the nipples.

It is not until about the seventeenth week of pregnancy that you will experience the most dramatic and conclusive sign that Baby is on the way. This sign is known as the "quickening"; it is *your* first perception of baby's movement within the womb. This is indeed a wonderful moment when first you feel Baby "taking exercise."

The sensation is that of a gentle flick with a finger, low down in the abdomen. It is as if Baby were giving you a little secret message: Here I am, mother, getting along fine.

As Baby grows bigger and stronger, the movements become more vigorous: but it is the *first* faint movement that you definitely feel that constitutes "quickening." Some mothers-to-be feel a little faint at this wonderful moment, or may actually faint, but this is of no consequence. Anyhow, you will feel terribly thrilled and happy to get this first definite sign of the separate life within you.

Of course, your doctor will have been able to assure you, long before quickening occurs, that your baby is on the way.

NATURALLY you want to know as soon as possible when to expect the birth of your baby.

You can easily calculate this for yourself. Write down the date of the first day of your last menstrual period. Now go back three months and add seven days to this date. This gives the most probable date of your baby's birthday. Let us take an example. If the first day of your last period was June 1, three months back brings you to March 1; seven days added brings you to March 8, the probable date of the confinement. So you must carefully note the first day of your last period.

This calculated date, however, is not the certain date of your baby's birthday. It is not possible to foretell this, unless the actual date of conception is known. So when you have worked out the probable date, you should allow for Baby to arrive at any time during the week preceding or the week following this date. This is the nearest you can get to an accurate calculation.

If you know definitely that you are pregnant by the end of the second month, you have seven more months to get everything ready for Baby. But if you are wise you will aim at completing all your preparation at least two months before Baby is due to arrive, because some babies are born as much as two months before the expected date.

So when, with your doctor's help, your early doubts are settled and you know that motherhood is certain, you should begin at once to make your plans and get busy with your knitting needles and work basket. As the little garments take shape under your hands, may your heart be filled with joy at the prospect of achieving the crown of womanhood, the thrilling happiness of motherhood.

34

4.5 *Mother*'s maternity doctor did not link earlier pregnancy diagnosis to earlier prenatal care or abortion, but rather to preparing the layette. *Mother*, Jul. 1940, 34, British Library LOU.LON 624, licensed by Future Publishing Ltd.

4.6 A young wife and her uniformed husband contemplate parenthood and the geopolitical future of Britain as *Mother*'s maternity doctor asks, "Dare you have a war baby?" *Mother*, Feb. 1941, 12–13, British Library LOU.LON 578, licensed by Future Publishing Ltd.

patient whose husband had been called to serve overseas (figure 4.7). With "wartime difficulties," she did not have the courage to "face it all" and had asked him for help. "Many people," he explained "have the mistaken idea that doctors are free to terminate a patient's pregnancy if they can be persuaded to do so. They do not appreciate that the doctor's legal powers, whatever his sympathies, are very definitely limited. Unless the pregnancy is likely to be fatal to the mother, or to cause serious injury to her health it is a criminal offence to attempt to procure an abortion." Instead, he attempted to reverse the meaning of a wartime pregnancy: if a lonely wife could take comfort in her baby, then the prospect of her husband leaving to fight became a "weighty argument in favour of having a baby."[128] Conjuring a "young wife who was working on munitions, standing at a bench," *Mother*'s maternity doctor explained that women could carry on working and doing "their bit" up to the seventh month of pregnancy as long as they were kept "under constant medical supervision."[129]

★ GREAT NEW SERIES

Motherhood ahead

If the idea of having a baby worries or frightens you, read this reassuring article by our Maternity Doctor

A PATIENT came to see me in great distress because she was pregnant. She had a horror of pain and of losing her figure. She was worried about the expense of having a baby; her husband had been called up and expected to go overseas shortly. Already, she said, it was difficult to manage on her separation allowance and she had been planning to get some war work. She assured me that they intended to have a family later on, after the war, as they were fond of children. She would have more courage to face it all when life was normal again, she said; but the wartime difficulties she could not face. Would I help her?

I had known this patient previously as a normal, healthy young woman, certainly not lacking in courage. I knew that life had not been easy for her since the war began; and I could see that she badly needed sympathy and reassurance. But I could see no reason why she should not have a healthy baby and herself benefit greatly in health and happiness. We had a long talk together; and I think you may be interested in what I told her.

Many people have the mistaken idea that doctors are free to terminate a patient's pregnancy if they can be persuaded to

4.7 The anonymous (faceless) maternity doctor imposingly leans over a young wife dressed for war work. In this carefully staged portrayal of the appropriately subordinate patient's perspective, their gaze locks as she stares up at him; for a discussion of this staging in 1960s women's magazines: Loughran 2020: 140. *Mother*, Jun. 1942, 71, British Library LOU.LON 564, licensed by Future Publishing Ltd.

RESISTING A NEW TECHNOLOGY

In April 1943, *Mother* revisited the diagnostic encounter, this time with an "eager and excited" fictionalized patient. "Mrs. Brown" expected her husband to be sent abroad "any day now" and wanted to "make quite sure" before "sending him the good news." Her periods were, however, "only about two weeks overdue," and like "many other young wives," she did not appreciate the "difficulties of making such an early diagnosis of pregnancy." The doctor informed her about a urine test, "which takes about a week." He could send her urine to a laboratory, where a "small quantity would be injected into a young female mouse. About four days later, the animal would be painlessly killed, and its ovaries examined for the definite changes which would have taken place if the urine were that of a pregnant woman." In the end, the doctor "arranged for Mrs. Brown to come for a medical examination at about the time when her second missed period was due, as she did not wish to go to the expense of having the urine test." So again, and despite the new emphasis on painlessness, a woman's magazine publicized the test only to reject it.[130]

Grantly Dick-Read's *Revelation of Childbirth: The Principles and Practice of Natural Childbirth*, written "when bananas were still available," explained that if the "menses are more than ten days overdue and accurate diagnosis difficult for any of the many reasons that may give rise to uncertainty of early diagnosis, an Aschheim-Zondek or some other similar test of the urine should be done."[131] But in practice, many doctors remained reluctant to diagnose early pregnancy and did not propose a test.

A rare firsthand account from the 1940s is that of Ruth Beck, a married secretary and Mass Observation diarist from Berkshire. Her concise narrative usefully corroborates and adds texture to other lines of evidence. On a day off from work in July 1942, at the age of twenty-seven, she did the ironing, had tea, and went to the doctor: "I screwed up my courage & asked if I was going to have a baby, but he wouldn't diagnose it definitely yet: he was awfully sweet about it all though."[132] In 1945, Nurse Crawford projected "hope" onto amenorrhea but cautioned against seeing a doctor before a second missed period "because, without special tests which are not normally made, no definite opinion can be given."[133] And Edgar Hope-Simpson, a GP who began practicing in rural Gloucestershire in 1946, later recalled using the test on a selective basis: "I can remember people coming and wanting to know if they were pregnant. And I would say 'I think you are' or 'I think you aren't.' Is it important that you should know before next month or whatever? And if there was some particular reason why they should know then we'd arrange an Aschheim–Zondek. It cost them a couple of quid."[134] Going to the doctor in the 1940s took courage and usually didn't result in a diagnosis, but a new culture of pregnancy and its early verification was under construction.

5

IMPORTED TOADS

In February 1938, the *Daily Mirror* reported that if a woman in South Africa "wants to know whether she is going to have a baby . . . she consults the common frog of the veldt" and that this animal would soon be imported in bulk to "answer the Great Question for Englishwomen."[1] Six months later, the same paper revealed that "huge consignments" of "one of the world's ugliest frogs" were "being shipped from South Africa to Britain and other countries at 4d. each" and that there were even "fears of a shortage."[2] The platanna ("flat hander") or African clawed frog or toad (the terms were used interchangeably), *Xenopus laevis*, would become the dominant animal in pregnancy diagnosis in postwar Britain.[3] In 1939, Francis Crew claimed the *Xenopus* test as a British invention by naming it after his friend, the socialist physiologist Lancelot Hogben.[4] In so doing, Crew touched off a priority dispute between Hogben and his South African colleagues who had locally reported the use of *Xenopus* in pregnancy testing a few months before Hogben's assistant announced the test in the prestigious journal *Nature*.

A well-networked critic of eugenics and bestselling author of *Mathematics for the Million* (1936), Hogben is a colorful figure in the history and philosophy of science.[5] Most historical writing about the "Hogben test" either places its eponym center stage or attempts to redistribute the credit.[6]

This chapter revisits the dispute, but in a way that decenters Hogben. It focuses less on who invented what first than on how *Xenopus* was made into a practical test animal and how it became institutionalized in Britain under the National Health Service. It also introduces the entrepreneurial GP and "failed" scientist, Edward Elkan, whose crucial role in domesticating *Xenopus* has gone largely unnoticed.

THE TOAD THAT HAS NOT TO BE SLAUGHTERED

Lancelot Hogben headed the Department of Zoology at the University of Cape Town from 1927 to 1930.[7] His research projects were mainly devoted to studying skin color change in amphibians, or what he called the "pigmentary effector system" (figure 5.1). In South Africa, he began using *Xenopus* as experimental material, studying the pigment cells visible in its

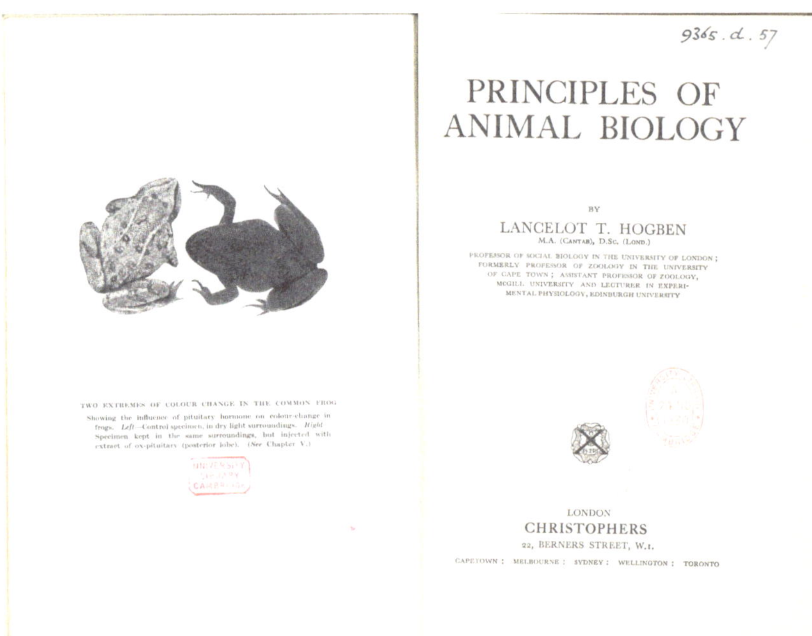

5.1 Frontispiece and title page of Hogben's popular textbook, *Principles of Animal Biology* (1930), showing color change in frogs. The animal to the left is a control and the animal to the right has been injected with ox pituitary extract. Hogben 1930, Cambridge University Library 9365.d.57.

webbed feet.[8] In March 1930, he reported to the Royal Society of South Africa that injections of ox pituitary restored ovulation in surgically altered females.[9] Hogben would later claim this 1929 finding as the basis of the pregnancy test, but at the time, he made no mention of any diagnostic application.[10] Later in the same year, possibly motivated by the shifting political climate, Hogben accepted a new chair in the Department of Social Biology at the London School of Economics.[11]

Hogben returned to London with some toads to establish a small research colony and hired Charles Bellerby to continue his investigations. Bellerby performed the first experiments, funded by the MRC's Sex Hormones Committee, on imported toads kept in a cold basement room; many failed to ovulate when injected with pituitary extracts obtained from local abattoirs.[12] Bellerby repeated the experiment but this time killed and examined the animals to find that unresponsive toads had atrophic ovaries. He next relocated the remaining toads to a warm, well-lit room and established a reliability of nearly 100 percent in subsequent experiments. Despite the inauspicious start, Bellerby concluded that a superior dose–response relation and reusability made *Xenopus* a more practical test animal than the rabbits he was more familiar with.

On March 31, 1934, he reported in a short letter to *Nature* encouraging results with pregnant women's urine, which was widely known to have many of the same properties as pituitary extract.[13] For each test, Bellerby injected ten toads and read a positive result if five of the ten ovulated within nine hours, an arbitrarily set end point. Bellerby claimed that *Xenopus* could be sourced "easily" and "cheaply" and maintained in colonies of several hundred without difficulty.[14] In London, they needed to be kept in clean water in a warm, well-lit room and fed a bit of raw meat once a week. Whereas mice and rabbits had to be dissected to reveal ovarian changes, *Xenopus* extruded numerous, large, visible eggs and so did not have to be killed. Females could be used repeatedly provided they were rested for about a week between injections. The reusability of *Xenopus* was its most obvious advantage over the reigning test animals. Other amphibian species were known to spawn spontaneously in captivity, making them unsuitable. Two months later, a lead article in the *BMJ* mentioned the new test but found insufficient data to assess its clinical value.[15] *Xenopus* was a promising but untested animal.

5.2 Photograph of Shapiro posing with pipette, microscope, and *Xenopus*, affectionately referred to on the reverse side of the print as his "coat of arms." Taken in the physiology laboratory of the Cape Town University medical school, 1936; courtesy of Roy Summerfield.

Bellerby's letter in *Nature* was not the first report of *Xenopus* in pregnancy diagnosis. In October 1933, Henri (Harry) Zwarenstein and his doctoral student Hillel Abbe Shapiro (figure 5.2), Hogben's former colleagues at the University of Cape Town, had presented preliminary results at a meeting of the Royal Society of South Africa.[16] News of the meeting, however, did not reach Britain until early March 1934.[17] Bellerby's letter, published four weeks later, credited Hogben but not Shapiro or Zwarenstein, thus preparing the ground for the dispute that was to follow. The pair responded with their own letter to *Nature*, which was published in May. In it, they reported an accuracy of 100 percent in ninety-seven tests undertaken in collaboration with the Cape Town gynecologist Ariel Goldberg.[18] They injected six toads per test and read a positive result if a single toad ovulated or if a postmortem examination revealed at least a single ovum in either or both oviducts. Their arbitrary end point was eighteen hours, compared to Bellerby's nine.[19]

In a lengthier article in the *South African Medical Journal*, Shapiro and Zwarenstein warned that Bellerby's definition of a positive result (five of ten toads) would result in false negatives.[20] They recommended using freshly collected toads because of ovarian atrophy in captivity, which would also lead to diagnostic errors. In their experience, the normal ovaries of freshly caught pond toads at peak breeding season were filled with large ova easily discerned with the naked eye. But after living for six months in a deep slate tank in a dimly lit animal house at the university, the ovaries of captive females often resembled "gelatinous masses" with indiscernible ova.[21] Shapiro and Zwarenstein attributed "ovarian retrogression" to insufficient sunlight and argued that the "captivity effect" would make the *Xenopus* test impractical in Britain.[22]

Because of these doubts, the test languished, and Bellerby went back to the drawing board. He experimentally injected and dissected freshly imported toads, toads that had lived in the London Zoological Gardens for years, toads that he irradiated with a 100-watt lamp at close range, and toads that had been kept in a cold, dark basement for several months. In the end, Bellerby failed to replicate the effect observed in Cape Town and concluded that reproductive activity was probably influenced not by light or temperature but by food supply. He speculated that, in the wild, females reabsorbed their ovaries seasonally (when the ponds dried up) and, in captivity, when their food ran out. Shapiro and Zwarenstein's toads, Bellerby claimed, had been overcrowded and underfed.[23]

As Crew put it in 1936 in the *BMJ*, "everyone" was still "waiting for the discovery of a new test animal" or a technique that would simplify pregnancy testing and reduce the time spent waiting for a result. *Xenopus* might be "ideal" in South Africa, where "fresh supplies" could be "quickly secured at regular intervals," but Crew despaired that the toad did not long tolerate laboratory conditions and so perceived its usefulness as limited in Britain. Test animals of any species would need to be "bred or bought, fed, housed, and cared for," and injections required a Home Office license and a "degree of surgical skill," so Crew looked forward less to a new bioassay than to a convenient and reliable test-tube reaction, which would remove the "necessity of maintaining and slaughtering thousands of animals [and] would surely replace them."[24] But all this changed in the following year.

Hogben, who never stayed in one place for very long, left the London School of Economics in 1937 to start a new job as Regius Professor of Natural History at the University of Aberdeen. He had some toads transported from London and hired Frank Walter Landgrebe as a research assistant to continue Bellerby's work there. The move crucially brought Hogben into closer proximity with his old friend Crew (they had worked together in Edinburgh in the early 1920s) and facilitated a productive collaboration between Hogben's research laboratory and Crew's diagnostic service. With Zwarenstein's help, Crew arranged for the importation of 1,500 large female toads, and with Landgrebe, he launched a large-scale comparative study of the use of mice, rabbits, and *Xenopus* as test animals in pregnancy diagnosis.[25]

In October 1937, Louis Bosman, another Cape Town gynecologist who collaborated with Zwarenstein, complained in the *BMJ* of the "almost universal ignorance of the method in vogue in South Africa."[26] Having given only six incorrect diagnoses out of 1,000 tests performed in five years, Bosman claimed the "frog test" as superior to all other methods reported in American and European journals. Its only drawback was that *Xenopus* did "not flourish in the northern hemisphere."[27] Crew responded that *Xenopus* was in fact well known in Britain and beyond. Laboratories in London and Aberdeen maintained a "considerable number of claw-toed frogs," and in Edinburgh, a large colony was "being extensively used in pregnancy diagnosis." The demand for *Xenopus* had lately "become so great" that the commercial exporters Crew dealt with warned of impending "restrictions." Yet Crew did not expect the use of *Xenopus* to become "at all widespread" until the toad could be bred and raised domestically. Setting up and maintaining a laboratory for 1,500 frogs was a "very much more serious matter" than establishing one for the use of mice or rabbits, which were available "locally at all times in considerable numbers."[28]

In 1939, adult rabbits cost about five shillings and young mice at least sixpence each. *Xenopus,* when imported in bulk, cost eightpence per toad. Setting up a large colony also involved significant overhead and maintenance costs. Crew kept his toads in galvanized metal tanks arranged in tiers on staging around the walls of a brightly lit room fitted with roof lights. A metal rim overhanging the water on all sides of a tank prevented escape. Each tank received one or two "heaping handfuls" of finely minced meat or liver per week, and electrical tubular heaters maintained the temperature of the toad room at a sweltering (for Scotland) 21°C.[29] Ideally, these investments in specialized equipment and maintenance costs would be offset by the reusability of a test animal that did not have to be killed. Beyond the clear economic advantage of reusability, Crew also factored in the emotional cost of killing laboratory animals, the brunt of which was borne by his all-female staff.[30]

Although Crew denied any compunction about the "killing of a 3-weeks-old mouse," he admitted that breaking the neck an adult rabbit, even one anesthetized with Nembutal, was "not a pleasant task," and his laboratory workers preferred to "deal with the toad that has not to be slaughtered."[31] Their knowledge of *Xenopus* was "still imperfect and incomplete," but they were getting to know the exotic animal and its needs "under artificial conditions." The reliability of the test depended on the "power of observation" of "human personnel." Laboratory workers did not grab toads randomly but "carefully selected" those "being to the eye and to the hand such as give the impression of possessing an ovary that will respond." They learned to avoid "flat" toads, which they removed from the reservoir for a period of extra rest and rations.[32]

After eighteen months of testing, Crew was ready to replace rabbits, although not yet mice, with toads, and in a prominent article in the *BMJ*, he proposed renaming the *Xenopus* test the "Hogben test" to bring it into line with the Aschheim–Zondek and Friedman tests.[33] John Gunn, the acting head of the University of Cape Town's physiology department, responded that there was "no justification whatever" for naming the test after Hogben. If it was to be named after anyone, it should be called the "Shapiro-Zwarenstein test."[34] Hogben countered that Zwarenstein and Shapiro's mistaken insistence on the "captivity effect" nullified their claim to priority and credited Bellerby and Landgrebe with working out the ideal "conditions of diet, density, pollution, temperature, and illumination" needed to maintain *Xenopus* for routine use.[35] In a final letter, Gunn explained that Shapiro and Zwarenstein had regarded themselves as merely extending the work of Aschheim and Zondek and so had modestly refrained from attaching their names to the test. In a conciliatory gesture, Gunn proposed crediting the "humble batrachian, which seems to give an invariably correct diagnosis, by calling this the 'xenopus test.'"[36]

5.3 Photograph of Zwarenstein, platanna in hand, taken on June 8, 1985, at his eighty-firth-birthday party. Held in the main lecture theatre of the University of Cape Town's Pharmacology Department, the highlight was a surprise guest: fifty-two-year-old Julian Mirvish, the "elderly foetus" responsible for the first positive *Xenopus* test on September 10, 1933: Zwarenstein 1986, 51; see also Phillips 1993; 326.

In Britain, where *Xenopus* would become the dominant pregnancy test animal after World War II, Crew's proposal mostly stuck.[37] But the priority dispute flared up for the second and final time when Hogben took issue with a London chemist's account of the "*Xenopus* pregnancy test."[38] Hogben, now at the University of Birmingham, attempted to set the record straight with his own "History of the Hogben test."[39] Shapiro and Zwarenstein responded to what they saw as "several gross misrepresentations of the true facts" and maintained their preference for "the *Xenopus* or frog test."[40] Hogben, in turn, blamed the "South African Press" for stirring up the controversy by boosting the test "as an indigenous South African discovery" and prompting "zeal for the credit of South African science."[41] In a final response before the *BMJ* editors formally closed the correspondence, Landgrebe, still in Aberdeen, insisted it was "beyond question that the test arose from Hogben's discovery in 1929" and that Crew had "very properly termed it the Hogben test" (figure 5.3).[42]

SEND 100 *XENOPUS*

At around the same time that Crew was installing his massive *Xenopus* colony in Edinburgh, a refugee doctor from Germany independently established his own more modest one in London. Born into a comfortably middle-class Jewish family in Hamburg in 1895, Rudolf Eduard Elkan studied medicine in Berlin, Freiburg, and Hamburg; served as a medic during World War I; and practiced in British Palestine before setting up a general practice back in Hamburg (figure 5.4).[43] The "Hamburg Nazis" harassed Elkan early on not only because he was Jewish but also because of his leftwing politics and activities in the birth control movement, which brought him into contact with the controversial socialist doctor and sex educator Max Hodann.[44] In 1933, some "hooligans" broke into Elkan's flat, stole his typewriter, arrested him, beat him, and dragged him into the street draped in red flags. Elkan found himself in a "rat-infested cellar" and then at a local hospital before being discharged. After some days recuperating with his in-laws, the police summoned him, gave him a passport, and escorted him to the SS *Manhattan*. His uncle Hans slipped him a ten-mark note to tuck under his hatband, and he departed for Le Havre and Southampton, where his ticket expired.[45]

5.4 Elkan as a young man in military uniform, posing with his prized Carl Zeiss microscope, a gift from his South African uncle. See Edward Elkan, "Sketches from my life," typescript, 1983, 15, Wellcome Collection MS.9151. Photo courtesy of Naomi Hull.

Elkan planned to join his friend and fellow birth control activist Elise Ottesen-Jensen in Stockholm, but the refugee support committee that received him at Russell Square insisted he spend "every penny" they gave him in Britain.[46] So Elkan contacted Helena Rosa Wright (née Lowenfeld), whom he had met at a Zürich birth control conference in 1929.[47] She took him to the headquarters of the recently formed National Birth Control Association (NBCA) and introduced him to the association's treasurer, Mrs. Gerda Guy, who "whisked" him off to her estate near Beaconsfield to recuperate. Elkan needed a British degree to practice medicine in London, but English medical schools were not accepting refugees. So he attended lectures in Glasgow, learned English from the radio, and passed an exam in Edinburgh on subjects he had "practiced for years."[48]

After requalifying, Elkan returned to London, where Wright put him in touch with Edgar Obermer, an endocrinologist from a wealthy family who drove an American car "as big as St Paul's" and ran a "large, peculiar but flourishing practice at Manchester Square."[49] Obermer had studied medicine in Lausanne and practiced at Papworth Village Settlement in Cambridgeshire, a "socio-medical experiment" in the treatment of tuberculosis, before settling in London. In 1933, he had applied for Ministry of Health funds to research the "individual's neuro-endocrine-circulatory-metabolic-adaptational mechanism" but was rejected possibly as too unorthodox.[50] Obermer promoted his individualistic approach to preventive medicine in numerous articles and two books.[51] Elkan cynically recalled that Obermer's "main cure consisted in bleeding the patients, subjecting the sample to procedures only known to himself and then re-injecting the product into any part of the patient's anatomy."[52] After working as Obermer's assistant for some time, he decided to set up his own somewhat more conventional practice.

With help from his third wife, Lotte "Maya" Lask, Elkan, who had by now changed his name to "Edward," set up a practice in a house overlooking Regent's Park. In 1937, he decided to try the Hogben test because the Aschheim–Zondek, "then en vogue, was cumbersome, expensive and needed hecatombs of young mice." Despite having to explain his cryptic telegraphic order, "Send 100 Xenopus," to suspicious authorities ("What kind of secret and probably dangerous war material was I ordering to the detriment of Old England?"), the first shipment reached him safely. In the

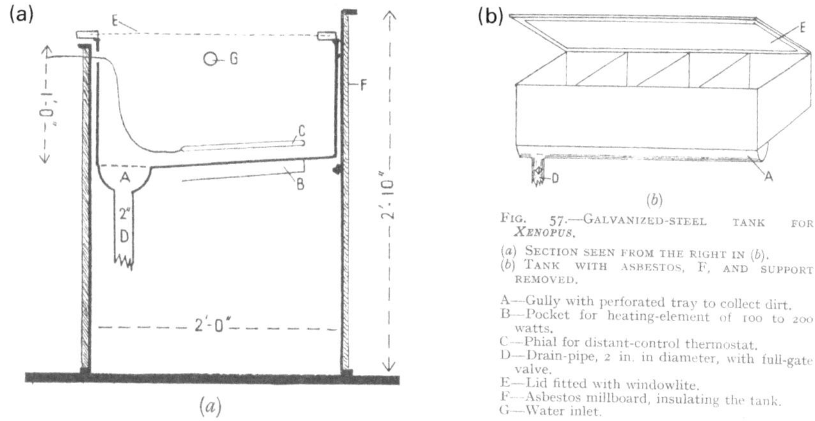

5.5 In 1938 Elkan constructed a galvanized steel tank with a sloping bottom and removable trays to facilitate cleaning. This (a) cross-sectional profile and (b) three-quarter view was published a decade later in the Universities Federation for Animal Welfare's *Handbook on the Care and Management of Laboratory Animals*. After World War II, Elkan obtained electrical heaters from the General Electric Company and thermostats from the British Thermostat Company: Elkan 1947a, 254, Cambridge University Library 9385.c.275.

summer, they lived in outdoor tanks on a balcony, where they became "comparatively tame." Females did not croak, so keeping them by the hundreds did not disturb the neighbors.[53] The rest of the year, Elkan kept them in a specially constructed tank (figure 5.5) and used "one of the practical aquarium heaters [then] on the market" to maintain the water temperature at around 25°C.[54]

Prior to widespread electrification, aquarium heaters had been rare and expensive. The Central Electricity Board had been established with monopoly powers in 1926, and by 1933, the national grid of high-voltage transmission lines, one of the most advanced in the world, was nearly complete. By 1939, the first wave of electrical domestic appliances (vacuum cleaners, cookers, radios, gramophones, irons) had entered many homes, two-thirds of which were wired for electricity.[55] Most pet shops stocked goldfish only, but fanciers could purchase "tropicals" from London dealers who imported and bred them in large numbers.[56] Tropical fish fancying had been pioneered in the United States in the 1910s, and by the early 1930s, American companies were making thermostatically controlled aquarium heaters that incorporated small electrical heating coils, the same technology found

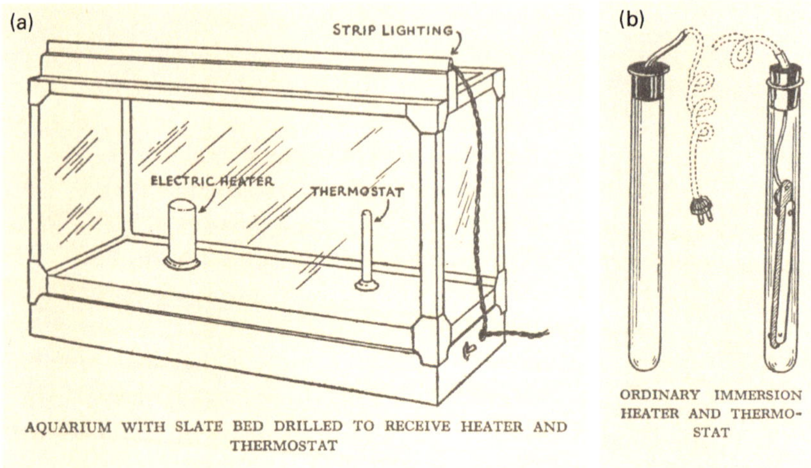

5.6 Line drawings of (a) a slate-bottom aquarium and (b) an immersion heater and thermostat from a manual on tropical aquariums: Wells 1937, 20, 44, Cambridge University Library 9385.d.92. Earlier manuals recommended using a Bunsen burner or an electrical bulb lamp to improvise a heating source: Hodge 1927, 25–26.

in increasingly familiar domestic appliances such as irons, toasters, and immersion heaters for warming soup or beverages.[57] So by the late 1930s, Elkan would have been able to choose from several efficient and inexpensive electrical heaters and thermostats on the market (figure 5.6).

Elkan, an amateur herpetologist with a passion for exotic reptiles and amphibians honed in the Palestine desert, later saw himself as an "obscure and very largely failed potential scientist" with "too many secondary interests" (figure 5.7).[58] He benefited from links to the birth control movement in London, but had no connections to Hogben's network of physiologists. This may have been disadvantageous in some ways, but it also liberated him from concerns about the "captivity effect," which turned out to be baseless. In the late 1930s, he energetically experimented with *Xenopus* husbandry and published several influential articles in British, French, and American medical journals.[59]

Elkan performed the Hogben test for his own patients and for other doctors. The postcode of his service was upscale, so he was able to charge one guinea per test, rather more expensive than the "modest fee" of five shillings set by Crew in 1929. Unlike the Edinburgh station, Elkan sometimes

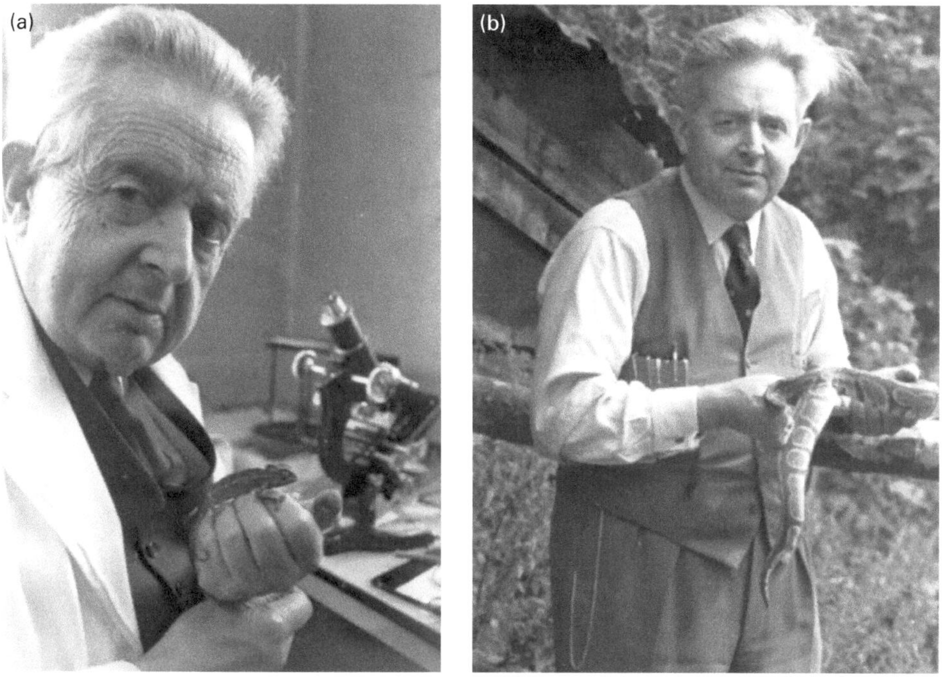

5.7 (a) An aging Elkan posing in lab coat with chameleon and microscope on the worktable in the background. (b) A somewhat younger Elkan with snake; courtesy of Kraig Adler.

posted and telephoned results directly to patients (figure 5.8). Most of the tests he performed were for women who had already missed one menstrual period.[60] Elkan's patients often drank tea or carbonated Vichy mineral water in the evening, which diluted the specimen of morning urine, making it worthless.[61] He later recalled that pregnancy tests were "required by three groups of people: those who *hope* they are going to have a child, those who *fear* they are going to have a child, and finally doctors who, faced with an arrangement of signs and symptoms that might, among others, be explained by pregnancy, do not feel that playing for time is what the patient expects of them." Bottles often labeled "URGENT" suggested to Elkan that "senders, whichever group they may belong to, do not think they are having a luxury test done."[62]

In the first year, Elkan performed nearly 300 tests using over 2,000 toads (counting the same ones multiple times), or about seven toads per test.[63] He operated a twenty-four-hour service. It took him an hour to prepare

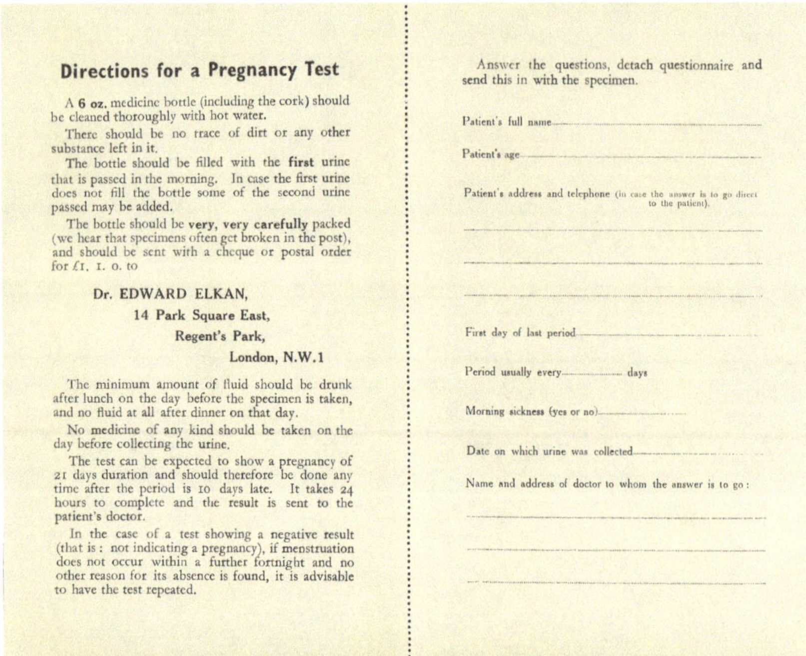

Directions for a Pregnancy Test

A **6 oz.** medicine bottle (including the cork) should be cleaned thoroughly with hot water.

There should be no trace of dirt or any other substance left in it.

The bottle should be filled with the **first** urine that is passed in the morning. In case the first urine does not fill the bottle some of the second urine passed may be added.

The bottle should be **very, very carefully** packed (we hear that specimens often get broken in the post), and should be sent with a cheque or postal order for £1. 1. 0. to

Dr. EDWARD ELKAN,
14 Park Square East,
Regent's Park,
London, N.W.1

The minimum amount of fluid should be drunk after lunch on the day before the specimen is taken, and no fluid at all after dinner on that day.

No medicine of any kind should be taken on the day before collecting the urine.

The test can be expected to show a pregnancy of 21 days duration and should therefore be done any time after the period is 10 days late. It takes 24 hours to complete and the result is sent to the patient's doctor.

In the case of a test showing a negative result (that is: not indicating a pregnancy), if menstruation does not occur within a further fortnight and no other reason for its absence is found, it is advisable to have the test repeated.

Answer the questions, detach questionnaire and send this in with the specimen.

Patient's full name..........

Patient's age..........

Patient's address and telephone (in case the answer is to go direct to the patient).

....

....

....

First day of last period..........

Period usually every.......... days

Morning sickness (yes or no)..........

Date on which urine was collected..........

Name and address of doctor to whom the answer is to go:

....

....

....

5.8 "Directions for a pregnancy test" instructs doctors on packing urine specimens "very, very carefully" and includes space for the patient's address and telephone "in case the answer is to go direct to the patient"; on the telephone in healthcare: Greene 2022. Wellcome Collection PP/EPR/A.1/1.

eight tests, including injections, and he read the results the next morning, "a fact usually appreciated by patients and doctors."[64] Elkan preferred to work with medium-sized animals: the smallest toads did not tolerate injections very well, and larger ones did not fit into an ordinary two-pound glass canning jar.[65]

Dissatisfied with his reliance on imported stock, Elkan also attempted to breed *Xenopus* in captivity. After "many unsuccessful attempts," he managed to hatch about two hundred tadpoles.[66] These resembled "young fish" standing vertically, head-down, to filter-feed on aquatic microorganisms in the "manner of whales."[67] Not knowing what to feed them, he first rather transgressively offered them emulsion of human blood, obtained from his patients' routine tests. This "worked perfectly," but the supply of "donors" was unpredictable, so he "tried emulsified butcher's liver instead."[68] This worked too, and although only one in five tadpoles survived to adulthood, Elkan predicted that "frog-farming" could become "as remunerative as mouse-farming" was already; he called on the "reptile specialist" to attempt the "difficult task" of breeding the "pathologist's pet."[69]

When, at the start of World War II, the Ministry of Food promoted rabbits for human consumption, Elkan argued that instead of discontinuing pregnancy testing altogether, pathologists should rather use *Xenopus*, which required only a "little scrap of meat once a week."[70] Elkan had performed over 800 tests on 5,000 toads by the time he was imprisoned as an enemy alien, first at Huyton Alien Internment Camp, near Liverpool, and then on the Isle of Man.[71] Although later released into duty, caring for civilians and wounded soldiers at an EMS hospital in Bishop Auckland, County Durham, Elkan did not resume pregnancy testing until after the war.[72] During the war, however, Elkan's former boss, Edgar Obermer, came to the attention of the Home Office for allegedly performing Aschheim–Zondek tests without license or registration.

THE FOREIGN SCIENTIFIC REFUGEE ELEMENT

In the winter of 1943, Obermer came under suspicion of the Ministry of Labour for carrying out "biochemical diagnostic procedures" and a "large scale survey of pregnancy" at his Institute for Medical Diagnosis.[73] Walter Kennedy, the medical intelligence officer assigned to the case, surmised that Obermer was performing Aschheim–Zondek tests, which required a license and a certificate.[74] Obermer's survey was in fact on "wartime rationing and nutrition in pregnancy." Carried out on pregnant women in the outpatient department of the City of London Lying-in Hospital, the results were not published until after the war, by which time Obermer had relocated to Italy.[75] Although Obermer was not actually injecting mice, Major James Alfred Giles, chief inspector at the Home Office under the Cruelty to Animals Act, worried that pregnancy testing, a potentially lucrative and possibly disreputable sideline, was attracting a "proportion of the foreign scientific refugee element."[76]

Ordinarily, Giles would have called on the local police to make inquiries, but in the case of pregnancy testing, his inspectors had long doubted whether an animal injection was "really an experiment within the meaning of the Act" and considered it unwise to "put legal machinery into motion and risk an adverse decision" in court until the question was settled. So, instead of pursuing Obermer, Giles decided to draft a memorandum to clarify the position of pregnancy tests under the act.[77] In it, he explained that a requalified medical refugee could apply for a license, and there was "no obvious reason" to refuse registering premises where pregnancy tests were one of several diagnostic services provided. Yet he also admitted that the "standard of these premises" varied "considerably" and contrasted Queen Charlotte's Hospital, where pregnancy tests were performed under a general registration, with a "small laboratory staffed entirely by foreigners and doing only pregnancy work." It was this second kind of establishment that troubled inspectors, even if there was "no direct evidence of irregularity or illegality."[78]

Giles dreaded the "public outcry" that would surely result if premises registered by the Home Secretary under the 1876 act were ever implicated in illegal abortion, say, if the counsel for the defense brought it up during a criminal trial. He could reasonably argue that was "not for the Secretary of State to trace the origin of specimens tested on any particular registered premises, and that his responsibility is fulfilled if those premises are conducted according to the requirements of the Act." But the position of the Home Office would be badly compromised if it came out that the Home Secretary had, "however unwittingly, registered a laboratory where material supplied by [Harley Street] abortionists is regularly examined." This is why Giles decided to reexamine whether pregnancy diagnosis came within the act.[79]

Pregnancy tests were "quite unknown" in 1876 and therefore, Giles reasoned, "could not have been in the minds either of the Home Office or of the [legal officers] at this time." After reviewing the meaning of "experiments calculated to cause pain" within section 2 of the act, he decided that there could be "no question at all that these tests do not, cannot, and are not intended to cause pain or disease." On the contrary, they seemed to "create a perfectly normal physiological reaction in the ovary," even a "sense of well-being." Giles recommended that pregnancy tests be "taken outside the Act." Not only would this remove all risk of public embarrassment for the Home Office, but it would also relieve inspectors of work that, in 1944, was taking up a "great deal of their time."[80] Although the reasons

(a)

CHAPTER 21

XENOPUS LÆVIS **DAUDIN**

By R. E. ELKAN,* M.D. (Frbg.), L.R.C.P. & S.
(*E.M.S. Hospital, Bishop Auckland, County Durham*)

1. INTRODUCTION. 2. HABITS. 3. NUTRITION. 4. BREEDING. 5. HANDLING. 6. TEST JARS. 7. DETERMINATION OF THE SEXES. 8. DISEASES.

1. Introduction

The South African claw-footed toad is an amphibian belonging to the group *Aglossa*. Morphologically it is a close relation of the Pipa toad, and the

FIG. 56.—*Xenopus lævis* DAUDIN IN COPULATION.
(Note eggs attached to bottom of tank.)

[Reproduced by the courtesy of the *Aquarist and Pondkeeper*.]

* Dr. Elkan's present address is: 62, Woodhall Gate, Pinner, Middlesex.

251

(b)

FIG. 3.—Test jar for xenopus pregnancy test. The centre of lid is perforated. The toad sits on a platform; the eggs fall to bottom of jar. Note typical attitude, holding nostrils above surface of water.

5.9 Two photographs of *Xenopus* in captivity with the distinctive marble-like eggs at the bottom of tank and jar. (a) First page of Elkan's chapter in the *UFAW Handbook*, showing *Xenopus* in amplexus: Elkan 1947a, 251. (b) The iconic *Xenopus* test jar in profile (naturalists and children had long kept frogs and toads in ordinary glass jam or canning jars): Elkan 1938. Cambridge University Library 9385.c.275, with permission by UFAW, and P300.b.111.188, with permission by the BMJ.

for reevaluating the position of pregnancy tests under the act were never made public, the decision to quietly exempt them would have far-reaching consequences, especially for the visibility and acceptance of *Xenopus*.

The removal of pregnancy tests from the ambit of the act was reported in 1947 in the first edition of the highly successful *UFAW Handbook on the Care and Management of Laboratory Animals*.[81] By then, Elkan had relocated his practice to Pinner, and his chapter in the handbook helped to establish *Xenopus* as a standard laboratory animal not only in pregnancy diagnosis but also in developmental biology (figure 5.9).[82] *Xenopus*, Elkan later recalled, had become "quite accepted by the medical profession," and he had assembled "quite a clientele."[83] Some enterprising medical students at

University College London launched their own private service using *Xenopus*, and the biologist Alan Parkes recommended the imported toad as a substitute for British frogs in schools.[84] By the early 1950s, hobbyists were enjoying the "strange and weird performances" of the "interesting species" in aquaria and garden pools.[85] The *Xenopus* test was also adopted on a large scale by the National Health Service, to which I now turn.

THE NATURAL DESIRE OF EVERY WOMAN

It is significant for pregnancy testing that on Monday, July 5, 1948, not one but two national health services came into being. The 1946 National Health Service Act created one service in England and Wales, administered from the Ministry of Health, Whitehall, London. It consolidated the wartime Emergency Laboratory Service (ELS) as the Public Health Laboratory Service (PHLS) with administrative headquarters at Westminster to coordinate a central laboratory at Colindale; four regional laboratories at Oxford, Cambridge, Cardiff, and Newcastle; several smaller "area" laboratories; and fewer reference laboratories for specialized examinations (figure 5.10).[86] In Scotland, a separate piece of legislation, the 1947 National Health Service (Scotland) Act, set the terms for an autonomous organization administered by the secretary of state for Scotland, St. Andrew's House, Edinburgh.[87] This act conserved the "general pattern" of Scotland's Emergency Bacteriological Service, administered by the secretary of state in Edinburgh.[88]

Crew was poised to begin using *Xenopus* routinely in 1939, but the war scuppered his plans. He was recalled by the Royal Army Medical Corps and had handed over control of his institute to Alan W. Greenwood, a zoologist from Melbourne who was running an autonomous subdepartment on poultry genetics.[89] Greenwood not only maintained the pregnancy diagnosis station in Crew's absence but also invested in new animal cages and general refurbishments.[90] It had been "exceedingly difficult" to import *Xenopus* during the war, and mice had "reigned supreme."[91]

Crew returned to the University of Edinburgh in 1944 to take up a chair in social medicine, and when Conrad H. Waddington took over the Animal Genetics Institute in 1947, he transferred the station from the suburban King's buildings site to the Usher Institute of Public Health in the more central area of Marchmont, where his chair was located.[92] Crew hired

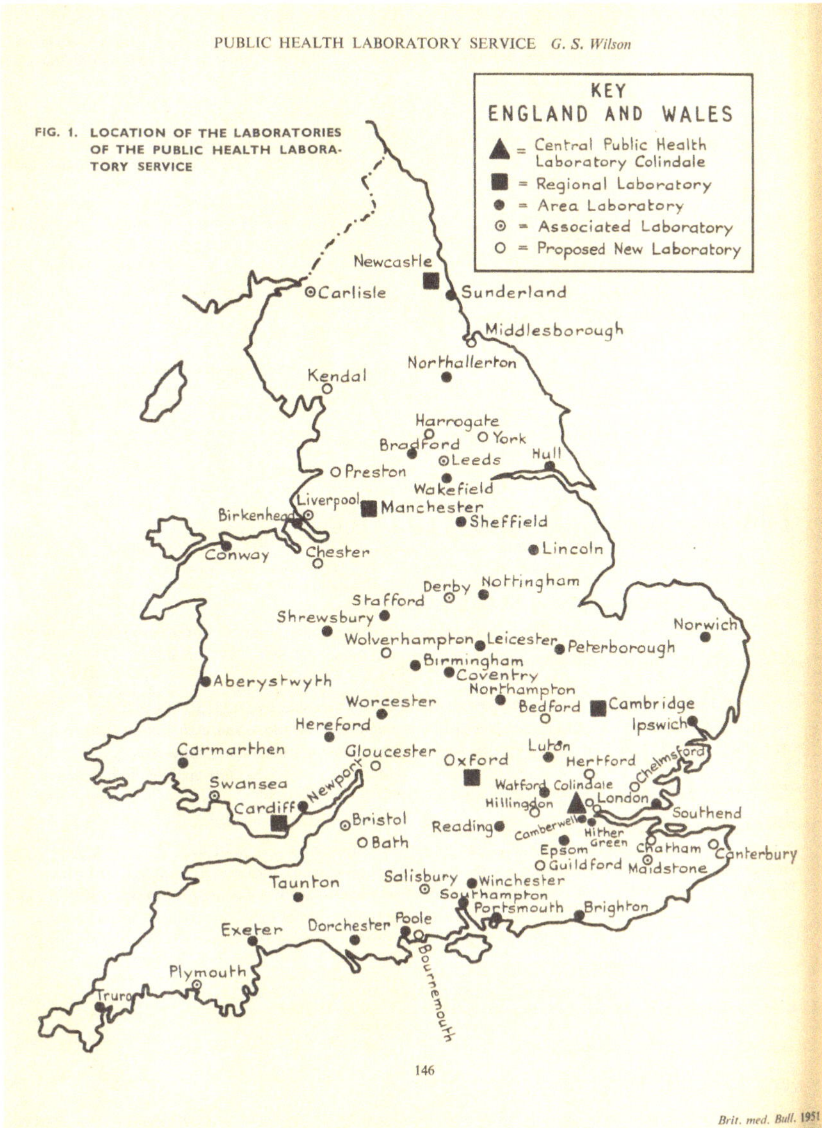

5.10 Map of PHLS labs in England and Wales. Wilson 1951, 146, Cambridge University Library P300.a.11.3, by permission of Oxford University Press.

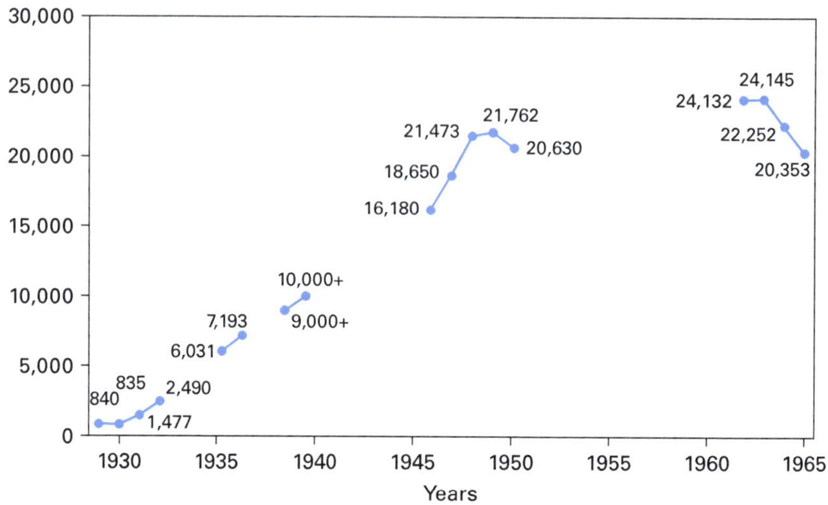

5.11 Annual number of tests performed by the pregnancy diagnosis station, Edinburgh, 1929–1965. Despite a significant gap in the record, the volume of work performed at the station seems to have levelled off at more than twenty thousand tests a year until the early 1960s when the proliferation of private labs using immunoassays caused decline (see chapter 8). In the 1950s the Watford and Sheffield centers (see below) would have contributed perhaps a further twenty thousand tests each, but it is unlikely the grand total (including numerous smaller pathology labs) would have surpassed eighty thousand or so per year. For perspective, married women in England and Wales the 1930s produced six hundred thousand livebirths each year, though the number of pregnancies (including those ending in miscarriage, abortion, or stillbirth) was undoubtedly higher: Szreter 1996, 428. This means that, in the 1930s, there was approximately one test per sixty livebirths (1.7 percent). Based on data from the National Records of Scotland HH 102/858.

Hogben's former assistant Landgrebe for a year to reestablish the *Xenopus* colony, and by 1948, the new laboratory was ready for business.[93] The number of tests had more than doubled from some 10,000 in 1939 to over 21,000 in 1948, and after three years of coexistence, the Hogben test was used exclusively from 1951. The two decades-long reign of mice had ended. By 1952, the stock of *Xenopus* had reached some 6,000 toads, four times as many as before the war (figure 5.11).[94]

During the war, Greenwood had employed a secretary at £7 a week to handle the large sums of money, three office clerks, and half a dozen laboratory technicians and animal attendants. Crew's enlarged station at the Usher employed Bruce Morris Hobson (BSc Aberdeen), as its new scientific director, a university assistant, three secretaries, nine laboratory technicians, a part-time laboratory worker, and a part-time animal attendant. Half of all specimens received in Edinburgh came from England, with half of those from south of Birmingham, with "heavy concentrations" in London and the Midlands. A test cost fifteen shillings for private clinics and seven shillings for hospitals, and in one ten-month period, the station received £6,700 in fees and spent £4,575 on wages, mice, laboratory supplies, telephone bills, stationary, stamps, and sundries to show a surplus of £2,125. But the increasingly voluminous "stream of urine" posed a "dilemma" for Crew. He faced the ruinous prospect of expanding his service at "considerable expense" to keep up with demand until the NHS in England and Wales began offering its own free service, at which point he would be left with an "organisation out of harmony with future needs." Crew endeavored to "steer a very careful course" between continuing to serve southern clients while preparing for the "time when the whole of this custom will collapse."[95]

Before the war, Crew had promoted the democratization of pregnancy testing. Postwar plans for a Scottish health service presented him with the irresistible opportunity to secure a place for his station within a nationalized system of laboratory services. All he needed was a block grant of £4,500 per year to cover equipment, staffing, and running costs.[96] In January 1946, he proposed a "scheme for the provision of pregnancy diagnosis facilities" to Sir Andrew Davidson, the chief medical officer of the Department of Health.[97] Centralization, Crew argued, would save the Scottish service money. As he knew from his wartime experience, when a pathologist performed the occasional test in a clinical laboratory, the cost soared from five shillings to the "absurd heights" of two guineas. Picking up from where he had left off in 1939, he also argued that a centralized service could provide facilities and material for research in human infertility, a vital aspect of social medicine. But, he admitted, there was "no point in . . . developing such a scheme" if it was "not to be used as the basis of a national pregnancy diagnosis service."[98]

But the five regional boards in Scotland had authority, and not all accepted Crew's proposal.[99] Wary of the "various defects of 'postal pathology,'" the Western Board preferred to rely on an "experienced pathologist" at the Glasgow Royal Maternity and Women's Hospital, who performed a steadily increasing number of Aschheim–Zondek tests every month, although Dumfries and other outlying areas continued to post specimens to Edinburgh.[100] When Landgrebe returned to Aberdeen University in 1948, he continued to perform *Xenopus* tests for the Northeastern Region, and from 1951, so too did the regional laboratory at the Aberdeen City Hospital. The Southeastern Board, responsible for Edinburgh, decided not to act on the block grant in the face of such uncertainty.[101]

Scotland's health department considered Crew's proposal, but Charlotte Ann Douglas, the department's adviser on maternity services, was "not too happy" about the prospect of a "central laboratory in Edinburgh [that] would have to be fed by postal services for about 90% of the population of Scotland." As with the specimens Elkan received in London, those sent to the Edinburgh station were "usually accompanied by a note stating that it is a matter of urgency," but Douglas wondered whether the urgency was medical or "emotional." The departmental committee on laboratory services concluded that those Scottish regions that did not rely on Crew's service were "quite happy" with "their alternative arrangements." Even the treasurer of Edinburgh University sided against Crew because of the financial risk posed by the centers that would be opening in England.[102]

During the war, pregnancy tests had been provided under the ELS and Sir Philip Panton, the recently knighted consultant in pathology to the Ministry of Health, expected the status quo to continue under the auspices of regional hospital boards. He favored the *Xenopus* test but saw no need to make a special arrangement for postal references to Crew's laboratory. In 1948, the Ministry began making plans to distribute dozens of toads to each of the major hospitals in England and Wales. But after discussing the matter with Landgrebe, the leading *Xenopus* expert on both sides of the border, Panton instead decided to follow the Scottish model by setting up a few large, specialized centers.[103] Herta Schwabacher (née Moos), a German-born, London-trained pathologist and one of the few women to hold a position of authority in this history, was placed in charge of the NHS pregnancy diagnosis center at Shrodells Hospital in the London suburb of Watford in 1949 with Elkan enlisted to assist with the *Xenopus* colony.[104]

After building the stock up to two hundred toads, the Watford center began testing the first specimens, collected locally in Hertfordshire. As the size of the colony grew, Schwabacher began accepting specimens from London and southern England. February 1949 saw a total of forty tests and

October, over five hundred. After one year in operation, nearly 4,000 tests had been performed, and the *Xenopus* colony had reached its full capacity of 3,000 toads. Doctors submitted urine specimens to Watford for all the usual medical reasons: threatened miscarriage, hormonal imbalance, hydatidiform mole, menopause, tuberculosis, heart disease, fibroids, tumors, and (rarely) testicular cancer in men. But Schwabacher also recognized a smaller portion of requests (around 4 percent) for unmarried girls and women, or what she called "anxiety" cases.

Although not strictly medical, Schwabacher accepted that early diagnosis in such cases would enable patients to "make readjustments in social and domestic life."[105] This might involve preparations for marriage or, if marriage was out of the question, adoption.[106] But a staggering 40 percent of doctors' requests were for married patients with no apparent pathology. Schwabacher feared these "curiosity cases" would eventually "swamp" the *Xenopus* colony to the "detriment" of legitimate "pathological" and "social" cases.[107] She formally discouraged tests from *"married women who are likely to have a normal pregnancy."*[108] A circular sent to London hospitals in 1950 clarified that the Watford center carried out tests within the "hospital pathological service where clinical or social conditions exist" and did "not accept specimens from a [married] woman who is likely to have a normal pregnancy."[109] Yet in practice, the proportion of "curiosity" cases increased slightly to 45 percent as the total number of specimens increased to 6,148 in 1950 (the proportion of "social" cases also increased to 6.2 percent).

A second center established in 1951 by the Ministry of Health at Sheffield City General Hospital to serve the Midlands and northern England had to "severely restrict" its service in 1954 on account of toad supply difficulties.[110] When the *Guardian* incorrectly reported that the Edinburgh station was similarly affected and that it "only dealt with special cases of women with abnormal conditions," Hobson contacted the paper to clarify that his was a "university venture" not affiliated with the NHS, had not "suspended operations," and accepted specimens not only for pathological but also for normal pregnancies.[111] Sheffield again suspended its service in 1958 when a bacterial epidemic nearly wiped out the colony.[112] Aside from these hiccups, the center maintained a stock of 4,000 toads and performed some 20,000 tests a year. Landgrebe, meanwhile, moved to the pharmacology department of the Welsh National School of Medicine in

Cardiff, where he established another *Xenopus* colony and, by 1960, was performing around 2,000 tests a year for more than 130 local GPs. Second only to blood tests for anemia, the Hogben test accounted for an impressive 22 percent of all laboratory investigations for the nearly 311,000 NHS patients in the Cardiff area.[113] Crew's prewar vision of state-led democratization was coming true.

In 1954, Robert Johnstone rated pregnancy tests as second only to X-rays in the "sweeping advances" in obstetrics in the first half of the twentieth century. Pregnancy tests had become "so generally adopted" as to have "largely replaced the skilled clinical methods that [had been] the pride of all previous generations of obstetricians."[114] In a letter to G. P. Wells, Landgrebe estimated that in Cardiff, Watford, and Edinburgh, some half a million or more tests were "done at the request of medical men for good medical reasons."[115] No laboratory could "satisfy the natural desire of every woman to know the truth as soon as possible," but, insisted Schwabacher, unmarried mothers "should have the advantage of laboratory diagnosis . . . in making re-adjustments in social and domestic life."[116] Despite statements to the contrary, the NHS accepted pregnancy testing for "social" reasons, a liberal approach that would be extended to contraception and abortion in the increasingly permissive decades to come.

6

FAMILY PLANNING

On Saturday, July 10, 1948, the novelist Mary Rose Alpers (née Coulton) and her husband, Anthony Alpers, began another day of work in the British Museum Reading Room by "reminding one another [to] telephone Dr. W.'s secretary at noon." When Mary Rose had telephoned the day before, the secretary had made no promises but suggested that the "results of the pregnancy test" might come in the next morning's post, in which case she would be able to tell the couple whether they were "positive or negative" before they went away for the weekend. The pair worked "till noon under the stifling dome" and then, on their way to lunch, Mary Rose telephoned from the "booth in the desolated entrance hall." The secretary answered in a "nice warm voice" that the post had "just this moment come in," that it contained the letter, and, on opening it, that the test result was "positive." "I do hope that's what you want," said the secretary in a "pleasant voice." It was, and the Alperses walked out into the "brilliant summer sunshine" to lunch under the plane tree and "happily" discuss "ways and means."[1]

The brief telephone conversation between Mary Rose Alpers and her doctor's secretary, documented in the book *National Baby* (1950), was the beginning but not the end of a protracted diagnostic experience. Endorsed in the *Guardian* as a "highly entertaining account" of prenatal care from the "consumer's end," *National Baby* was Alpers's "personal chronicle of what it is like to have a first baby under the National Health Scheme."[2]

Although pregnancy testing was by no means routine in the late 1940s, Alpers, who wrote under the pseudonym of Sarah Campion, qualified for a pregnancy test under the new system because she was over forty and "Philip" was her first child. Her unusually detailed account is particularly instructive for pregnancy diagnosis, not only in one woman's experience but also in the broader culture of the early NHS.

This chapter starts by examining firsthand accounts, beginning with Alpers's. It proceeds to recover the proliferation and diversification of postwar services as smaller clinics embraced the "male toad test" and Argentinian innovation, and the Family Planning Association (FPA) established its own "frog lab" and controversially advertised to chemists. It takes an excursion to the western cape of South Africa, to follow platanna collectors on their ever-widening search for the lucrative animal. And it thematizes the publicity drummed up by the Family Planning Association. By the end, a conservative health minister will have visited the association's premises at Sloane Street and posed for press photographs with *Xenopus*, a remarkable milestone in the history of pregnancy testing and more evidence of the permissive trend that would intensify in the 1960s but was already evident in earlier decades.

NATIONAL BABIES

On Monday morning, Alpers trekked to a suburban hospital (she does not say which one) bearing a doctor's letter "in one hand" and the test result "in the other," only to be told to return at the "end of next month" for a booking. She returned home "fulminating against the official mind, feeling like that grotesque old Viscountess in [William Makepeace Thackeray's 1852 novel *The History of Henry*] *Esmond* who was always thinking herself pregnant and never had anything to show for it in the end." Although worried that she might be "undergoing only a hysterical pregnancy," morning sickness soon restored her self-confidence.[3] She welcomed the "butterflies" in her stomach as evidence of the hospital authorities' overcautiousness. Perhaps, she speculated, they were "cautious only because miscarriages [were] so common in the early months, and they themselves so desperately busy."[4] Only some weeks later, when her first prenatal clinic was finally booked for August and a maternity bed for the following March, did she feel "as if the words 'Now officially pregnant' were blazoned on [her] bosom."[5]

In August, when a hospital nurse finally granted her a chit for an expectant mother's ration book (rationing continued until 1954), Alpers declined because she already had been given one by her Health Visitor friend and had been "drawing the correct rations" for the past six weeks. Reflecting on her protracted diagnostic experience, Alpers found it "odd" that the hospital authorities expected her to "wait one and a half months from the declaration of pregnancy, and some two and a half from its inception, for the rations which are supposed to support the infant's growing demands on the body." But in the end, she charitably rationalized, as before, that there are "so many falls by the wayside in the first three months that they have found this the better way."[6]

Perhaps the most relevant aspect of Alpers's account here is the irrelevance of the pregnancy test from the perspective of the hospital authorities. Made by arrangement between Alpers, her GP, and a diagnostic laboratory, probably the Edinburgh station, the positive result, initially so meaningful to Alpers and her husband, was disregarded when it came to hospital protocols regarding prenatal care, the maternity bed, and the ration book. This disjuncture between general practice, on the one hand, and hospitals, on the other, sheds light on the marginal place of pregnancy testing in the early NHS.

Although firsthand accounts are scarce, the evidence suggests that women's diagnostic experiences varied significantly in the 1950s and early 1960s. For a young unmarried girl, the choice was not necessarily hers to make. In her memoir, *Bad Blood*, novelist Lorna Sage recalled her obstinate denial of pregnancy at the age of sixteen in 1959: "It was so unthinkable that when I felt ill, bloated, headachy, nauseous and, oh yes, my period hadn't come, I stayed in bed and called out our new doctor, a pale, prim man in his thirties, Dr. Clayton. After taking my temperature, asking about bowel movements and looking at my tongue, he looked out of the window at the copper beach tree, cleared his throat and asked could I be—um—pregnant? No, I said, feeling hot suddenly, No. He recommended a urine test anyway." Sage was kept home from school for days and took aspirins for her persistent aches until the "embarrassed and puzzled" doctor returned to confront her about the positive test result, at which point she "knew it was true, just as absolutely as until that moment I knew it couldn't be."[7]

Sheila Walker, an Edinburgh-born woman interviewed for the Millennium Memory Bank oral history project, was living in the Surrey countryside and expected to marry her boyfriend when she "got pregnant," also in 1959, but at the somewhat older age of nineteen. She "went to the doctor's and of course in those days you didn't get early pregnancy tests. You waited till you missed your second period. And it was then that you might start worrying. And I went to the doctor and of course the doctor would take a test and it would take a week before you got, you know, the result. So that's what it was like in those days." When she told her boyfriend she had "missed [her] second period" and the doctor had "confirmed" she was "having a child," her boyfriend, who had been cheating on her, "just went completely cold" and refused to marry her; her father convinced her to give up the child for adoption.[8]

For married women, a pregnancy test could still be seen as an expensive luxury. Hope, a London-born woman had worked as a lab technician for British Drug Houses before getting married and moving to Oxford. In an interview with historian Angela Davis, she later recalled visiting her doctor at the age of twenty-five in 1955:

> I went to the doctors. Erm, and we hadn't actually planned to have one quite that soon, so I was moderately upset erm, and pregnancy testing was only just starting then, and I went and you know until the doctor could actually feel something you couldn't be certain that you were pregnant. I said "What about a pregnancy test?" which in those days I think they injected some of your urine into a frog and it ovulated or something, and he said, "Well yes, we can arrange that but it would cost I don't know how much," and I said, "Oh we can't afford that we need that money for the baby," so erm anyway I was pregnant and it duly arrived at the appropriate time.[9]

Ovulating toads occasionally featured in women's diagnostic narratives, including that of Claire Rayner, a prolific broadcaster, agony aunt, and novelist.[10] In the early 1960s, Rayner's doctor broke the news by telephone: "'Claire,' she said, in as delighted a voice as I had ever heard her use, 'that specimen you left with me—I have to tell you I was right and there are, as I've just heard from the laboratory, a couple of toads there are in a state of *great* excitement. You are pregnant, my dear, most definitely pregnant. You did say you were planning another baby, didn't you? How lovely for you!'"[11] As a "young wife" in Manchester in the 1960s, Maureen Symons was "desperate for a family" but would "miscarry and then be told a few days later that indeed I had been pregnant, my toad had laid eggs." She eventually gave birth to a "beautiful daughter, Kate, and then a son, Daniel." Decades later, she would still smile upon seeing a toad in her garden, remembering the "efforts they made on my behalf (even if they aren't the right species!)."[12]

Audrey Peattie, a former laboratory technician I interviewed in 2011, remembered "quite a few" specimens from "unmarried mothers" at the Watford center in the 1950s (figure 6.1). But otherwise, she did not know "anything much" about the lives of the women whose urine specimens she centrifuged and injected into toads. Although "quite scared" of Schwabacher, a "formidable lady," she fondly recalled Elkan as a "funny old boy that came down a couple of times a week and just fiddled about in the lab." She expressed an interest in his work, and he often called her over to "look at things" and "explain a bit," which is why she also remembered *Bufo bufo*, "quite little ordinary toads like the kind you see in the garden."[13]

A DELEGATION OF TOADS

In December 1948, a *Lancet* editorial surveyed twenty years of pregnancy diagnosis since the invention of the Aschheim–Zondek test. "The day has not yet arrived," it concluded, "when the doctor can tell his patient that she is pregnant by pouring her urine into a tank of fish and watching their bellies become red; though this was the great expectation which the male bitterling at one time held out."[14] The editorial was, however, cautiously optimistic about the Galli Mainini test, on which Magnus Haines, director of pathology at the Chelsea Hospital for Women, had recently reported. In 1947, the Argentine physiologist Carlos Galli Mainini determined that pregnant women's urine injected into the lymph sac of the male South American toad, *Bufo arenarum* (today usually reclassified as *Rhinella*), caused it to ejaculate.[15] After two or three hours, its urine could be pipetted from the cloaca and inspected under a microscope for the presence or absence of sperm (figure 6.2).

Haines had been driven to the male toad test by a sense of dissatisfaction with the others: the supply of infant female mice, he complained, was

6.1 Photograph by Elkan, an avid amateur photographer, of Peattie (front) injecting *Xenopus* with urine while her colleague Marion (back) prepares the next syringe. Note the prominent test jar on the work surface and the holding tanks on the shelving units in the background of the warehouse-like room; courtesy of Peattie.

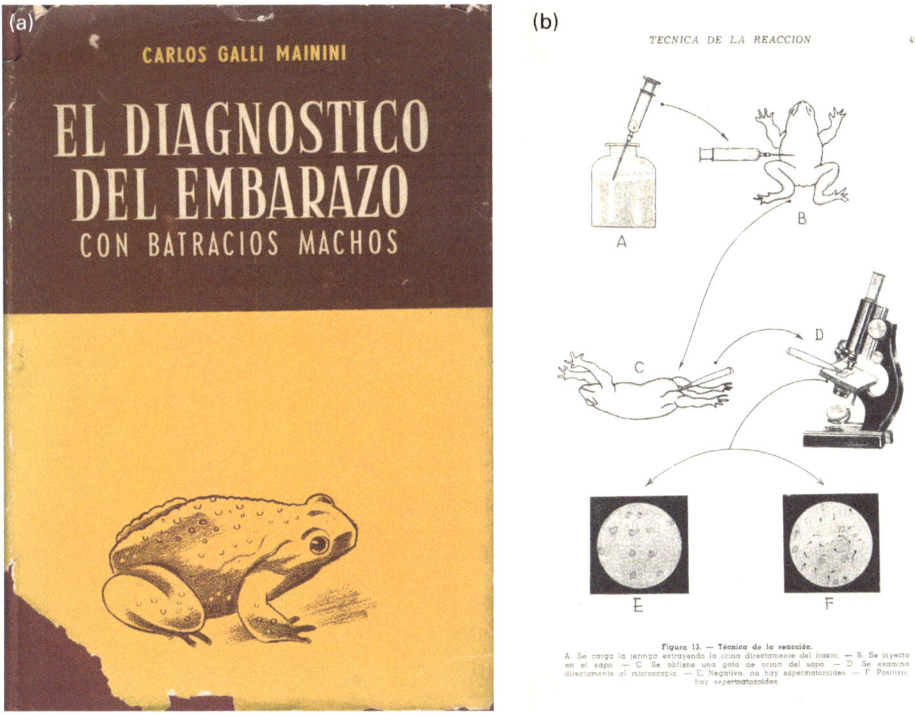

6.2 (a) Dust jacket from Galli Mainini's monograph on the male toad pregnancy diagnosis test. (b) Diagram from the same book depicting the steps involved and the two possible outcomes. Gailli Mainini 1948a, 45.

unreliable; rabbits had become "costly and difficult to obtain"; the "early enthusiasm" for rats had waned; and *Xenopus* had "not been adopted generally."[16] He had written to Galli Mainini in Buenos Aires, who had arranged for the transport of 136 toads to London as part of a traveler's luggage. Haines praised the male toad test on the grounds that it only required three hours to complete, reading the result did not require any "special skill," and the "whole operation of pipetting off the urine and its microscopical examination" lasted only a few seconds. The toads required "no special tanks," but importing them was inconvenient.[17] Researchers in Paris had recently reported "equally good results" with "one of the common frogs found in France," the edible *Rana esculenta*.[18] So, with help from the keeper of zoology at the British Museum and the curator of reptiles at the London Zoological Gardens, Haines began using the locally abundant *Bufo bufo*.

In the 1950s, it was still "more convenient and economical" to collect dogs, cats, monkeys, and amphibians "from the wild" than to breed these "slowly maturing" species in captivity.[19] *Bufo bufo* could be purchased from a dealer for anything from a few pence to one shilling depending on the season.[20] Haines and others rapidly determined that British toads worked just as well as their South American counterparts in the Galli Mainini test (ironically, *Xenopus* was one of the only species found to be unsuitable). Clinical pathologists soon began using *Bufo bufo* at St. Mary's Hospital Medical School, London.[21] As pathologists at Chase Farm Hospital, Enfield, became familiar with the "fascinating variety of protozoal life to be found in a toad's cloaca, a cursory low-power examination became sufficient to establish a diagnosis."[22]

In the second edition of his pocket-sized handbook, *Gynaecological Endocrinology for the Practitioner* (1951), Peter Bishop predicted that the male toad test would "in time supersede the other tests."[23] Gwen Barton, a pathologist at the Salisbury Infirmary, praised the test for making pregnancy diagnosis available to even the "smallest laboratory."[24] Grace Jeffree of Bristol predicted that the "cheapness, simplicity, and speed of the toad test are such that it may be expected to replace the Friedman test in many laboratories."[25] And a review of pregnancy tests in the *Postgraduate Medical Journal* recommended *Bufo bufo* over *Xenopus* for smaller laboratories (figure 6.3).[26]

By the mid-1950s Galli Mainini's test had been "carried out all over the world using nearly every species of male toad and frog with almost universal success."[27] At the end of the decade, a UFAW report ascribed the 3,802 Friedman tests performed in 1952 to "inertia" and the rise of amphibia to "their increasing popularity for pregnancy diagnosis, in which they are tending to displace mice and rabbits."[28] Although hospitals were unaccustomed to housing toads, pathologists were able to improvise with available materials. Rhoda M. Allison, an assistant pathologist at the Huddersfield Royal Infirmary in Yorkshire, converted standard metal guinea pig trays into residential boxes for toads, ordinary fish tanks into feeding tanks, photographic developing dishes and enamel surgical trays into water pots, and pathology specimen jars into test jars (figure 6.4).[29] Although *Bufo bufo* was "adequate and satisfactory for routine work," she preferred to rely on a "number of foreign toads and frogs" during the winter months.[30]

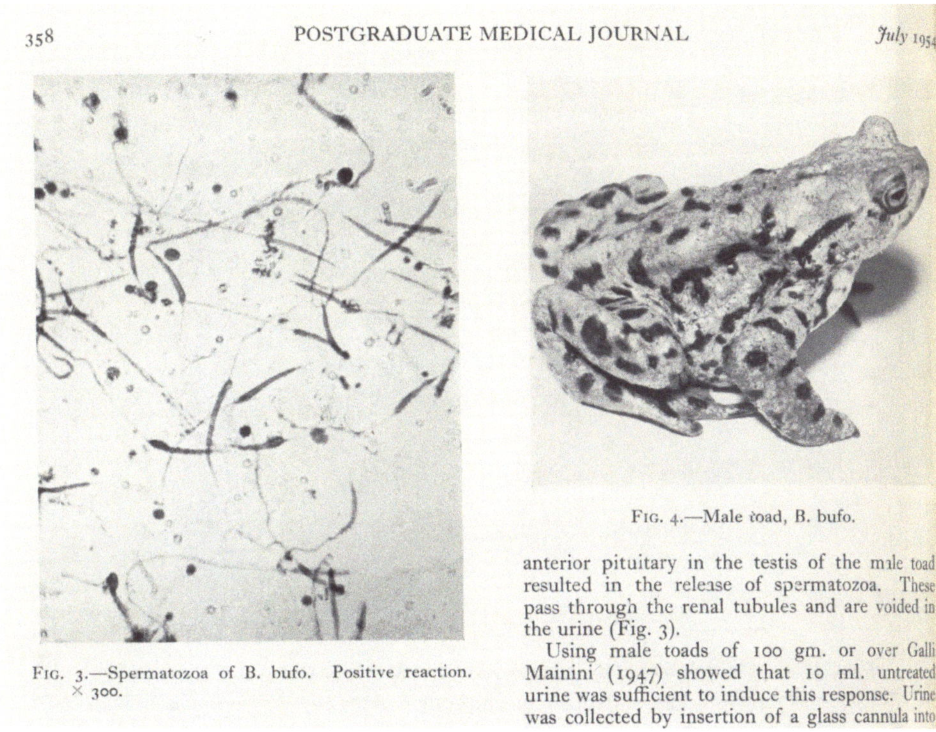
358 POSTGRADUATE MEDICAL JOURNAL *July* 1954

FIG. 3.—Spermatozoa of B. bufo. Positive reaction. × 300.

FIG. 4.—Male toad, B. bufo.

anterior pituitary in the testis of the male toad resulted in the release of spermatozoa. These pass through the renal tubules and are voided in the urine (Fig. 3).

Using male toads of 100 gm. or over Galli Mainini (1947) showed that 10 ml. untreated urine was sufficient to induce this response. Urine was collected by insertion of a glass cannula into

6.3 Photographs of the male British toad, *Bufo bufo*, and his magnified sperm. Ferreira 1954, Cambridge University Library P300.b.285.30, with permission by the BMJ.

Although generally considered more convenient than *Xenopus* for small-scale laboratories, *Bufo* stubbornly resisted scaling up. Elkan lost about half his experimental stock at Watford from unknown causes every year. The situation was paradoxical. Imported toads thrived in captivity while native ones starved to death in the presence of abundant flies and mealworms. In solitary confinement, a British toad could survive for many years, but it fared less well when kept in a larger group, even when provided with abundant water and food and moss for shelter. *Bufo* evidently lost its appetite in a crowd and, unable to overcome its inhibitions, starved to death. Elkan concluded that British toads were "rigidly conditioned animals and even if it costs them their life they cannot learn. The whole picture might change if we could adjust our laboratories so that each toad had its own cage," but that suggestion was "too uneconomical to deserve much consideration."[31]

(a)

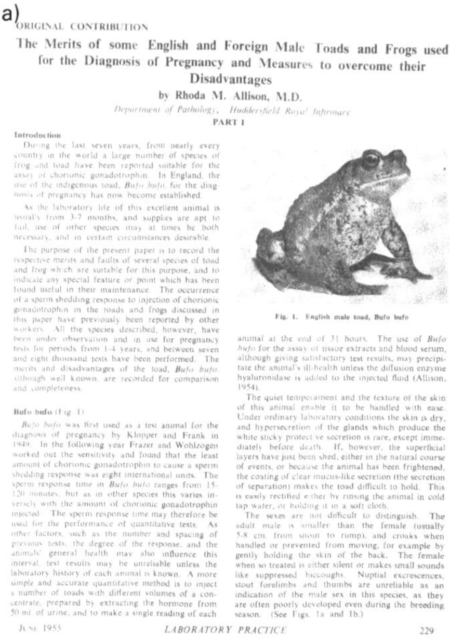

ORIGINAL CONTRIBUTION

The Merits of some English and Foreign Male Toads and Frogs used for the Diagnosis of Pregnancy and Measures to overcome their Disadvantages

by Rhoda M. Allison, M.D.

Department of Pathology, Huddersfield Royal Infirmary

PART I

Introduction

During the last seven years, from nearly every country in the world a large number of species of frog and toad have been reported suitable for the assay of chorionic gonadotrophin. In England, the use of the indigenous toad, *Bufo bufo*, for the diagnosis of pregnancy has now become established.

As the laboratory life of this excellent animal is usually from 3-7 months, and supplies are apt to fail, use of other species may at times be both necessary, and in certain circumstances desirable.

The purpose of the present paper is to record the respective merits and faults of several species of toad and frog which are suitable for this purpose, and to indicate any special feature or point which has been found useful in their maintenance. The occurrence of a sperm shedding response to injection of chorionic gonadotrophin in the toads and frogs discussed in this paper have previously been reported by other workers. All the species described, however, have been under observation and in use for pregnancy tests for periods from 1-4 years, and between seven and eight thousand tests have been performed. The merits and disadvantages of the toad, *Bufo bufo*, although well known, are recorded for comparison and completeness.

Bufo bufo (Fig. 1)

Bufo bufo was first used as a test animal for the diagnosis of pregnancy by Klopper and Frank in 1949. In the following year Frazer and Wohlzogen worked out the sensitivity and found that the least amount of chorionic gonadotrophin to cause a sperm shedding response was eight international units. The sperm response time in *Bufo bufo* ranges from 15-120 minutes, but as in other species this varies inversely with the amount of chorionic gonadotrophin injected. The sperm response time may therefore be used for the performance of quantitative tests. As other factors, such as the number and spacing of previous tests, the degree of the response, and the animals' general health may also influence this interval, test results may be unreliable unless the laboratory history of each animal is known. A more simple and accurate quantitative method is to inject a number of toads with different volumes of a concentrate, prepared by extracting the hormone from 50 ml. of urine, and to make a single reading of each

Fig. 1. English male toad, Bufo bufo

animal at the end of 3½ hours. The use of *Bufo bufo* for the assay of tissue extracts and blood serum, although giving satisfactory test results, may precipitate the animal's ill-health unless the diffusion enzyme hyaluronidase is added to the injected fluid (Allison, 1954).

The quiet temperament and the texture of the skin of this animal enable it to be handled with ease. Under ordinary laboratory conditions the skin is dry, and hypersecretion of the glands which produce the white sticky protective secretion is rare, except immediately before death. If, however, the superficial layers have just been shed, either in the natural course of events, or because the animal has been frightened, the coating of clear mucus-like secretion (the secretion of separation) makes the toad difficult to hold. This is easily rectified either by rinsing the animal in cold tap water, or holding it in a soft cloth.

The sexes are not difficult to distinguish. The adult male is smaller than the female (usually 5-8 cm. from snout to rump), and croaks when handled or prevented from moving, for example by gently holding the skin of the back. The female when so treated is either silent or makes small sounds like suppressed hiccoughs. Nuptial excrescences, stout forelimbs and thumbs are unreliable as an indication of the male sex in this species, as they are often poorly developed even during the breeding season. (See Figs. 1a and 1b.)

JUNE 1955 *LABORATORY PRACTICE* 229

(b)

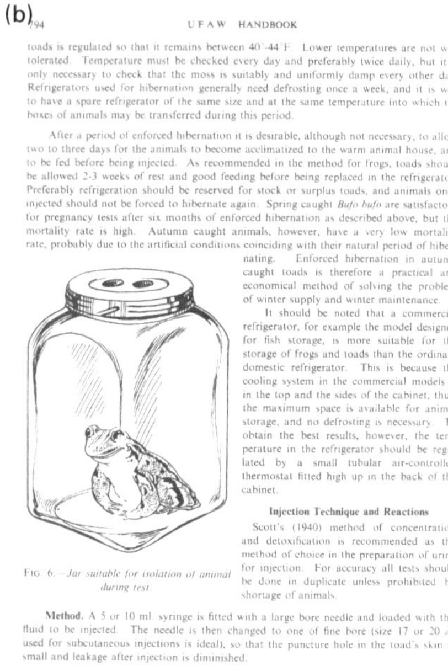

794 UFAW HANDBOOK

toads is regulated so that it remains between 40°-44°F. Lower temperatures are not well tolerated. Temperature must be checked every day and preferably twice daily, but it is only necessary to check that the moss is suitably and uniformly damp every other day. Refrigerators used for hibernation generally need defrosting once a week, and it is well to have a spare refrigerator of the same size and at the same temperature into which the boxes of animals may be transferred during this period.

After a period of enforced hibernation it is desirable, although not necessary, to allow two to three days for the animals to become acclimatized to the warm animal house, and to be fed before being injected. As recommended in the method for frogs, toads should be allowed 2-3 weeks of rest and good feeding before being replaced in the refrigerator. Preferably refrigeration should be reserved for stock or surplus toads, and animals once injected should not be forced to hibernate again. Spring caught *Bufo bufo* are satisfactory for pregnancy tests after six months of enforced hibernation as described above, but the mortality rate is high. Autumn caught animals, however, have a very low mortality rate, probably due to the artificial conditions coinciding with their natural period of hibernating. Enforced hibernation in autumn caught toads is therefore a practical and economical method of solving the problem of winter supply and winter maintenance.

It should be noted that a commercial refrigerator, for example the model designed for fish storage, is more suitable for the storage of frogs and toads than the ordinary domestic refrigerator. This is because the cooling system in the commercial models is in the top and the sides of the cabinet, thus, the maximum space is available for animal storage, and no defrosting is necessary. To obtain the best results, however, the temperature in the refrigerator should be regulated by a small tubular air-controlled thermostat fitted high up in the back of the cabinet.

FIG. 6.—*Jar suitable for isolation of animal during test.*

Injection Technique and Reactions

Scott's (1940) method of concentration and detoxification is recommended as the method of choice in the preparation of urine for injection. For accuracy all tests should be done in duplicate unless prohibited by shortage of animals.

Method. A 5 or 10 ml. syringe is fitted with a large bore needle and loaded with the fluid to be injected. The needle is then changed to one of fine bore (size 17 or 20 as used for subcutaneous injections is ideal), so that the puncture hole in the toad's skin is small and leakage after injection is diminished.

6.4 Allison's lengthy and detailed reviews of the male toad test in (a) the journal *Laboratory Practice* (1955), 229, and (b) the second edition of the *UFAW Handbook* (1957), 794, established standard guidelines for the use of the male toad test, much as Elkan's articles had done for *Xenopus* a decade earlier. Cambridge University Library L340.1.c.175.4 and 384:2.b.95.1.

Pat Fincham, who began using the Aschheim–Zondek test as a junior technician at the pathology laboratory of the Royal Northern Hospital at Holloway Road, London, in 1949, soon switched to male toads. At eighty years, she still remembered "being shocked at the thought of killing 3 mice to do a pregnancy test and much happier to see the toads sitting in their jars and knowing they would survive the ordeal."[32] The Group Laboratory at St. Stephen's Hospital, Chelsea, used male toads in the early 1960s.[33] And a technician at Chester city hospital remembered bringing them "outside to play on the grass" as late as 1966.[34] The use of toads made diagnostic work less unpleasant for technicians, and the decision by the Home Office to exempt pregnancy tests from its definition of "experiments calculated to cause pain" made it possible for medical writers to promote the Hogben and Galli Mainini tests as a benign, even pleasurable, experience for the animal.

Within the NHS, local authorities took charge of health education in the 1950s. Health reporting, meanwhile, was concentrated in women's magazines.[35] Inspired by the success of the American Medical Association's mass-market magazine *Today's Health*, the British Medical Association (BMA) launched *Family Doctor* in 1951. Available from bookstalls and newsstands around Britain, *Family Doctor* communicated medical knowledge and heath advice to general readers, especially mothers.[36] Before the end of the decade, it had provided Liverpool physician and medical writer Robert Kemp with a forum to promote toads as "friendly creatures . . . doing a most valuable job in pregnancy diagnosis."[37]

In lively, humorous prose, Kemp argued that it was "most unfair" to think of toads as "ugly, slimy, and repulsive" even if they were unlikely to "turn into a handsome young prince as the frog does in the fairly tale." The toad was "actually a creature of deep thoughtful character who might be quite friendly if only he could express himself." Kemp told the story of how his "obviously upset" pathologist friend had recently led him "rather tragically to his animal house," where the previous day a "small consignment" of "eagerly awaited" female *Xenopus* had been "carefully" placed in "warmed water in a deep porcelain basin." At night, they "jumped very easily out of their white pond and wandered off in all directions, possibly in search of male company. They made for all the grids of the neighbourhood and several never came back."[38]

Kemp was also shown some dry tanks containing "English toads" with the "lovely name," *Bufo bufo*, and skin the "colour of autumn leaves." He returned to his friend's laboratory the next day with a sample of blood serum taken from one of his hospital patients (most likely without her knowledge), a girl he "thought might be pregnant."[39] She was, and Kemp witnessed the positive end point of both the Hogben and Galli Mainini tests. In a glass jar assigned to Kemp's patient, one of the remaining female *Xenopus* was "busy laying long streams of black dotted eggs."[40] A laboratory assistant then showed Kemp the "swarms" of sperm "put out by the male toad in response to some mysterious message given by the pregnancy hormones circulated in [his] patient's blood." Although there were "many occasions when it is really important [for a gynecologist] to know at the earliest possible moment whether someone is pregnant or not," as a generalist, Kemp "did not use this sort of test much."[41] So why did he welcome these "simple" and "convincing" tests with so much verve?

One new reason that had not been available in the 1930s or 1940s had to do with concerns over a possible link between diagnostic radiology in utero and childhood leukemia and other cancers. In 1956, Alice Stewart, the assistant director of the Institute of Social Medicine at Oxford, had reported in the *Lancet* that mothers of deceased children were twice as likely to have been X-rayed while pregnant than mothers of living children. The full report, published in the *BMJ* in 1958, confirmed her preliminary findings.[42] In 1959, while the medical profession debated the "possible harm" of fetal irradiation, Kemp preferred to "keep mothers away from any *x*-rays that [were] not absolutely essential." Although "not very new," the value of pregnancy testing was "now becoming more widely recognised and many more laboratories [were] providing the service." He also disavowed the "'let nature take its course' school": "There are so many social and medical reasons for knowing whether the first missed period does mean a pregnancy. It might even be said that a woman has the inherent right to know where she stands on this very important matter. I am sure that most mothers would find plenty of reasons for supporting me on this vital point."[43]

Kemp's appeal to a woman's "inherent right to know" was precocious in 1959; it would gain ground a few years hence, when newspapers and magazines openly debated the rise of direct-to-consumer pregnancy testing (see chapter 8). The toads, Kemp concluded, were "certainly doing a very useful friendly job," but what did they "think of it all?" His pathologist friend had "every reason to believe that it is a very pleasurable sensation and occasion for both male and female toads. They really seem to enjoy it, and they certainly seem quite happy living here in these tanks." Kemp ended his article with the fanciful image of a "delegation of toads seeking admission to his [friend's] animal house because they had heard that what Nature had decreed to be an irregular event was there taking place under the plushiest conditions every month in a toad's life" (figure 6.5).[44] A far cry from the gruesome dissections that shrouded laboratory practice in the 1930s, pregnancy testing with toads presented a softer, more humorous image and new opportunities for public visibility. These were seized on by the FPA, to which I next turn.

Toads have been given a bad name for too long and with no reason. Some families of toads are friendly creatures doing a most valuable job in pregnancy diagnosis

The friendly bufo bufo and his chums

by ROBERT KEMP, M.D., M.R.C.P.

6.5 Kemp's article, including photographs of *Xenopus* and *Bufo* by Colonel Basel Blewitt, M.D. The captions, from left to right, read: "Hard at work on a pregnancy test. This lady comes from South Africa and sits on an elegant glass tray with water right up to her neck"; "This gentleman is not suffering. He was injected with the serum of a woman through to be pregnant and now samples of his water are being taken through a glass tube to be examined. In this particular case the test was positive"; "And here is what *Bufo bufo* was producing. A swarm of male sperms are seen under the microscope, looking exactly like baby tadpoles." *Family Doctor*, May 1959, 280–281, Cambridge University Library L300.b.142.9, with permission by the BMA.

THE FAMOUS TOADS OF SLOANE STREET

In 1949, Margaret Pumphrey, a twenty-eight-year-old newlywed from Birmingham, wrote to the FPA, "I am very irregular with my periods. They vary from four to six weeks, but never have I been beyond the six week mark. I am now well on into my 7th week and I do not know whether I am extremely late or pregnant . . . I had always been under the impression that a water test could be taken within a few days after conception. . . . Could you please enlighten me on these points." Pumphrey, who was looking forward to becoming a mother for the first time, also asked for the "names of one or two reliable books on the subject of pregnancy and motherhood" and enclosed a postal order of two pounds. Although the FPA did not have any relevant books, it did have a "special Centre in London for the diagnosis of pregnancy," and Stephanie Robinson, the general secretary of the FPA, enclosed "instructions as to the way in which a specimen can be sent."[45]

Birth control was not championed in the "reforming zeal" to create the NHS; negotiations leading up to the 1946 act made no mention of family planning and left underlying contraceptive legislation unamended.[46] In 1949, the Royal Commission of Population recommended the NHS take charge of clinics providing infertility services and contraceptive advice, but the Chancellor of the Exchequer rejected the proposal as too costly for taxpayers.[47] Politicians generally shied away from reproductive policies in peacetime. The NHS enabled local health authorities to contribute to voluntary organizations like the FPA, but it was unclear whether GPs could charge a fee for contraceptive advice.[48] For guidance, local authorities looked to a 1930 Ministry of Health circular (153/MCW) that permitted prenatal clinics to dispense contraceptive advice to a married woman but only if her health was seriously threatened by pregnancy. In 1946, the FPA's sixty clinics could not keep up with demand, and by 1950, just over a third of some 145 local health authorities in England and Wales dispensed contraceptive advice (often liberally interpreted as medically indicated) at special clinics for married women. Other authorities referred patients to a local FPA clinic or hired premises to an FPA branch, while still others provided their own services.[49]

Toward the end of World War II, the FPA extended its range of activities to include infertility treatment. With money donated by Gerda Guy, Hans Adolf Davidson set up a seminological laboratory in 1945 at 33 Wimpole Street, near Harley Street.[50] Some seventy patients attended the clinic each month for semen analysis, a postcoital test, or testicular biopsy in the first year, and the number more than doubled in the second.[51] Hospitals, infertility clinics, private consultants, and GPs referred men, mostly recently married husbands, to Davidson's laboratory for testing.[52] And by June 1948, he was struggling to "keep up with the demand for microscopic investigation."[53] In 1949, the FPA purchased larger premises at 64 Sloane Street, just south of Hyde Park, and—at the suggestion of Helena Wright's son, Beric Wright—opened a pregnancy diagnosis center in the basement.[54] The association would come to regard the "famous" toads of Sloane Street as a "much-prized asset."[55] Getting to see them was a memorable experience for ladies and schoolboys alike.[56]

Beric Wright had learned to perform Hogben tests during the war while working with Schwabacher in the EMS.[57] He expected the initial investment to be quickly offset if the service were adequately promoted and argued that the FPA would benefit financially from the income generated by pregnancy testing.[58] In his letters to potential benefactors and clients, he anticipated a "steady profit" for the association even after taking a "small fee" for his own time and involvement. The FPA was in a "desperate financial situation," and Wright promised that pregnancy testing would "balance the Association's budget," liberating it from dependence on "private subscriptions and the usual methods by which voluntary organizations are always fighting to balance their overdrafts."[59]

The FPA agreed to provide Wright with £100 to cover startup costs, and Wright planned to open the laboratory in March 1949.[60] He prosed two rates: twenty-five shillings for private patients and twelve shillings and sixpence for hospital patients. This was more expensive than the Edinburgh station, but Wright claimed that there was a "shortage of laboratory animals and . . . difficulties in getting a pregnancy test done rapidly."[61] Harley Street doctors and other local clients would no longer have to wait an entire week for a urine specimen and test result to work its way through the postal service. He even anticipated setting up monthly accounts for regular users. And when he spent twice as much as planned on equipment, including glass bottles and a bespoke galvanized steel "frog tank," Guy agreed to donate a further £100.[62]

Wright first approached his wartime supervisor Schwabacher about whether she might be able to refer cases to his laboratory or if anyone else at the NHS would discuss the matter and arrange fees. Schwabacher informed Wright that, unfortunately for him, the Ministry of Health was already in the process of establishing its own center in Watford to serve southern England and offered him the use of her toads when his were "overworked." Wright next consulted Robert Forbes, secretary of the Medical Defence Union, to find out whether he would be within his rights to distribute a promotional leaflet about the laboratory. Forbes explained that Wright's leaflet would not be "objectionable" because a BMA resolution adopted in 1932 stipulated that a "practitioner who wishes to draw the attention of his colleagues in the profession to the fact that he has recently commenced or intends to practise any particular branch of medical or surgical work, may do so . . . by calling upon practitioners already established in the area & giving a personal explanation of his arrangements."[63]

Wright sent his leaflet to doctors and pathologists in and around London who were already known to the FPA. And general secretary Robinson contacted the Ministry of Health to request the names and addresses of the Group Pathological Laboratories around Britain to inform them of Sloane Street. George Godber responded that the Ministry was in the process of establishing its own pregnancy diagnosis service at Watford and would soon be opening a second one at Sheffield. He clarified, however, that these centers would provide an "essentially medical service within the hospital scheme" and would not be available in "any early pregnancy but only when this is really necessary on medical grounds." Watford was "not yet able to cope with all requirements," but information about it and Sheffield had already circulated within the hospital service, so Godber doubted the Ministry would also agree to publicizing Sloane Street.[64]

Robinson pressed Godber for a list of addresses. She suggested that group pathology laboratory directors might be "grateful for [the FPA] to do some tests for them" until the NHS was able to meet the demand. Furthermore, although laboratories within the health service were "unable to perform [tests] from doctors for their private patients," they might still be able to refer such cases to the association. But after consulting with Philip Panton, Godber informed Robinson that the Ministry would not be able to encourage a "redundant" laboratory. There might be plenty of demand "outside the scope of the Health Service," but the association could not count on the NHS for support.[65]

Wright nevertheless managed to secure contracts with hospitals not only in London but also in East Surrey and as far north as Hull, Salford, and Newcastle. He finally turned to Francis Crew, his principal competitor, with whom he had become acquainted during the war. Wright had heard from a colleague that the Edinburgh station was "overworked" and "anxious to try and reduce the number of tests coming in." Furthermore, hospitals in southern England had to "wait ten days or so" for a test result from Edinburgh. Because of this delay and because Wright was "anxious to build up" the Sloane Street service, he cautiously asked if Crew would be willing to pass on "some of the work from the South." But, as we saw in the previous chapter, Crew was in fact worried about losing his clients from England to the NHS, and so he rejected Wright's offer.[66]

COLLECTING PLATANNAS

Wright's first shipment of *Xenopus* arrived in April 1949. Their passage was arranged by the travel agency Thomas Cook, which in turn obtained them from the Department of Inland Fisheries of the Cape Provincial Administration (CPA) (figure 6.6).[67] The toads, colloquially known as platannas, were initially "obtained from coloured [*sic*] collectors, who caught them mainly in small waters on the Cape Flats, or from one or more European dealers who also did a small amount of export before the war."[68] But supplies were "intermittent," and it was not always possible to obtain the mature females needed for pregnancy testing. Land drainage for agriculture and the introduction of largemouth bass for fishing made matters worse.[69] With demand increasing and fears of a shortage looming, Louis Bosman, a colleague of Zwarenstein and enthusiast of the *Xenopus* test, successfully petitioned the CPA to introduce protective legislation and plan for artificial cultivation at the provincial trout hatchery in Jonkershoek.[70] In August 1941, the CPA amended the Inland Fisheries Ordinance to cover "aquatic fauna generally" and authorized the hatchery to construct concrete holding tanks, use hatchery personnel and transport to collect *Xenopus* from farm dams, and breed them for medical and scientific purposes.[71]

Collecting was hard work. Douglas Hey, the hatchery's resident biologist, would set off with three assistants in an "ex-army three-ton Chevrolet truck loaded with containers, nets and traps." They visited several farms and learned how to "assess the potential of a dam almost at a glance." At first, they foraged in the vicinity of Stellenbosh but were gradually forced "further afield," to Paarl, Caledon, Malmesbury, and at last Piketberg (figure 6.7). They paid regular visits to one farm that employed a team of Italian prisoners of war who "objected strongly" when they started catching toads. The prisoners had recently identified platannas as a "delicacy and were catching them regularly for food." Decapitated, skinned, eviscerated, salted, peppered, rolled in flour, "fried to a crisp brown in boiling

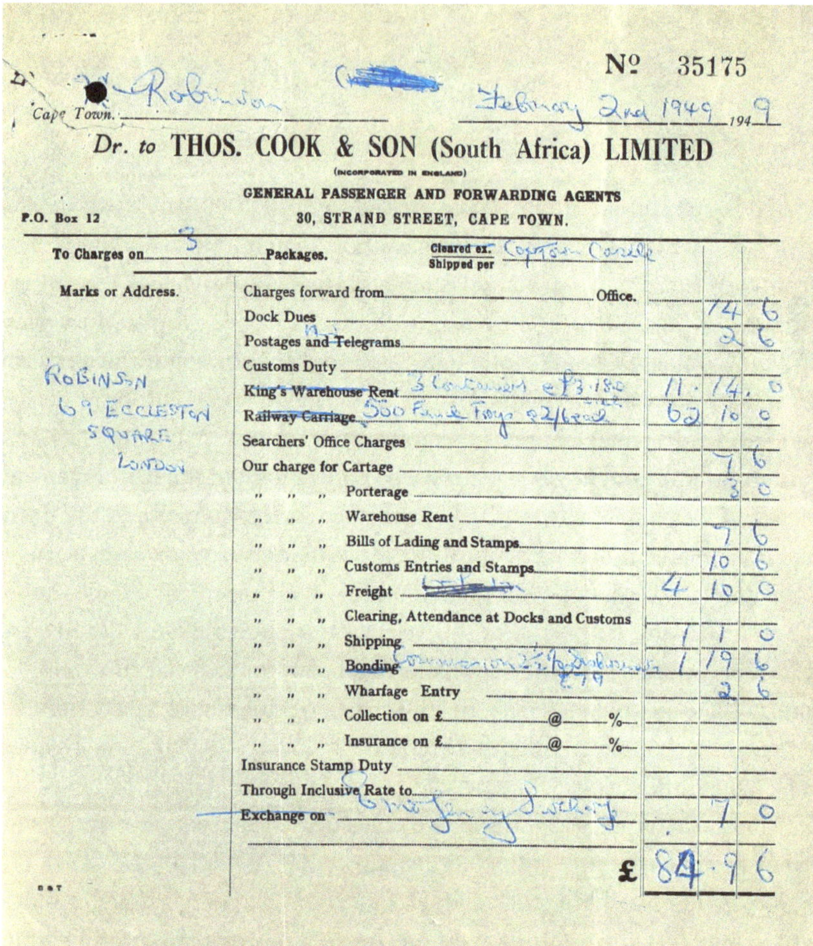

No. 35175

Cape Town. Robinson February 2nd 1949 194 9

Dr. to THOS. COOK & SON (South Africa) LIMITED

(INCORPORATED IN ENGLAND)

GENERAL PASSENGER AND FORWARDING AGENTS

P.O. Box 12 30, STRAND STREET, CAPE TOWN.

To Charges on 3 Packages. Cleared ex. / Shipped per Capetown Castle

Marks or Address.

ROBINSON
69 ECCLESTON
SQUARE
LONDON

Charges forward from ... Office.			
Dock Dues		14	6
Postages and Telegrams		2	6
Customs Duty			
King's Warehouse Rent 3 containers @ £3.18.0	11	14	0
Railway Carriage 500 Frogs @ 2/6 each	62	10	0
Searchers' Office Charges			
Our charge for Cartage		7	6
" " " Porterage		3	0
" " " Warehouse Rent			
" " " Bills of Lading and Stamps		7	6
" " " Customs Entries and Stamps		10	6
" " " Freight	4	10	0
" " " Clearing, Attendance at Docks and Customs			
" " " Shipping	1	1	0
" " " Bonding	1	19	6
" " " Wharfage Entry		2	6
" " " Collection on £ @ %			
" " " Insurance on £ @ %			
Insurance Stamp Duty			
Through Inclusive Rate to			
Exchange on		7	0
£	84	9	6

6.6 Documentation for three packages of five hundred toads shipped by Cook & Son from Cape Town to the FPA headquarters in London in 1949. Wellcome Collection SA/FPA/A3/13.

oil," and served with "bread and a glass of cold white wine," platanna was a "meal for a gourmet."[72]

The hatchery supplied domestic hospitals and doctors by rail, sending toads in batches of three hundred in standard five-gallon carboys insulated against temperature fluctuation and painted inside to prevent metal poisoning. The toads were starved for up to three days before embarking on rail journeys of up to five days. For overseas shipments, 125 toads were

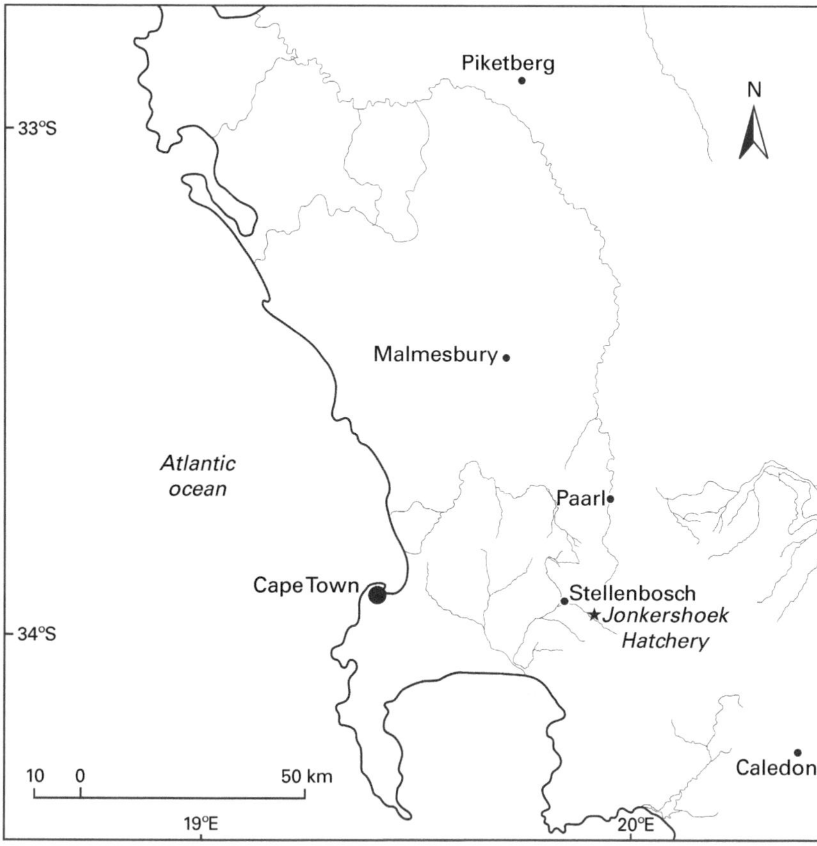

6.7 Map of the Western Cape, *c.*1940s, showing the Hatchery and successively more distant collecting spots; map by Wendy Phillips.

placed in a flatter container with a larger surface area. Casualties resulted from exposure to "sun and salt spray" as consignments were usually carried on deck, and during the war, ships could be delayed for weeks or even months in a tropical African port.[73] Survival significantly improved when a technical officer accompanying the shipment regularly removed corpses, changed the water, and fed the passengers.[74] The hatchery supplied 350 toads in 1941, 2,700 in 1942, and 4,300 in 1943. In that year, the CPA established the Inland Fisheries Department with Hey as its first superintendent (later director), consolidating the government's control of the trade.

In 1949, the hatchery distributed 10,866 toads, of which 3,803 (35 percent) were exported.[75] But the supply was "erratic" and Hey looked forward to the day when universities and hospitals would be able to order disease-free hatchery-cultured toads of any size and quantity.[76] Despite the optimistic prediction that platanna would soon be "cultured as easily as fish," the hatchery did not become a breeding center but rather a clearinghouse for harvested farm toads. Toads collected from farmers in exchange for a token payment to the landlord were brought to Jonkershoek, where they were sorted by sex and size before shipping. The hatchery charged slightly above cost on a sliding scale from local to international delivery and only supplied teaching, research, and medical institutions. This effectively allowed consumers to reliably purchase somewhat standardized toads directly from a wholesaler, stabilizing the market.[77]

UNDESIRABLE INTENTIONS AND ILLICIT DESIGNS

In April 1949, the *BMJ* and *Lancet* reported that the FPA's seminological center had relocated to Sloane Street and now included a twenty-four-hour pregnancy diagnosis service.[78] In July, a notice in the *Chemist & Druggist* alerted retail pharmacists to a discount on pregnancy testing and invited them to contact the FPA for details.[79] Two weeks later, the *Pharmaceutical Journal* published a letter by Wright, which explained that the association would charge pharmacists a reduced rate of twenty-five shillings per test and suggested they could turn a profit by charging customers thirty.[80] Wright claimed that as a voluntary organization, the FPA was "not primarily interested in profits" and so was "able to do pregnancy tests at a somewhat lower rate than the majority of commercial laboratories" (figure 6.8).[81] This approach marked a major departure from Crew's policy of dealing exclusively with the medical profession. Where Crew refused to deal with chemists, Wright courted them.

Wright's letter in the *Pharmaceutical Journal* provoked Norman Jones van Abbé, a Muswell Hill chemist, to "question the desirability" of placing pregnancy tests in chemists' hands. In his own letter, van Abbé suggested that a test was justified only in cases of "real urgency," namely, "to indicate the clinical necessity for the interruption of pregnancy or special

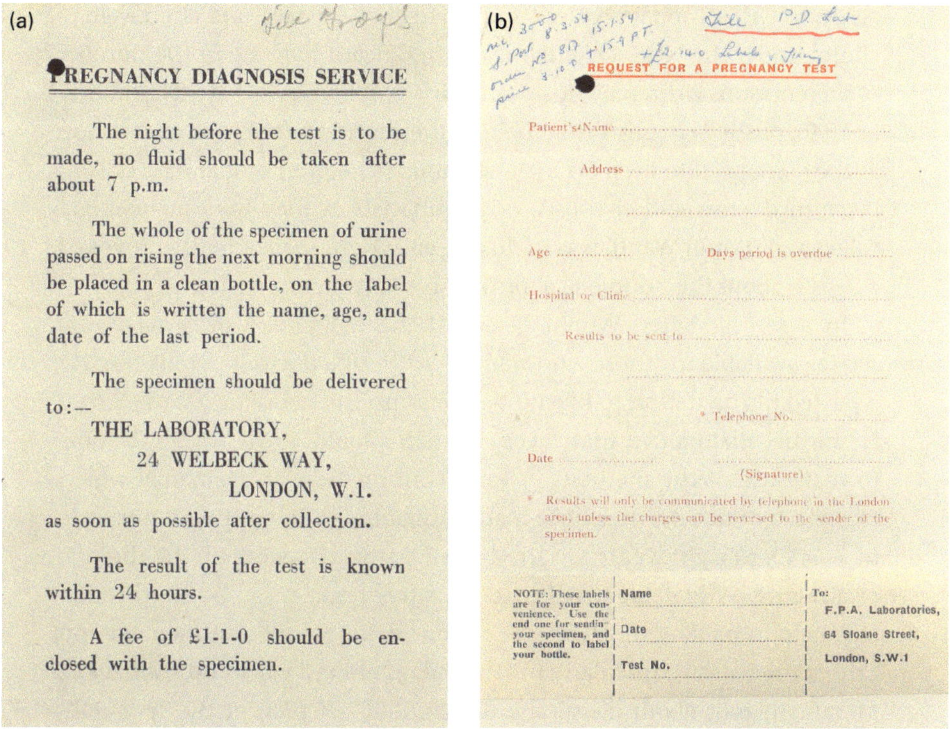
(a)

PREGNANCY DIAGNOSIS SERVICE

The night before the test is to be made, no fluid should be taken after about 7 p.m.

The whole of the specimen of urine passed on rising the next morning should be placed in a clean bottle, on the label of which is written the name, age, and date of the last period.

The specimen should be delivered to:—

THE LABORATORY,
24 WELBECK WAY,
LONDON, W.1.

as soon as possible after collection.

The result of the test is known within 24 hours.

A fee of £1-1-0 should be enclosed with the specimen.

(b)

REQUEST FOR A PREGNANCY TEST

Patient's Name ....

Address ....

Age .... Days period is overdue ....

Hospital or Clinic ....

Results to be sent to ....

* Telephone No. ....

Date .... (Signature)

* Results will only be communicated by telephone in the London area, unless the charges can be reversed to the sender of the specimen.

NOTE: These labels are for your convenience. Use the end one for sending your specimen, and the second to label your bottle.

Name ....
Date ....
Test No. ....

To:
F.P.A. Laboratories,
64 Sloane Street,
London, S.W.1

6.8 (a) Flyer for Welbeck laboratory, one of the more prominent commercial pregnancy diagnosis services in the postwar years. (b) Request form for an FPA pregnancy test including labels. Wellcome Collection SA/FPA/A3/11 and SA/FPA/A3/12.

ante-natal care, or to establish legal evidence." But a test might also be requested to "satisfy ordinary curiosity" or "establish grounds for illegal interference with pregnancy." Better not to encourage such cases, van Abbé argued, "even . . . unwittingly, especially as the pharmacist is then likely to be pestered by these people with undesirable intentions." Finally, van Abbé objected to the "uncivilised" use of animals to "gratify the curious or the person with illicit designs." He acknowledged the necessity of vivisection for "instructional purposes or genuine research" but regarded pregnancy tests as an "abuse of the Home Office licence."[82]

Chemists had long discretely dispensed sexual and contraceptive advice, condoms, and abortifacients. They also provided dental care, first aid, and postpartum care for mothers and babies, activities that diminished with free access to GPs, hospital accident and emergency services, and child welfare clinics under the NHS. Overnight, the significant increase in the number of prescriptions written by doctors, from 70 million in 1947 to 241 million in 1948, pushed chemists from behind the counter to the back of the shop as they struggled to keep up with demand.[83] The professional relationship between doctors and chemists, or pharmacists as they were increasingly called after World War II, was in flux when Wright inadvertently provoked a debate about the nonmedical provision of pregnancy testing.

In a second letter, Wright defended the desirability of a diagnostic service available to "those who want it" without a doctor's permission. He repeated his invitation to likeminded pharmacists to call on Sloane Street. He further maintained that "every woman should, if she wishes, be able to make use of the methods of early confirmation of pregnancy which recent scientific advances have made available." Women should not have to wait six to eight weeks as van Abbé suggested. Drawing on the rhetoric of austerity, Wright suggested that a positive result could be a "considerable help in making the necessary arrangements for the confinement and care of the baby," particularly in "these days of hard work and shortages." Moral concerns about illegal abortion should not prevent the "possibilities for good which the service offers." According to Wright, women had been able to obtain pregnancy tests "through pharmacists for a number of years with . . . no undesirable results." As for allegations of "vivisection," Wright countered that his toads were "merely subjected to a subcutaneous injection, isolated for a short period and then returned to the aquarium," a simple procedure that no longer required a license.[84]

Around the same time, fellows of the Royal College of Obstetricians and Gynaecologists (RCOG) met to discuss their own misgivings about Wright's initiative, which they objected to on "ethical grounds" and because his first letter possibly constituted medical advertising, which was prohibited. Several fellows sat on FPA committees, and others had been appointed vice presidents, so the association could not afford to offend such a prestigious and politically powerful group of supporters. For reasons of political allegiance, if not legal obligation, the FPA decided to act in accordance with the wishes of the RCOG.

In a conciliatory letter, the FPA explained that because the NHS offered a strictly medical service, a woman with "domestic reasons for wishing to know as soon as possible" might prefer to go through a chemist and that, in the event of a positive result, she would be advised to "consult her doctor immediately." Demand from FPA clinics and affiliated doctors was too small to sustain the running cost of Wright's laboratory, the association's "smallest and newest venture." This was why a medical subcommittee had approved the service for "all medical practitioners, hospitals, clinics and chemists," unaware of the offense it would cause.[85] Negotiations, also with the BMA, eventually led to a compromise whereby Wright continued accepting tests from chemists and women but communicated results exclusively to doctors. Women no longer needed medical permission to have the test done, but doctors remained firmly in control of the result.[86]

THE ADVERTISING ANGLE

Wright ceased all activities that could be interpreted as inappropriate or unethical advertising, but articles could not count as such if they did not mention him by name, and general secretary Robinson was fair game because she had no medical qualifications. So medical journalism constituted a new and valuable source of publicity. In August 1949, the FPA took out a classified ad in the *Pharmaceutical Journal*, and an article in *Reveille*, a small weekly tabloid, reported that the FPA's new twenty-four-hour service carried out thirty-five tests a week for twenty-five shillings a test and was available to "any woman who wishes to apply [to her doctor] for it."[87] The article, which appeared under the headline "'Radar' for the stork," emphasized the advantages for expectant mothers "anxious to book a bed in a nursing home" and quoted Robinson as saying that "many couples" had "urgent reasons for wanting the information as quickly as possible so that they can plan their domestic arrangements accordingly," for instance, a "serviceman about to leave for overseas can make proper arrangements for his wife's confinement."[88]

The *South London Advertiser* carried essentially the same story, this time with the headline, "New Science Aids Family Planning: Tells if Baby Is on the Way."[89] In December, the medicine and health section of the *News Review* carried a more substantial article ("Trial by Toads") by an anonymous staff reporter who had visited Sloane Street to see the "four, metal-lined tanks," where 1,000 "greyish-brown" toads were kept in "thermostatically heated water." This article also adopted the usual tactic of precluding the possibility of abortion by clarifying the toads did an "important job in connection with childbirth." It described the "Hogben test," praised its accuracy, and explained that the toads did not have to be killed and could be used "every two or three weeks for three or four years." Wartime meat rationing continued well into the 1950s, so the sympathetic article went out of its way to mention that the toads were fed on a "special grade of minced liver unfit for any other use" (figure 6.9).

The *News Review* article elaborated that, although most tests were currently performed for "hospital patients and those attending clinics," every woman would soon be able to procure a test from "her local chemist." It explained the arrangement whereby results were communicated not to women or chemists but to doctors, a "compromise on an earlier plan" that had been "open to misuse." And it commended Sloane Street for its value to not only patients booking a maternity bed but also businesswomen and women planning holidays, sources of demand that seemed to herald a new era.[90] The remainder of the article generally praised the FPA and promoted its broad range of activities beyond birth control, including work on male infertility. Because of the expenses incurred by these activities and the move to Sloane Street, the association was now "some £1500 into the red." Although its relations with the Ministry of Health were "cordial" and many doctors reportedly favored integrating contraceptive services with the NHS, the FPA had recently been rejected by the BBC's "Good Cause" charity appeal.[91]

Wright, meanwhile, was struggling to keep up with demand, so he enlisted his secretary, Mrs. Northgate, as "full-time frog queen" and began training her in doing tests. Margaret Pyke, a founding member of the association that became the FPA, agreed that the "splendid rise in the number of P.D. tests" made it difficult for Wright to manage without a technician but reminded him that the FPA would have to approve his budget first. Frustrated with bureaucratic constraints, Wright complained that the association would not become "really energetic" unless it was "prepared to delegate." Publicity was a "case in point." The FPA had already lost about three months' worth of trade while committees "gently wrangled over" advertising. Meanwhile, the toads continued to "flourish" and Wright had "already banked about £80 to £90."[92]

(a)

"Radar" for the stork

By a STAFF REPORTER

EXPECTANT mothers anxious to book a bed in a nursing home, and other married couples who want to know urgently whether a baby is on the way, can now find out within two days

to us. Some hospitals, too, refer them to us.

"The scheme is well under way. We are carrying out tests at the rate of 35 a week."

The charge for the test is 25s.

(b)

New Science Aids Family Planning

TELLS IF BABY IS ON THE WAY

Test Can Now Be Arranged

(c)

Pharm. Journal—Classified—Sept. 20

PREGNANCY DIAGNOSIS
(Hogben Test).
Family Planning Association Laboratories
64, Sloane St., London, S.W.1 (Sloane 9112)
Specimens of urine accepted for testing from Pharmacists. The result, available within 24 hours of receipt of specimen, will be communicated to patients' doctor. Terms and details on request.

(d)

NEWS REVIEW, *December 22, 1949*

Medicine and Health

TRIAL BY TOADS

IN four, metal-lined tanks in the Family Planning Association's basement laboratories in Sloane Street, London, 1000 greyish-brown female *Xenopus lævis* toads from South Africa are kept in thermostatically heated water. They do an important job in connection with childbirth.

When a woman is pregnant, certain hormones develop. These, when injected into a toad, cause it to lay eggs a few hours later. In a glass jar, the toad is made to sit on a grid, which prevents it from touching the eggs.

Mistakes are very rare, and the test (called the Hogben test) is considered a shade more accurate than the similar Ascheim-Zendek test, in which mice are injected. But whereas the mice have to be killed for examination, the toads are not, and each can be used every two or three weeks for three or four years. They are fed twice a week on a special grade of minced liver unfit for any other use.

Most of the F.P.A.'s tests are made for hospital patients and those attending clinics, but a scheme is now coming into operation under which a woman will be able to consult the F.P.A. through the medium of her local chemist. The result will be communicated to her doctor, who will pass the verdict on to her within 24 hours. This is a compromise on an earlier plan for the chemist to tell the woman the result direct. It was felt that without the intervention of a doctor the service might be open to misuse.

6.9 Classified ads in trade journals and headlines in tabloid newspapers provided valuable sources of non-medical advertising for the FPA's pregnancy diagnosis service and significantly increased the public visibility and acceptability of pregnancy testing: (a) "'Radar' for the stork," *Reveille*, August 27, 1949; (b) "New science aids family planning," *South London Advertiser*, September 16, 1949; (c) "Pregnancy diagnosis," *Pharmaceutical Journal*, September 20, 1949; (d) "Trial by toads," *News Review*, December 22, 1949. Wellcome Collection SA/FPA/A3/11.

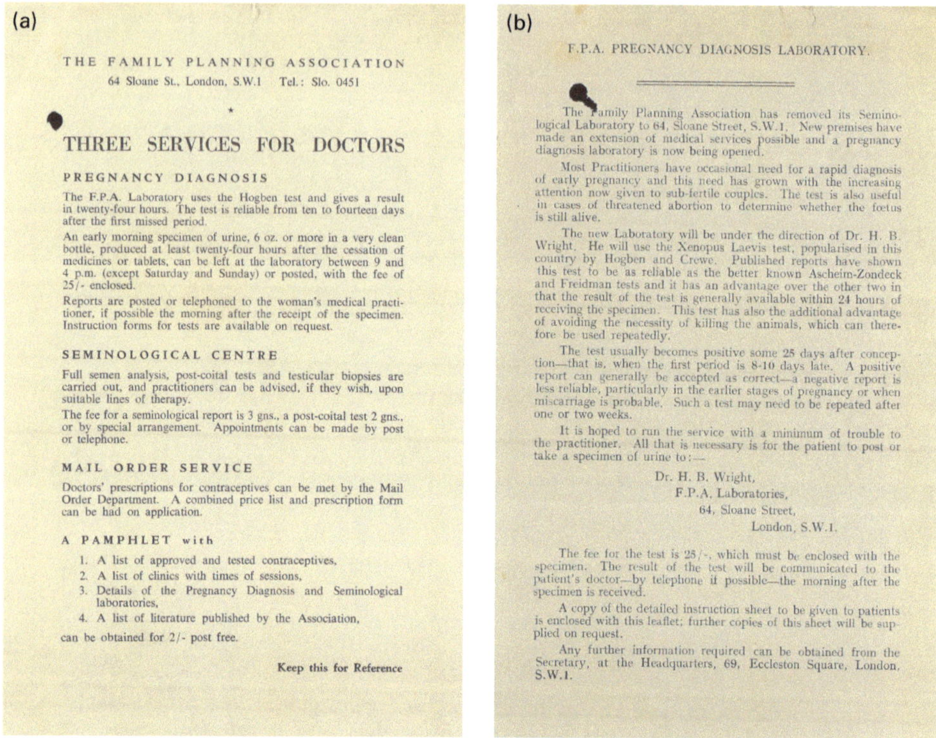

(a)

THE FAMILY PLANNING ASSOCIATION

64 Sloane St., London, S.W.1 Tel.: Slo. 0451

*

THREE SERVICES FOR DOCTORS

PREGNANCY DIAGNOSIS

The F.P.A. Laboratory uses the Hogben test and gives a result in twenty-four hours. The test is reliable from ten to fourteen days after the first missed period.

An early morning specimen of urine, 6 oz. or more in a very clean bottle, produced at least twenty-four hours after the cessation of medicines or tablets, can be left at the laboratory between 9 and 4 p.m. (except Saturday and Sunday) or posted, with the fee of 25/- enclosed.

Reports are posted or telephoned to the woman's medical practitioner, if possible the morning after the receipt of the specimen. Instruction forms for tests are available on request.

SEMINOLOGICAL CENTRE

Full semen analysis, post-coital tests and testicular biopsies are carried out, and practitioners can be advised, if they wish, upon suitable lines of therapy.

The fee for a seminological report is 3 gns., a post-coital test 2 gns., or by special arrangement. Appointments can be made by post or telephone.

MAIL ORDER SERVICE

Doctors' prescriptions for contraceptives can be met by the Mail Order Department. A combined price list and prescription form can be had on application.

A PAMPHLET with

1. A list of approved and tested contraceptives,
2. A list of clinics with times of sessions,
3. Details of the Pregnancy Diagnosis and Seminological laboratories,
4. A list of literature published by the Association,

can be obtained for 2/- post free.

Keep this for Reference

(b)

F.P.A. PREGNANCY DIAGNOSIS LABORATORY.

The Family Planning Association has removed its Seminological Laboratory to 64, Sloane Street, S.W.1. New premises have made an extension of medical services possible and a pregnancy diagnosis laboratory is now being opened.

Most Practitioners have occasional need for a rapid diagnosis of early pregnancy and this need has grown with the increasing attention now given to sub-fertile couples. The test is also useful in cases of threatened abortion to determine whether the foetus is still alive.

The new Laboratory will be under the direction of Dr. H. B. Wright. He will use the Xenopus Laevis test, popularised in this country by Hogben and Crewe. Published reports have shown this test to be as reliable as the better known Ascheim-Zondeck and Freidman tests and it has an advantage over the other two in that the result of the test is generally available within 24 hours of receiving the specimen. This test has also the additional advantage of avoiding the necessity of killing the animals, which can therefore be used repeatedly.

The test usually becomes positive some 25 days after conception—that is, when the first period is 8-10 days late. A positive report can generally be accepted as correct—a negative report is less reliable, particularly in the earlier stages of pregnancy or when miscarriage is probable. Such a test may need to be repeated after one or two weeks.

It is hoped to run the service with a minimum of trouble to the practitioner. All that is necessary is for the patient to post or take a specimen of urine to:—

Dr. H. B. Wright,
F.P.A. Laboratories,
64, Sloane Street,
London, S.W.1.

The fee for the test is 25/-, which must be enclosed with the specimen. The result of the test will be communicated to the patient's doctor—by telephone if possible—the morning after the specimen is received.

A copy of the detailed instruction sheet to be given to patients is enclosed with this leaflet; further copies of this sheet will be supplied on request.

Any further information required can be obtained from the Secretary, at the Headquarters, 69, Eccleston Square, London, S.W.1.

6.10 Information sheet produced by the FPA publicized pregnancy diagnosis alongside semen analysis as well as mail order contraceptives and information pamphlets. It attributes the increased demand for early pregnancy diagnosis to the "increasing attention now given to sub-fertile couples" and points out that the Hogben test "avoided the necessity of killing the animals." Wellcome Collection SA/FPA/A3/11.

In January 1950, Wright had a thousand revised information sheets printed and distributed. In a letter to chemists who already used the service, he suggested the new arrangement would offer the "best chances of protecting the Pharmacist and the Laboratory" and serve the "best interests of the patient concerned" (figure 6.10). A second letter assured new customers that the method of notifying results did not signal "any lack of confidence" in pharmacists but had been recommended by medical authorities as the only way to protect them and laboratories "from occasional abuse." These terms were the only ones under which Wright was permitted to receive tests from "non-medical sources."[93]

By the end of March 1950, Wright had performed over 1,500 tests. He had cleared a small profit (£165) after spending around £1,200 on overhead, running costs, and salaries. He budgeted £1,400 for the following year and anticipated a monthly income of around £160, or about £2,000, for the year. For £50 a month, an advertisement in the *BMJ* was "still bringing in a steady flow of new doctors, at the rate of 6 to 10 a week." Wright expected continued expansion to necessitate hiring a "bottle washer" and "general cleaner up" to assist the "frog queen." He also invested in a new tank with an immersion heater, thermostat, and Sunvic control. Wright's test animals were becoming a "little bit overcrowded," and he planned to order another batch come summer.[94]

The FPA approved Wright's plans for expansion, and he commissioned new tanks and electrical equipment. He also asked ordered a further six hundred toads at about five shillings each from Thomas Cook. Wright requested approval for cardboard bottle containers to compete with commercial firms, which were sending "far more beautiful containers." His workload was "still expanding" "pretty fast," and with "judicious propaganda," he expected it to stabilize at around three hundred tests per month. By October 1950, five or six hundred doctors were using the laboratory. Davidson's wife Victoria prepared a "semi-humorous design," based on the work of seminological and the pregnancy diagnosis laboratories, and Wright sent out Christmas cards to generate "good will" and remind his "more wayward clients" of the service.[95]

In 1951, the Post Office complained that the glass bottles Wright used did "not meet their very stringent regulations for sending specimens through the post." So he contacted Industrial Appliances to receive samples of plastic bottles. Subject to approval by the Post Office, Wright planned to buy 250 "completely unbreakable" bottles for one shilling each. The high cost of plastic was offset by what Wright called the "advertising angle." In other words, "by having something which is entirely up to date we are leading the field and are likely to create a good impression by so doing." Wright suspected that plastic would soon "replace glassware" in medical packing. Although self-conscious of the image projected to clients, his basement laboratory was less impressive in other ways. It lacked heating, and the glass roof over the frog tank leaked and made it impossible to work when it rained without getting get. Wright had no filing cabinet and so folders were kept in an "untidy" partition in a cupboard. But clients never saw any of this, and his service continued to expand. The number of tests he performed in 1951–1952 increased to over 3,300 and the income to over £2,500.[96]

Meanwhile, women's magazines began promoting Sloane Street as they had not done with Edinburgh. *Woman's World* published a letter attributed to a "very worried" aunt from Glasgow whose niece, a "single girl," had "made a terrible mistake." "How soon is a pregnancy certain?" she asked. The magazine's agony aunt, "sorry to hear" that the girl was "in such trouble," explained there was "now a urine test for pregnancy which can be performed when the period is a fortnight late" and that would "give the answer in a few days." "If a doctor thinks it is necessary," she continued, "this test can be done for free under the National Health Service. If performed privately, through the Family Planning Association there is a fee of twenty-five shillings."[97]

Mother similarly published a letter attributed to a woman who wished she "knew whether or not [she] was going to have a baby." She hadn't had a period for "nearly three months," and she was certain her "tender" bust was "fuller." Her doctor, however, refused to examine her because she had already miscarried twice, and he was "afraid of doing anything which may start a period off, just in case [she] had conceived." "Would it be any good having a pregnancy test, do you think? How does one go about it?" she asked. "By all means ask your doctor about a pregnancy test," encouraged *Mother*'s agony aunt, "It simply involves sending a specimen of your urine to The Family Planning Association, 64 Sloane Street, London, S.W.1."[98]

On November 29, 1955, the conservative Minister of Health Iain Macleod visited FPA premises in formal recognition of the association's Silver Jubilee. He posed for photographs outside the North Kensington clinic as well as at a "microscope in the sub-fertility laboratory." This was the first time a minister for health had publicly endorsed any voluntary organization promoting birth control, a publicity stunt that generated headlines in the national newspapers, an interview with Margaret Pyke on BBC television, and a talk on *Woman's Hour*.[99] In June 1958, *Family Doctor* published a photograph of Macleod being shown a positive *Xenopus* test by an obliging lab technician at Sloane Street (figure 6.11). Publicity didn't get much better.

Wright later recalled having started the "frog lab" to "make money for the FPA and it would have made a lot more if they had let us report direct to the patients."[100] Although he had been frustrated by the association's bureaucracy and compromises over the involvement of nonmedical clients, a progress report praised Wright's service for generating "useful revenue"

6.11 Beric Wright looks on as a woman wearing a labcoat, probably Mrs. Northgate, holds up a test jar for the health minister to see. The caption read: "The Rt. Hon. Iain Macleod sees a positive result of the Hogben test. In the bottle is a toad and the eggs she has laid. She only lays these eggs if the urine injected into her was an 'early morning specimen' from a pregnant woman. Within twenty-four hours the results are absolutely certain. Every week over a hundred couples use this service at the Family Planning Association's Pregnancy Diagnosis Laboratory at 64 Sloane Street, London, S.W.1." Capel 1958, 381, Cambridge University Library, L300.b.142.8, with permission by the BMA.

and bringing in new (medical) clients.[101] His energetic campaigning, along with the generally softer image of reusable toads and frogs, significantly increased the public visibility of pregnancy testing in the 1950s. Extending the theme of visibility through marketing, the next chapter significantly complicates the historical understanding of pregnancy testing as a progressive narrative of technological improvement by examining what happened in the 1950s and 1960s, when pharmaceutical companies successfully marketed hormone injections and tablets as diagnostic tests for early pregnancy that were cheaper and more convenient than toads or frogs.

7

RISKY HORMONES

In *Sex and Destiny: The Politics of Human Fertility* (1984), Australian feminist Germaine Greer speculates whether the "marketing of high doses of sex hormones as pregnancy tests was not a disguised way of selling do-it-yourself abortion kits. The instructions for the use of Primodos are simply too good to be true." These were to take "1 tablet on each of two consecutive days. Bleeding follows in 3–6 (rarely as long as ten) days, if there is no pregnancy. An existing pregnancy is unaffected by Primodos." Greer goes on to recount how, in 1967, Dr. Isabel Gal, a researcher at Queen Mary's Hospital for Children at Carshalton in Surrey, reported a correlation between children with spina bifida and mothers who had "used a hormone pregnancy test." Despite mounting concerns within British Schering, the makers of Primodos, and even after the Royal College of General Practitioners produced more evidence of harm, the German parent company "refused to take any action." Crucially, the tests "performed no useful function, because there were already more practical urine tests available."[1]

This chapter investigates the peculiar and troubling history of Primodos and other "hormone pregnancy tests" up to and including Gal's report in 1967. It concentrates on the circulation and reception of these products in Britain and interrogates Greer's speculation that they were disguised abortifacients. I conclude that Greer was half-right. From the vantage of the

1980s, Primodos looked suspiciously like an abortion pill, but the evidence presented in this chapter suggests that many women (including those longing for pregnancy) and their doctors (including those helping couples with infertility) took claims about these unusual drugs at face value, even as others collaborated in using them transgressively to induce miscarriage. The implications contained in the history of Primodos and related products are far-reaching, and I will have more to say about them elsewhere. Here, I ask how the market for hormone tablets and the market for urine tests mutually shaped one another in the 1960s. But first, where did Primodos come from, and how did it become thinkable to market hormone tablets as a pregnancy test?

ZONDEK'S METHOD OF MEDICAL CURETTAGE

Today, most medicines are proscribed in pregnancy to protect the developing fetus, so it may seem strange that doctors ever prescribed tablets to diagnose pregnancy. Yet attempts to develop a "therapeutic" or "clinical" test for pregnancy that act on the woman herself go back to the early twentieth century.[2] These included sugar tolerance tests, eye drops, and skin reactions inspired by those for diabetes, diphtheria, scarlet fever, tuberculosis, hay fever, and other allergies.[3] As discussed in previous chapters, clinicians further gained access to proprietary injectables, including Schering's Maturin and Parke, Davis's Antuitrin S in the 1920s and 1930s, respectively. In 1940, shortly after the short-lived enthusiasm for Antuitrin S had waned, Samuel Soskin and colleagues of Michael Reese Hospital in Chicago announced a new kind of test in the *Journal of the American Medical Association*. They induced menstrual bleeding in nonpregnant women by injecting patients with Hoffmann-La Roche's prostigmine methylsulfate and claimed the injection as a treatment for amenorrhea and, crucially, also as a test for early pregnancy.[4]

The simplicity of the "prostigmine test" appealed to some clinicians, but seasoned pregnancy testers greeted it with skepticism. Abner Weisman, a prominent New York gynecologist and fertility specialist who had recently visited Edinburgh to learn the *Xenopus* method from Crew, dismissed prostigmine as "practically worthless."[5] Others reported favorable results with small groups of patients, but larger and more controlled studies tended to confirm Weisman's assessment.[6] Like countless other bids to make animals redundant, the prostigmine test did not live up to expectations. The principle, however, of an injection that would confirm pregnancy by failure to induce menstruation had staying power. It would soon be revived, indirectly, inadvertently, and with far-reaching consequences, by Bernhard Zondek.

Following his dismissal in 1933, Zondek spent a year in Stockholm before settling in Jerusalem as head of gynecology at the Rothchild-Hadassah University Hospital. There he developed a successful research program to simulate menstruation if not restore ovulation in predominantly religious patients for whom infertility was a significant source of anxiety. This work in many ways culminated with his announcement in *JAMA* in 1942 of a "simplified hormonal treatment of amenorrhea" using compounds furnished by Roche-Organon of New Jersey.[7] The standard treatment for amenorrhea involved thirteen injections of estrogen and progestogen over a month to simulate the menstrual cycle. Following the experimental induction of midcycle bleeding in "normally menstruating women," Zondek proposed a regimen of just two to five injections in as many days. This was reduced in 1948 to a single injection of a specially prepared estrogen–progesterone mixture produced locally for Zondek by Teva, Middle East Pharmaceutical and Chemical Works Ltd.[8]

Although Zondek did not propose the treatment as a pregnancy test, others soon did. In 1950, Schering, the Berlin company that had marketed Maturin before the war, launched Duogynon in West Germany. As suggested in the name, the new drug innovatively combined two of its proprietary hormones: Proluton, a progestin, and Progynon, an estrogen. Schering initially marketed Duogynon primarily as a treatment for amenorrhea, but its secondary function as a pregnancy test was there from the start and increased in prominence to become the dominant indication a few years later. New orally active forms of progestogen enabled the crucial shift from injections to better-tolerated tablets, and major pharmaceutical companies responded to the commercial success of Duogynon by launching their own dual-purpose products. In Britain, Roussel Laboratories established a pedigree for Amenorone Forte by describing it as "Zondek's method of medical curettage," a treatment for menstrual disorders that doubled as a "simple and safe" pregnancy test (figure 7.1).

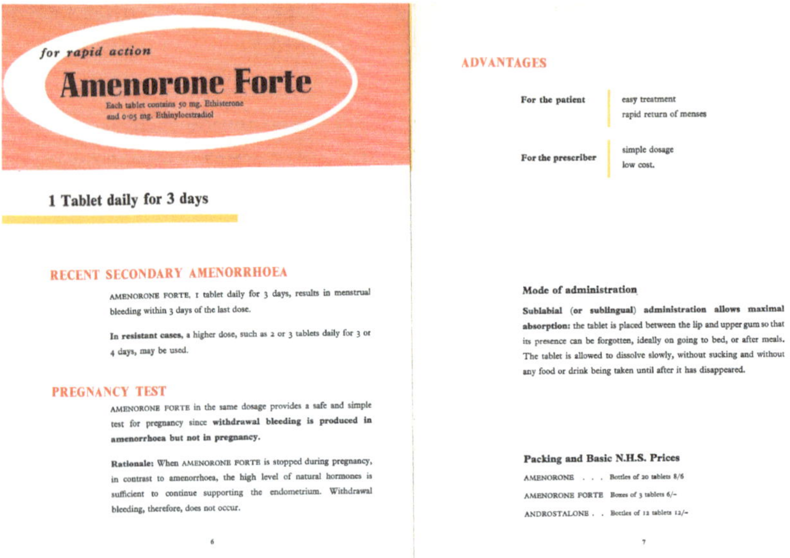

7.1 Two inside pages of a brochure promoting Amenorone Forte, a combination of ethisterone (50 mg) and ethinyl-oestradiol (0.05 mg), as a treatment for recent secondary amenorrhea and as a "simple and safe" pregnancy test that would produce withdrawal bleeding in cases of non-pregnancy only. It was given to historian Naomi Pfeffer as an abortifacient on two separate occasions in the 1960s by her family doctor, a refugee from Germany, who "didn't let on" to her parents. Pfeffer later recalled dissolving the tablets under the tongue on the 52 bus to Chelsea Art School and for "weeks afterwards every time I got on the bus I could taste them—a Pavlovian response!" Pfeffer to author, Nov. 1, 2011. *Amenorone and Amenorone Forte* (London: Roussel Laboratories, 1957), The National Archives MH 149/1105.

PREGNANCY, LIKE MURDER, WILL OUT

As the new products began circulating in British clinics, laboratory workers began testing them against the known standard of *Xenopus*. In 1953, George Matthew, an infertility specialist at the University of Edinburgh, teamed up with Bruce Hobson of the pregnancy diagnosis station to compare the Hogben test to two consecutive daily injections of Disecron, a combined progestin–estrogen manufactured by British Schering Ltd.[9] Although apparently more reliable than *Xenopus* in the early weeks of pregnancy, Matthew disliked the "discomfort" injections caused his patients, so he conducted a second trial in which he administered ninety-four women with ten Orasecron tablets for two days (figure 7.2). Of the sixty-two who did not experience

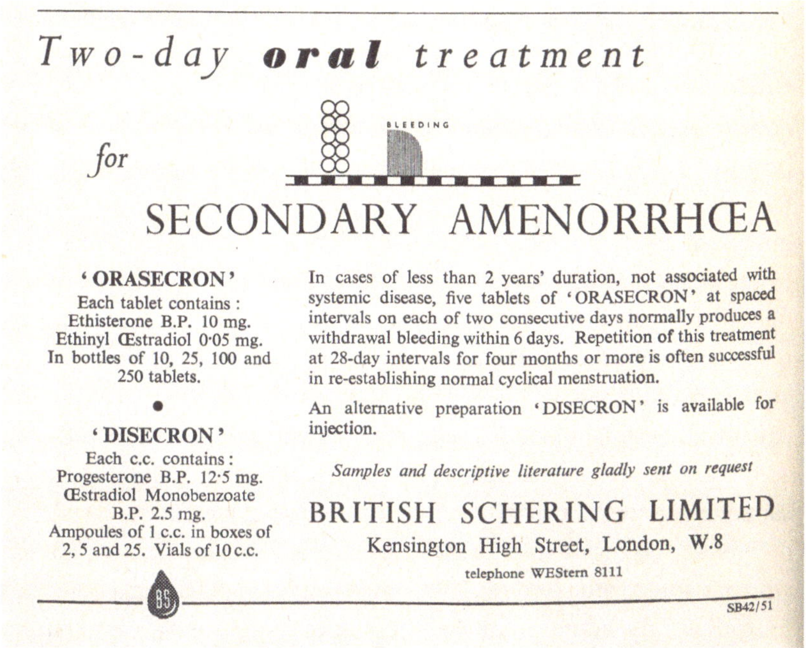

7.2 British Schering's "Disecron" and "Orasecron" were advertised as treatments for amenorrhea, not as pregnancy tests. *Proceedings of the Royal Society of Medicine*, 45, Nov. 1952, vi, British Library (P) GP 00-E(103).

any bleeding, all were confirmed with the Hogben test to be pregnant. Follow-up examinations excluded pregnancy in the remaining thirty-two who did bleed after a week or two. Matthew enthusiastically reported the oral administration of Orasecron in the *BMJ* as a "reliable clinical method of diagnosing early pregnancy."[10] As a gynecologist specializing in fertility care, it seems unlikely he would have knowingly prescribed abortifacients to his patients, many of whom would have been trying to become pregnant.

But not all doctors shared Matthew's optimism. A cautious *BMJ* editorial explained that, considering the well-known drawbacks of bioassays and the absence of a reliable biochemical alternative, it seemed "rational as well as economical to use the patient herself as the test animal." Researchers had investigated the potential of histology, the microscopic examination of vaginal smears and cervical mucus in pregnancy diagnosis. Although considered reliable, histological techniques required "specialist interpretation" and so were not widely adopted in general practice. Because patients usually sought medical advice on account of a missed period, researchers had also investigated methods that would "produce uterine bleeding only if the patient were not pregnant and which would not harm the pregnancy if present." The editorial agreed that the commendable "simplicity and relative economy" of the test would endear it to GPs but also spelled out some qualifications.[11]

Although Matthew had not noted any "untoward effects on the pregnancy," the possibility remained that the treatment had provoked miscarriages in one or more of the thirty-two women who experienced bleeding. To be sure of the safety of Orasecron, Matthew would have had to examine patients' menstrual blood for clots and conduct endometrial biopsies. The editorial suggested repeating Matthew's clinical trial with these added precautions. Another drawback was that the test was of little use in cases where differential diagnosis was "perhaps most often sought," that of patients who presented with bleeding. For women suspected of having a threatened miscarriage, the administration of hormones to induce bleeding could be of no assistance. Rather conservatively, the editorial concluded there was "usually no urgency for a certain diagnosis of early pregnancy, and if the family doctor cannot make a confident diagnosis, re-examination of the patient in three or four weeks' time is nearly always conclusive. Pregnancy, like murder, will out."[12]

A few months earlier, Hubert Britton, a London physiologist, had written, "It is with dismay that I have received this morning a brochure from a drug firm describing a test for differentiating between pregnancy and amenorrhoea by the administration of a mixture of synthetic hormones, and the induction of withdrawal bleeding in those with amenorrhoea." Britton worried that drugs administered in the "first few weeks of pregnancy . . . when the embryo is most susceptible to noxious influences . . . will upset the delicate hormonal balance of the mother and the foetus." He condemned clinical trials of the safety of such drugs on the grounds that a "continued pregnancy and an apparently normal child is no guarantee that no harm is being done."[13]

Britton reiterated his concerns following the *BMJ* editorial, this time drawing on a recent report of the birth of a "malformed foetus" following the administration of a hormone pregnancy test, despite the writer's view that the "association was coincidental." He criticized the editorial for raising the risk of miscarriage but not teratogenicity and hoped the "widespread use of these tests [would] not lead to a repetition of the story of X-ray pelvimetry, for a procedure of no therapeutic value."[14] But concerns that prescribing drugs in pregnancy caused fetal anomalies were not widespread in the years before thalidomide, the West German sedative that caused a global epidemic of "birth defects" in the late 1950s and early 1960s. For example, even as the BMA's *Refresher Course for General Practitioners* (1956) warned of the "danger" posed by "repeated exposures of radiation . . . to the fetal gonads," it remarked on how "how seldom drugs harm the foetus."[15]

David Lambert of Ruislip had often used Disecron as a pregnancy test but switched to Orasecron because his patients appreciated "not having to be injected and not having to come to the surgery twice for that purpose." He agreed with Matthew that Orasecron was a "reliable clinical method of diagnosing early pregnancy." He also took issue with the view, recently expressed in the *BMJ*, that time would tell:

> Leaving aside those several cases seen each year in general practice where one must consider advising a therapeutic abortion, the general practitioner will at any given time in a large practice have anything up to half-a-dozen cases on his hands of patients who are terribly anxious to know early on whether they are pregnant. I have precisely six such cases in my practice at this minute. For example, recently a patient consulted me when one week overdue. She was in a terrible state of anxiety because she has four children. I believe that she is becoming menopausal. With the aid of orasecron I shall be able to advise her one way or the other within seven to fourteen days.

The *BMJ*'s "annotator," Lambert complained, had "little knowledge of this problem in general practice. It is a very real one, and orasecron is a very real help. Eight to twelve weeks of anxiety are harder to bear than one to two weeks. The grateful patient will tell you that it is so."[16] This was anxiety-driven demand, an increasingly important feature of the clinical encounter that many consultants and pathologists failed to appreciate. The previously standard practice of telling a patient to come back in a month was becoming untenable; women expected more, sooner. If a urine test was too slow, expensive, inconvenient, or unavailable, then perhaps Orasecron could help.

A MODERN SCIENTIFIC ACHIEVEMENT

Organon Laboratories, a Dutch company that had formed a European hormone cartel with Schering and Ciba in the 1930s and would soon pioneer immunological test kits for pregnancy (see next chapter), brought out Menstrogen, its own withdrawal bleeding test and treatment for amenorrhea (figure 7.3).[17] *An Introduction to Endocrinology*, a handbook published by Organon's British subsidiary in 1957, explained that Menstrogen was a "safe, simple and effective" alternative to laboratory animals that produced "cyclic bleeding in cases of amenorrhoea due to endocrine dysfunction" but not pregnancy. The "failure to induce menstruation after four

(a)

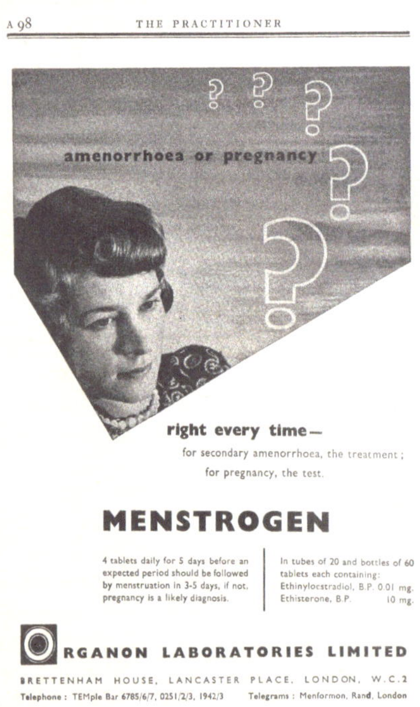

(b)

7.3 Menstrogen marketed as (a) a test for pregnancy (*Practitioner*, 184, Apr. 1960, A98); and (b) a treatment for amenorrhea (*Practitioner*, Jun. 1960, A80). The question mark motif was a favorite across brands; the metronome, less common. Cambridge University Library P300.c.155.217.

tablets . . . have been given daily for five days indicates a diagnosis of pregnancy." Menstrogen was considered safe because the "addition and withdrawal" of the hormones it contained did not "interfere with the existing hormonal balance and have no effect on the pregnant uterus."[18] But Schering's products had a stronger hold on the market.

Douglas Hogg, a Newcastle GP, favored Orasecron because of the high cost, waiting period (at least a week), and "trouble" of collecting, packaging, and posting the urine for a Hogben test and because overworked laboratories often requested that GPs "ask for such tests only when absolutely necessary." Hogg prescribed Orasecron, which he judged simple, cheap, rapid, and reliable, to women who suspected pregnancy on the grounds of a missed period but showed no other clinical symptoms. One of Hogg's patients made the "veiled suggestion that the drug had produced an abortion" and so he warned the practitioner to be "guarded in the wording of his instructions to a patient" to avoid misunderstandings. Hogg further commended Orasecron as a "most useful drug when it is necessary for a woman to regulate her periods to prevent menstruation at awkward times such as examinations or sporting events."[19] Mary Bew, a Belfast practitioner, found Orasecron "particularly useful as an aid to diagnosis when pregnancy is possible in an unmarried girl" and did not "suspect that it had interfered with the course of pregnancy in those women who were pregnant."[20]

D. H. Forster, a Bristol GP, disliked the "cumbersome" Hogben test. Urine had to be "collected, packed and posted, sometimes to a very considerable distance." The container might break, spill, or leak in transit, and even if it did get there "intact," there might not be enough urine for the test. The results were "not always accurate," and they were slow: "ten days or even longer" from the patient's first appointment. In the past few years, Forster had administered "hormone tests for pregnancy" to forty-six patients using an "oily injection" of Disecron. In view of the "distraught state of mind" in many of his patients, he preferred injections "rather than risk an incorrect diagnosis through a misunderstanding by the patient over the dosage of the tablets." He considered Disecron to be "at least as accurate" as *Xenopus* and had not "heard of any foetal abnormalities resulting from its use." Finally, the basic NHS cost of six shillings for two injections compared "favourably" with the cost of a bioassay.[21]

Bruce Hobson, Britain's leading proponent of the Hogben test, had reservations about Disecron as an alternative. He maintained that the Edinburgh station routinely provided reliable results within twenty-four to forty-eight hours (except on weekends) and that urine specimens packed in polyethylene bottles would not leak. Hobson conceded that Disecron "might be more convenient for some general practitioners" but argued that few women with the option of preparing a urine specimen would choose the "discomfort" of injections. Finally, his strongest objection to any "pregnancy test involving the injection of steroid material when other adequate tests [were] available" was not the risk of birth defects but possibility that the "resulting hormonal imbalance, however small," could induce miscarriage in "susceptible women."[22]

Before thalidomide, consumers and experts associated congenital anomalies more with heredity or radiation than with the prenatal use of prescription drugs.[23] Progesterone therapy was widespread in the 1950s to prevent miscarriage, so (Hobson's reservations notwithstanding) it was not a stretch to imagine that hormone pregnancy tests, far from causing harm, might even "help implant the ovum properly."[24] Anticipating to some extent the greater privacy commercial laboratories would soon confer on women by serving them directly as clients, they were also a test the woman took home to perform privately, the result known to her alone.[25] An anxious patient may have taken greater comfort in the visceral certainty of bleeding than in a laboratory report, if her period had still not come.[26] And tablets generally appealed to a generation increasingly accustomed to the convenience of prescription medicines, heavily subsidized on the NHS.[27] Perhaps most important from the patient's perspective, she did not have to wait two weeks from a missed period for the hormone pregnancy test, like she did for *Xenopus*.

But patients didn't always have a choice. A working-class woman was less likely to know about laboratory tests or to challenge her doctor, who still had sufficient authority to decide which test she would get. And the convenience of tablets enticed the busy practitioner. A Glasgow GP preferred them because they demanded less of his time than injections or urine tests. He routinely prescribed a course of four tablets of Primodos, "one tablet night and morning on each of two consecutive days," to produce "withdrawal bleeding (no pregnancy) or no bleeding (indicating

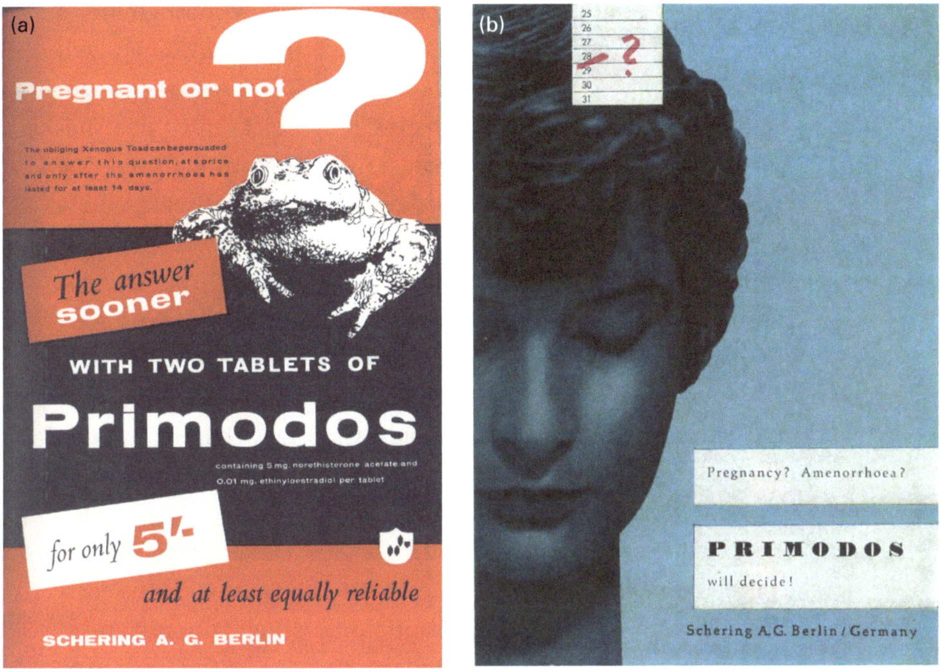

7.4 Pair of Schering adverts displaying different marketing tactics while making prominent use of the question mark. (a) A direct attack on the "slow" toad in fiery red also emphasizes the low cost of five shillings (5/-): *Practitioner*, 187, Jul. 1960, A49, Cambridge University Library P300.c.155.217. (b) A subdued blue ad plays on the patient's quiet anxiety over a missed period, Schering Archives, Bayer AG, S1 166.

pregnancy) within 3–6 days." As with Disecron, the cost to the NHS of five shillings compared favorably with that of bioassays (figure 7.4).[28]

Albert Davis compared injections of Organon's pregnancy test ampoule (PTA) to *Xenopus* in one hundred patients from outpatient gynecological clinics in ethnically diverse London suburbs. Each was given a routine examination and a single injection, instructed to bring a urine specimen the next day, and seen a week later to verify whether the "presence or absence of bleeding correlated with the Hogben test," which was repeated in cases of disagreement. All patients were seen later "either for artificial reinstitution of menstruation, or for supervision of their pregnancy if pregnant." Davis reported in the *Practitioner* that PTA had been correct in all cases, that it was "utilizable at an earlier stage" than the Hogben test, and that there had been "no adverse effect in cases of established pregnancy."[29]

Gabriel Jaffé, a Bournemouth practitioner, used pyridostigmine, a prostigmine analogue, as a pregnancy test in one hundred women with amenorrhea. He reported in the *Lancet* an overall accuracy of 97 percent for the "simple, accurate, and inexpensive" test, which cost three shillings on the NHS.[30] Drs. Higgins and Sadler, who provided prenatal care to 7,500 patients in Bristol, an industrial city of 500,000, disparaged the Hogben test as "cumbersome and lengthy" and the circulation of urine as a "considerable inconvenience to an already busy person." They decided to give Primodos to all amenorrheic women, excluding those who were "clearly pregnant," "after explaining the nature of, and the reasons for, the test."[31] Yet cautious views continued to be expressed.

Republished from the *Practitioner*, Ursula M. Lister's chapter on pregnancy diagnosis in *Calling the Laboratory* (1962) warned that in "susceptible cases," the tablets could "upset" the hormonal balance of pregnancy and cause "bleeding." The patient's desire to know earlier notwithstanding, she praised reexamination (after a few weeks had passed) as the "best test of all."[32] This view represented the cautious noninterventionist end of the spectrum. But anxiety-driven demand was only increasing, and many GPs felt pressured by their patients to do something. The unknown risks of tablets and injections, on the one hand, and the increasing demand for pregnancy testing, on the other, contributed to an even greater positive presence of toads in women's magazines. In June 1961, an article by Joan Seaward in *Woman* promoted the Hogben test, not Primodos, as a "modern scientific achievement." A full-page story conveyed the pros and cons of different tests in the form of a fictionalized encounter between "Mrs. Berry" and her doctor (figure 7.5).

Three years ago, Mrs. Berry had miscarried in the third month of her first pregnancy. She and her husband had been "bitterly disappointed at the loss of what they hoped would be their first child." Subsequently, Mrs. Berry's periods had been regular, but they were now a fortnight overdue. She suspected pregnancy, but her doctor would not risk an internal examination, which could provoke another miscarriage. "But doctor," she implored, "how much longer must I wait before knowing for certain? It means so much to my husband and me. Couldn't I have one of those pregnancy tests I've heard about?" Mrs. Berry's doctor informed her that the most popular tests in Britain cost one guinea ("but I take it you think

"WOMAN" W/E 24TH JUNE 1961

PREGNANCY TEST ★ ★ ★ a modern scientific achievement

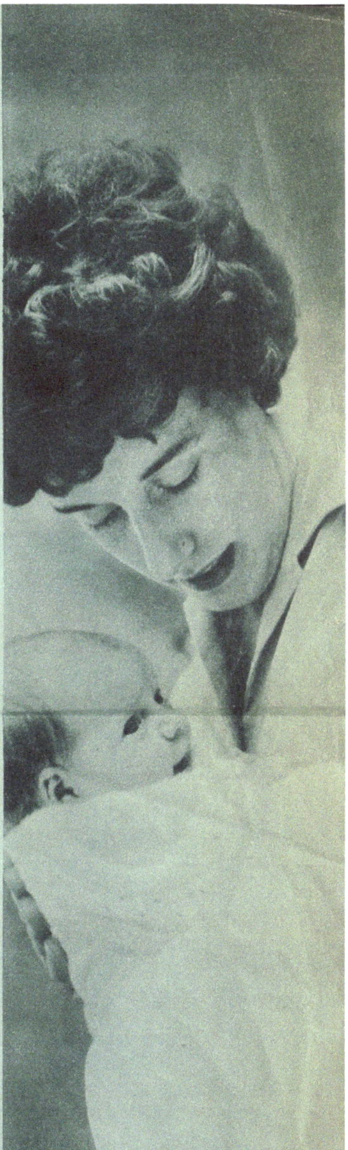

GLANCING through the records on his desk, the doctor noted that the last time he had seen the young woman in front of him was when he had attended her for a miscarriage just three years ago.

Both she and her husband had been bitterly disappointed at the loss of what they had hoped would be their first child. And when Mrs. Berry had completely recovered, he had advised her to become pregnant again just as soon as she could.

"Doctor," she began, "once I got over my miscarriage, I never missed a single period—not even a day late. But now I'm a whole fortnight overdue. That could mean I'm pregnant again at last, couldn't it?"

"It could," he answered. "But I won't be able to give you a definite diagnosis today."

Her face fell. "Do you mean you can't tell by examining me?"

"Not yet," he replied. "In any case, it might be unwise for me to do an internal examination very early in your pregnancy. We have to remember that last time you had a miscarriage in the third month, so it's best to take no risks."

What does a pregnancy test involve?

"But doctor," she said, "how much longer must I wait before knowing for certain? It means so much to my husband and me. Couldn't I have one of those pregnancy tests I've heard about?"

"Well yes, Mrs. Berry," he answered, "as your period is two weeks overdue that could be arranged. The laboratory charges a guinea for making the test, but I take it you think it's worth that?"

"Oh yes," she said fervently. "What happens?"

"Well, in the first place, a carefully prepared specimen of your urine is needed and it's essential that your period is at least two weeks overdue, because it's the presence of what we call gonadotrophic hormones in the specimen of urine which gives proof that conception has occurred. And it takes twenty-five to thirty days from time of conception for these hormones to be first manufactured and then excreted in sufficient strength in the urine."

"Can you do the test right here in the surgery?"

The doctor shook his head. "It has to be done in a properly equipped laboratory. You see, apart from anything else, the test we use most in this country involves the co-operation of toads!"

"How perfectly extraordinary. How on earth do toads help?"

"In the test, which is called the Hogben test, the urine is injected under the skin of a female toad. If the urine contains gonadotrophic hormones, twelve to twenty-four hours later the toad will begin to produce streams of eggs.

"In another test, called the Galli-Mainini, male toads are used. If the urine injected into them is definitely from a pregnant woman, they produce sperm as the female produced eggs.

"Two other tests—the Ascheim Zondeik test for which mice were used and the Friedman test which involved rabbits—have been largely discarded nowadays as although they (like the Hogben and Galli-Mainini tests) are practically a hundred per cent reliable, they take longer to get results and also involve killing the animals concerned, whereas the toads can be used over and over again with no ill effects."

"How amazing," said Mrs. Berry. "Thank you for explaining. Could I ask just one more question? Aren't there some tablets that act like a pregnancy test?"

"The tablets you mean," the doctor answered, "are a combination of two of the ovarian hormones, oestrogen and progesterone. A woman can start taking them when her period is just one week overdue and continue for four to five days. If she is *not* pregnant, then four to five days after this her period will commence. If she *is* pregnant, there'll be no bleeding.

"A similar test can be given by means of a hormone injection when the period is one week overdue. Again it's a combination of the same two hormones. And again the period will start after a five day interval if the patient is not pregnant, while there'll be no bleeding if she is.

"But like most doctors I prefer my patients to have the Hogben test. There is still much we have to learn about hormones—although the pregnancy tests are reliable enough.

"And a hormone test wouldn't have got you the result any quicker. You see, for five days of this past week you would have been taking the necessary tablets (for it's these I would have prescribed). Then you would have to wait another five days to see if your period started. Which brings you up to the day after tomorrow.

"As it is, if you take your specimen of urine round to the laboratory tomorrow, which you can do as you live in London, we'll have the result from the Hogben test just twenty-four hours later. So you see you haven't lost time by not coming earlier!"

Hopes confirmed in forty-eight hours

"Here's the necessary pregnancy test form," he said. "I've filled in my part. When you get home, fill in your name, address, age and the number of days your period is overdue. And do adhere strictly to the instructions on the form, won't you?"

"Nothing to drink after your evening meal today so that the morning specimen of urine will be really concentrated. And no aspirin, indigestion mixture or any kind of drug, because that would contaminate the specimen and might possibly harm the toad into which it is injected.

"And finally, use the urine you pass first thing in the morning. Put at least six ounces into a clean glass bottle or jar. See your name is on the bottle. Put the filled-in form, specimen of urine and the fee together, and hand them in at the address on the front of the form. As you'll see, the laboratory is open for specimens from 9 a.m. to 4 p.m. daily except at the weekends. And luckily for you this is only Tuesday!

"Now away you go and try to possess your soul in patience!"

Forty-eight hours later, an ecstatic Mrs. Berry was able to tell her husband a telephone call from the doctor had confirmed she was pregnant.

And just seven months after that she declared herself to be the happiest woman in the world. For she had been safely delivered of a beautiful baby boy.

JOAN SEAWARD

The pregnancy test proved positive—and now the baby they wanted so much is safely in her arms

27

7.5 The caption reads, "The pregnancy test proved positive—and now the baby they wanted so much is safely in her arms." Joan Seaward, "Pregnancy test . . . a modern scientific achievement," *Woman*, Jun. 24, 1961, 27. On *Woman*'s health page in the 1960s: Loughran 2020: 138–142. Wellcome Collection SA/FPA/A3/12, licensed by Future Publishing Ltd.

it's worth that") and "involved the co-operation of toads!" "How perfectly extraordinary," Mrs. Berry replied, "How on earth do toads help?" The doctor explained how the Hogben and Galli Mainini tests worked as well as the now "largely discarded" Aschheim–Zondek and Friedman tests. "How amazing," explained Mrs. Berry, before asking one "just more question" about "tablets" she had heard of that "act like a pregnant test?"

The tablets, the doctor explained, are a combination of the ovarian hormones, estrogen and progesterone, and a woman "can start taking them when her period is just one week overdue and continue for four to five days." If she is not pregnant, then "her period will commence" four or five days after taking them. If she is pregnant, there will "be no bleeding." There was also an injection that worked the same way but, "like most doctors," he preferred the Hogben test. He did not doubt the reliability of tablets and injections, but there was "still much we have to learn about hormones." Furthermore, he elaborated, the "hormone test wouldn't have got you the result any quicker." The doctor calculated that the days spent taking the tablets and waiting to see if bleeding started added up to "the day after tomorrow." Because she lived in London, Mrs. Berry could take her specimen "round to the laboratory tomorrow" for a twenty-four-hour test, so she hadn't "lost time by not coming earlier!"

Mrs. Berry's doctor provided the necessary form with his part already completed for her to fill in her name, address, age, and the number of days her period was overdue and instructed her not to drink after her evening meal, to take no aspirin or other drugs that might harm the toad (not the fetus), and to deliver the form, fee, and at least 6 ounces of (concentrated) morning urine in a clean glass bottle or jar labeled with her name to the address on the form. "Forty-eight hours later," the narrative concluded, a "telephone call from the doctor" (not the lab) "confirmed she was pregnant" and "just seven months after that she declared herself to be the happiest woman in the world. For she had been safely delivered of a beautiful baby boy."[33] This strong endorsement of the Hogben test in Britain's most prominent women's magazine was a direct response to concerns about hormone tablets and injections. The medical debate continued in professional journals and, soon after thalidomide, expanded to include the risk of birth defects.

ANOTHER THALIDOMIDE STORY

In the immediate aftermath of thalidomide, Victor Dubowitz, a South African–born pediatrician at the Children's Hospital in Sheffield, warned in the *Lancet* of a "possible association" between the use of Amenorone in pregnancy diagnosis and "virilisation in the female infant." The case involved a thirty-four-year-old woman who had become pregnant for the first time after six years of marriage. After a second missed period, she had consulted her GP, who prescribed one tablet of Amenorone daily for three consecutive days. She did not bleed and, following an "uneventful pregnancy," gave birth to twins: one "apparently normal male" and one with "ambiguous" genitalia. The latter was transferred to the children's hospital, where a buccal smear and karyotyping disclosed a genetically female "pseudohermaphrodite." Dubowitz could "only speculate" whether a "small dose" of Amenorone had masculinized the infant.[34] This planted a new seed of doubt regarding a practice already suspected of causing miscarriage.

Richard Smithells, a Liverpool pediatrician who had recently established a fetal malformation register in the Merseyside area, explained in the *Practitioner* in 1965 that for the "first two weeks of embryonic life pregnancy is usually unsuspected and there is a natural anxiety that during this unguarded fortnight drugs may be taken, anaesthetics administered or x-ray exposures made which would have been avoided had pregnancy been recognized." But what of hormone pregnancy tests? Smithells surveyed 189 women who had been prescribed Amenorone Forte or Primodos in the first twelve weeks of pregnancies that lasted longer than twenty-eight weeks. Although the "small group" provided "no evidence to support [Dubowitz's] suggestion that pregnancy-test drugs are teratogenic," he nevertheless warned that a "heavy responsibility lies on the shoulders of every practitioner who orders the administration of any drug to a woman in the first twelve weeks of pregnancy."[35]

The question of teratogenicity became acute in 1967 because of Isabel Gal's research on spina bifida. Born into a middle-class Jewish family in 1925 in Hungary, Gal (née Gunsberger) survived Auschwitz with her mother and sisters and returned to Hungary with them after the war. After qualifying in medicine at the University of Budapest, Gal worked as a pediatrician at the Bókay János Children's Hospital.[36] In 1953, she married Endre, a

7.6 Photograph of Isabel Gal in Hungary, probably Budapest, c.1953, around the date she and Endre were married; courtesy of Kathy Gal.

mathematician and the son of a timber merchant for whom Gal's father had worked (figure 7.6). She fled the 1956 revolution with her mother, her husband, and their newborn daughter Katinka (Kathy), first across the border to Austria, where her mother was from, and then to England, where a sister had settled. After requalifying at the University of Edinburgh, she found work at Great Ormond Street Hospital and at Queen Mary's Hospital, Carshalton.[37]

Spina bifida and other neural tube defects are painful, debilitating, and often fatal. Innovations in surgical treatment and management improved survival but engendered new kinds of "personal and family problems, as well as complicated medical, socio-economic and legal issues" in the 1960s.[38] So medical attention turned to surveillance and prevention.[39] In 1964, Gal received funding from the Society of Mentally Handicapped Children to investigate, in collaboration with Westminster Children's Hospital, the genetic and environmental causes of central nervous malformations. For a period of twelve months, Gal personally interviewed the parents of 131 children with spina bifida and hydrocephalus and selected as controls the mothers of "healthy babies" at another hospital in Surrey, matching them as closely as possible by age, reproductive history, and so on.[40] Careful analysis of the interviews, she hoped, would disclose potential causal agents that medical notes failed to capture, for example, the use of over-the-counter drugs.

Gal had planned to use the "customary edge-punched card system" but was soon overwhelmed by the amount of data she had collected. Her husband, Endre, suggested using a computer to salvage her project. At first, Gal rejected the idea that a "machine could replace a very important part of human work" and worried that her "carefully gathered" data would get "mixed or messed up." Endre persisted, explaining the machine and programming to her, but Gal remained skeptical until she saw the computer in "no time" produce results that had taken her "about a fortnight of hard work to arrive at."[41] So finally she agreed to send the questionnaire for processing by her husband on the Atlas Computer at the University of London's Institute of Computer Science, Gordon Square (figure 7.7).[42]

Among other questions, Gal asked about drugs taken during pregnancy and (probably in ignorance of hormone pregnancy tests) about how pregnancies had been diagnosed. One difference between the "index" and "control" groups especially jumped out. Nineteen of the one hundred index mothers finally selected for analysis but only four of the same number of controls "reported having received oral tablets for the diagnosis of pregnancy," a surprising and disturbing ratio. Concerned that hormone pregnancy tests "could be another thalidomide story," Gal and her colleagues Brian Kirman, a physician, and Jan Stern, a biochemist, drafted a letter for *Nature*.[43] Before submitting it, Kirman sent the draft to the Committee on the Safety of Drugs (CSD) with a note stating that, "if soundly based," their findings appeared to be "important" to not only hormone

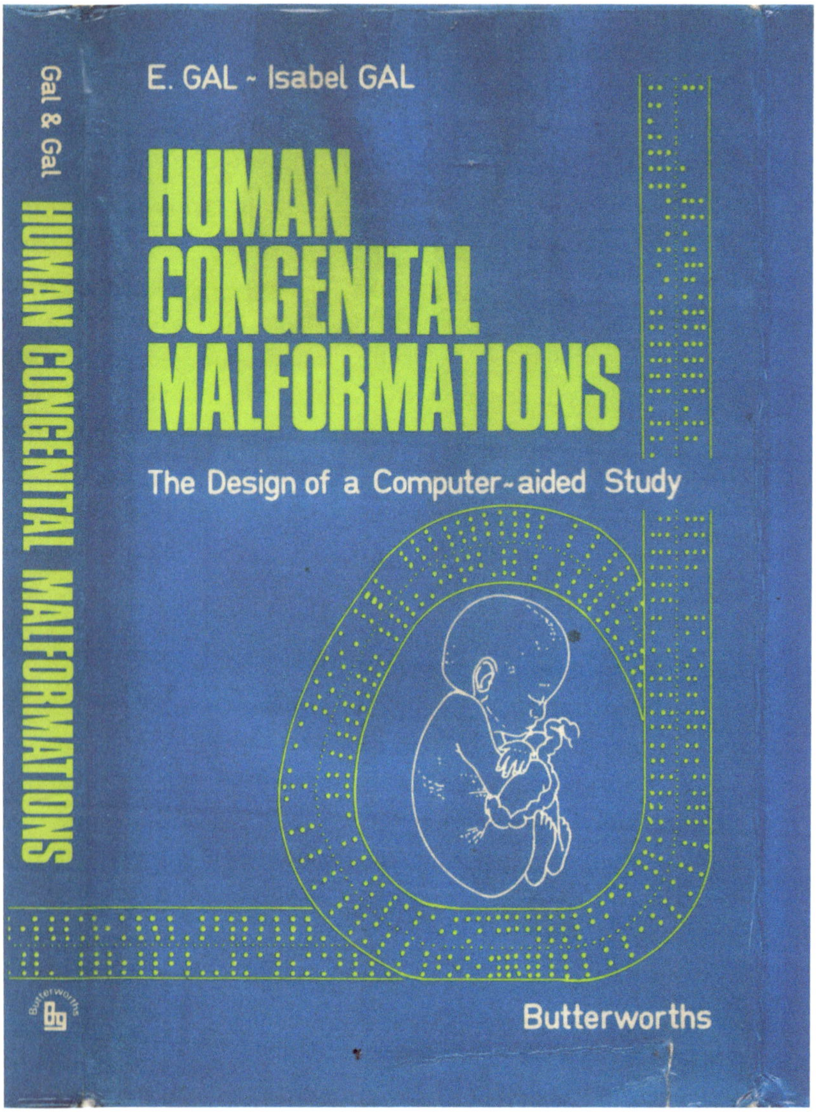

7.7 Dust jacket of Gal's "computer-aided" study, belatedly published as part of the Computers in Medicine series edited by D. W. Hill: Gal and Gal 1975. The striking image, of tickertape protectively wrapped around the fetus, was designed by Gal's daughter, Kathy; courtesy of Kathy Gal.

pregnancy tests but also "contraceptive medication, and also possibly in regard to the use of hormones for maintenance of pregnancy."[44]

THE EARLY WARNING SYSTEM

In 1963, amid the fallout from thalidomide, the government established the CSD, known colloquially as the "Dunlop Committee" after its first chair Sir Derrick Dunlop, a prominent Scottish physician and pharmacologist. The CSD reviewed data submitted by pharmaceutical companies and advised manufacturers and the government on whether new drugs had been adequately tested for market. By the end of 1965, the CSD consisted of a small team of six doctors, three pharmacists, and a modest administrative staff. It had no legal powers and depended on voluntary cooperation from industry. Its members were not employed by pharmaceutical companies but were permitted to have financial interests, such as shareholdings or research grants. The confidential review process was designed to be rapid to avoid delaying the introduction of potentially beneficial drugs.[45]

The CSD was divided into three subcommittees—on Toxicity, Clinical Trials, and Adverse Reactions, the latter of which was first chaired by Oxford professor Leslie Witts. William "Bill" Inman joined the CSD as senior medical officer and medical assessor for the Witts subcommittee in 1964. Previously, he had battled polio as a medical student at Cambridge before acting as medical adviser to Imperial Chemical Industries, the company his father had cofounded in 1926. He was later promoted to principal medical officer and is today remembered as "father of the mini-pill" for his role in reducing the estrogenic content of oral contraception.[46] Starting in 1967, he was the government advisor chiefly responsible for deciding what action, if any, to take on hormone pregnancy tests.

Inman oversaw the "yellow card" early warning system of monitoring adverse reactions, so named for the distinctively colored post–free business reply cards the CSD periodically issued to clinicians for reporting a suspected reaction to a new drug or a serious reaction to any drug, new or old. In 1964, the CSD received around 1,000 cards every month.[47] Inman did not have access to a computer. His "statistical calculations were worked with a slide-rule or log-tables and a hand-cranked 'Facit' adding machine."[48]

In the absence of robust baseline data, he developed a comparative method of assessing reactions caused by chemically similar drugs.[49]

The first drug that came to Inman's attention was a vasodilator that "quite obviously caused jaundice." The company, Inman later recalled, was "persuaded to remove that voluntarily without any pressure. It was kept under wraps. There wasn't much publicity. We didn't seek any publicity."[50] This was the model outcome for Inman, who preferred to quietly resolve any potential safety issues without involving doctors, the media, or the public. Owen Wade, deputy chair of the Adverse Reactions subcommittee in the mid-1960s, lamented this policy, which was intended to protect cooperative companies from bad press. Publicity, he later argued, would have "shown doctors the value of reporting adverse drug reactions and our reporting system would have got off the ground much quicker than it did."[51] In reality, doctors voluntarily reported only a small fraction of the suspected adverse reactions they observed in clinical practice, perhaps only 10 percent for serious reactions.[52] The fraction was even smaller for minor reactions and for birth defects.

Inman later described the system's "inability to detect teratogenic drug effects" as one of "several fundamental defects" that had been "obvious" from the start.[53] This is surprising for a system set up in response to thalidomide, a teratogenic drug, but several factors militated against detection. For one, gestation slowed down the already imperfect process of voluntary reporting. Noticing an adverse reaction nine months after a drug had been prescribed—and in another patient, the child, who had not been given the drug directly—was a particular challenge; women went to gynecologists but took their children to pediatricians, and first-time mothers often moved house and changed doctors. Finally, the CSD insisted on the premarket testing of new drugs on pregnant animals, but this did not affect hormone pregnancy tests or other "old drugs."

PRIMA FACIE EVIDENCE

Even before the draft of Gal's article crossed his desk, Inman had "picked up about a dozen reports of congenital abnormalities following the use of oestrogen-progestogen mixtures, either for control of menstrual irregularities or conception, or as pregnancy tests." Because of the large number of women using these drugs—"well over a million" each year—he did not consider these cases to "constitute evidence of teratogenicity." And from the start, he doubted Gal's methodology. Inman did not doubt the "statistical significance" of Gal's data. He did, however, suspect that sampling bias could explain the pattern. Were the groups "precisely comparable"? Gal's findings, Inman speculated, would be "invalidated" if the index mothers "had had pregnancy tests [because] they were more than normally worried about a further pregnancy following the birth of a deformed baby" or because of "some medical illness such as diabetes or hypertension which might make their doctors anxious to detect pregnancy at an earlier stage."[54]

The CSD considered attempting to delay Gal's publication, but Witts advised against it. "The most useful thing now," he wrote to Inman, would be for the findings to "become known" so that others could "confirm or refute them." Compared to Inman, Witts was less charitably disposed to hormone pregnancy tests, which he described as a uniquely "big dose" of progesterone "when the embryo is most vulnerable." Nor was he convinced that Inman could so easily explain away the large difference between the groups.[55]

On Witts's instruction, Inman contacted Norman Jeffcoate, a professor of obstetrics and gynecology at the University of Liverpool, for expert advice. Jeffcoate shared Inman's benign outlook on "pregnancy diagnosis pills." He did not know of any controlled studies, but clinical experience—the empirical use of comparatively "massive" doses of sex hormones in pregnancy to treat infertility and prevent miscarriage—suggested harmlessness. The only caveat Jeffcoate allowed was that norethisterone, the progestogen found in Primodos, had become "unpopular" because it was known to "virilise" the female embryo. Cases of virilization were, however, "exceptionally rare," and as far as Jeffcoate knew, "no other harm" had been reported. He suspected (with Inman and against Witts) that Gal's findings had "some explanation other than a direct teratogenic effect"—that it was not the test but the "circumstances calling for the test" that were to blame.[56]

Jeffcoate's reasoning was simple. Hormone pregnancy tests were "generally unnecessary" and typically "reserved for those cases in which the diagnosis of pregnancy [was] doubtful or of unusual importance." Hence, they were "most likely to be used, not in the woman of normal fertility, but in one who has long been sterile, has had irregular menstruation or recurrent abortion (or even previously malformed babies) or is threatening to abort." Jeffocate recommended investigating the matter further, controlling the cases not with "unselected women producing normal babies" but with women "matched for age, parity, abortions and previously malformed babies," and he referred Inman to Smithells for further expertise.[57]

Following a meeting at Queen Mary's, Inman wrote to Gal, thanking her on behalf of the CSD for the heads-up: "So often, the first indication we get that an important hazard is suspected, is the sudden appearance of a paper in one of the medical journals." Inman did not doubt that Gal had produced "prima facie evidence" of teratogenicity, and he encouraged her to publish. He further reassured her that, even if she was later "shown to be wrong," no harm would come of communicating her findings; hormone pregnancy tests were "not essential," and it would "not be a disaster" if Gal's paper reduced the "frequency of their use." On the other hand, the "rights and wrongs of using these hormones for diagnostic purposes [was] an entirely separate issue from the scientific interpretation of a possible cause/effect relationship between the tests and congenital deformities." And there were "major difficulties" in the "purely scientific evaluation" of Gal's data, including the possibility of selection bias and the "higher incidence of congenital abnormality among parents and siblings in the survey cases than in the controls." On balance, however, the CSD "fully endorsed" Inman's advice that the "correct thing to do would be to publish this work so that the profession may be alerted to the possibility that these tests may be dangerous and in the hope that further work will be stimulated."[58]

Inman reported back to Jeffcoate on his meeting with "these people at Queen Mary's Hospital." Gal, Kirman, and Stern were "very charming" and had done a "great deal" of "slightly misguided" work. They had "gone into the minutest detail" but, Inman claimed, "selected their cases badly, and on looking at their data it was also apparent that there were rather more congenital abnormalities among parents and siblings on the affected group than the unaffected controls." Inman doubted they would "have much success convincing the more scientific members of the profession that bias [had] been avoided in these areas." On the other hand, he was impressed by Stern's record as a biochemist of congenital abnormality and not prepared to dismiss his work "simply on the grounds" of selection bias. Ultimately, Inman accepted that the group had produced a "prima facie case against the pregnancy tests" and that "further work should be done to sort this problem out." The CSD concurred.[59]

Smithells, meanwhile, independently met Gal at a conference in London and, according to Jeffcoate, "was quite unconvinced as to the validity of the statistics, especially because they were based on retrospective studies."[60] For Inman, this was becoming a "rather awkward problem." Privately, he claimed to be "as unconvinced" as Jeffcoate and Smithells about the "validity" of Gal's data (on the grounds of selection bias), but he remained unwilling to "rule out the possibility" entirely. His doubts stemmed from uncertainty about the prevalence of hormone pregnancy tests: "If we assume that there is no teratogenic effect," he confided in Jeffcoate, "Kirman's 19 infants with hydrocephalus would have to have been born to at least 10,000 women who had had the [hormone] pregnancy test. This is a very large number indeed and I rather doubt whether the test is used all that frequently. I think there is a case for further investigation and I hope Dr. Smithells may be able to help us."[61]

In other words, if hormone pregnancy tests were very widely prescribed, then the nineteen hydrocephalic infants might just be coincidental. If the tests were uncommon, as Jeffcoate seemed to think, then it would be harder to explain them away. Lending his weight to the interpretation of the data as coincidental, Smithells claimed that Merseyside GPs prescribed hormone pregnancy tests on a "fantastic scale" and that Merseyside women regarded them as abortifacients and obtained them under false pretense, "by giving a misleading history."[62]

Gal's report, meanwhile, was published in *Nature* on October 7, 1967. It conceded that the observed difference between the two groups of mothers might have emerged "purely by chance" and called for more research into the "role of hormonal preparations in the causation of congenital malformations, particularly when taken in the organogenic stages of pregnancy."[63] The CSD issued a statement that it had seen "no evidence" to corroborate Gal's findings, but *Medical News* compared them to those of a Mexican survey that disclosed a higher incidence of birth defects in babies of mothers who kept taking the pill following conception.[64] Gal continued her correspondence with Inman, writing to him in December, "It may be that our finding is pure coincidence, as I have expressed on many occasions, but I still feel that to use hormones in the most sensitive stage of the pregnancy for diagnostic purposes is unnecessary when other good methods are available."[65] Inman agreed: "My personal view about the value of pregnancy tests is identical to yours, I frankly do not think that they are sufficiently useful when compared with other biological methods to justify even the slightest risk of teratogenicity."[66] By then, bioassays had given way to immunoassays, a change that in the 1960s enabled the decentralization, demedicalization, and commercialization of pregnancy diagnosis.

8

DIRECT TO CONSUMER

Delia Davin was a twenty-four-year-old doctoral candidate at the University of Leeds when she began to "think of having a baby." She was married and, having successfully used contraception for eight years, "slightly nervous" about having a "fertility problem." Because of this and to "fit any pregnancy in with research," she "decided to spend the equivalent of a week's rent on a commercial pregnancy test as soon as [she] noted signs of pregnancy." Davin had probably read about such tests in *New Statesman* or seen an advertisement in *Peace News* or *Private Eye*, all of which were "easily available" in West Yorkshire. Her result "came back promptly as positive," and she went "at once to the Student Health Service to request ante-natal care." When Davin "began to bleed" three weeks later, the "positive test" helped to persuade her doctor that she was not just having a late period, and he "immediately arranged" to drive her to Student Health, where she was "monitored" and "nursed for several days," causing her to miss the historic Vietnam Solidarity Campaign demonstration in London on October 27, 1968.[1]

Commercial pregnancy testing services of the kind Davin used were an innovation of the mid-1960s. Widely advertised, fiercely debated, and vigorously resisted by the medical establishment, commercialization restructured the power dynamics between doctors and women. It was enabled by a material shift away from bioassays. As we shall see, immunoassays were a highly disruptive technology. In less than a decade, they completely replaced bioassays in pregnancy testing. But this does not mean that test kits were vastly superior to toads. As historian David Edgerton has noted, "To become widely used, a thing does not have to be massively better than what preceded it; it need only be *marginally* better than alternatives."[2] In any case, technological change is never the whole story. This chapter reconstructs the changing practices and social relations of pregnancy testing in an eventful (and frequently mythologized) decade of protest and revolution.[3] It is also about how new forms of journalism drew on the increasingly available language of consumer choice, public interest, and reproductive freedoms to sustain a damaging critique of medical paternalism and, in so doing, forged a liberal consensus on pregnancy testing as a "woman's right."[4]

I SEE A RING, GIRL, GET YOUR RING

Pregnancy, made hormonal in the heyday of endocrinology, was made immunological in a brave new world of postwar biomedicine.[5] New tools and networks played an important role, but so too did older links among farm, lab, and clinic. Stephen Boyden, an Australia-born veterinarian, developed a method of binding protein antigens to the surface of sheep erythrocytes (red blood cells) treated with tannic acid while visiting the Rockefeller Institute in New York on a Wellcome Trust fellowship in 1950.[6] Using tuberculin preparations, sera from human patients, and rabbit antisera, he noticed that antigen-coated cells formed a mat pattern when they sedimented in a test tube in the presence of antibodies, whereas uncoated cells formed a visible ring or dot in the center. The addition of free antigen neutralized the antibodies and inhibited the mat pattern formation, which made it possible to detect the antigen in a solution.[7] Immunologists used Boyden's test to detect insulin in a buffer solution. And Leif Wide, a Swedish medical student, applied it to pregnancy testing in 1960 (figure 8.1).[8]

In the autumn of 1959, Wide approached Carl Gemzell, his teacher at the Karolinska Institute in Stockholm, to start some research in parallel with his studies. Gemzell, a gynecologist specializing in reproductive endocrinology and infertility treatment, directed Wide toward a recently reported immunoassay for the human growth hormone in blood.[9] Wide had little success

8.1 Portrait photograph by Gösta Glase of Leif Wide (c.1960) looking smart as a young doctoral student in lab coat, slicked back hair, horn-rim glasses, and bowtie, demonstrating a draft of a figure from his doctoral thesis; courtesy of Wide.

measuring the minute concentrations of "hGH" in serum or plasma, so in early 1960, he decided to apply the method to detecting hCG in pregnant women's urine. Gemzell arranged for specimens to be rerouted from his gynecology ward for testing. It was too soon to say, but the immunological technique seemed to work where countless attempts to develop a biochemical reaction had failed; only pregnant women tested positive.[10]

Wide treated sheep erythrocytes with formalin and tannic acid and then coated them with Pregnyl, a commercial hCG product marketed by the Dutch multinational Organon. With a view toward a test kit that would be suitable for clinical or pharmacy use, he next attempted to freeze-dry the hCG-coated cells and the antiserum in separate bottles. Developed in the 1930s by biochemists at the University of Pennsylvania, the technique of freeze-drying (lyophilization) was used to preserve human plasma and penicillin during the war and orange juice and other foodstuffs in peacetime.[11] Following encouraging results, Wide began using ampoules of the two freeze-dried reagents combined in a round-bottomed test tube. To perform a test, he first added a single drop of woman's urine and half a milliliter of buffer solution to an ampoule containing the reagents and then waited ninety minutes before inspecting a mirror beneath the test tube. In a positive reaction, the hCG bound the antibodies, and the hCG-coated cells slid down the glass wall to settle as a "sharp ring or disc." In a negative reaction, the antibodies reacted with and covered the cells, which adhered to the wall and formed a "mat pattern" as they sedimented (figure 8.2).[12]

By May 1960, Wide had tested over three hundred specimens from Gemzell's ward and had not obtained a single incorrect result. But the pair needed a commercial partner to manufacture the standardized reagents on a large scale, so Gemzell approached Organon, a company he had previously dealt with. Marius Tausk, the managing director in Oss, was "deeply impressed" and proposed the somewhat presumptive mnemonic, "I see a ring, girl, get your ring."[13] On his return flight, he drafted a contract, which granted Organon a few weeks to apply for a patent before the Swedes submitted their manuscript for publication.[14] Tausk anticipated launching the test kits by Christmas but had "seriously underestimated" the technical difficulties of scaling up the production of freeze-dried reagents.[15] With no guarantee the venture would pay dividends, the delay nearly provoked Organon to abandon the project.

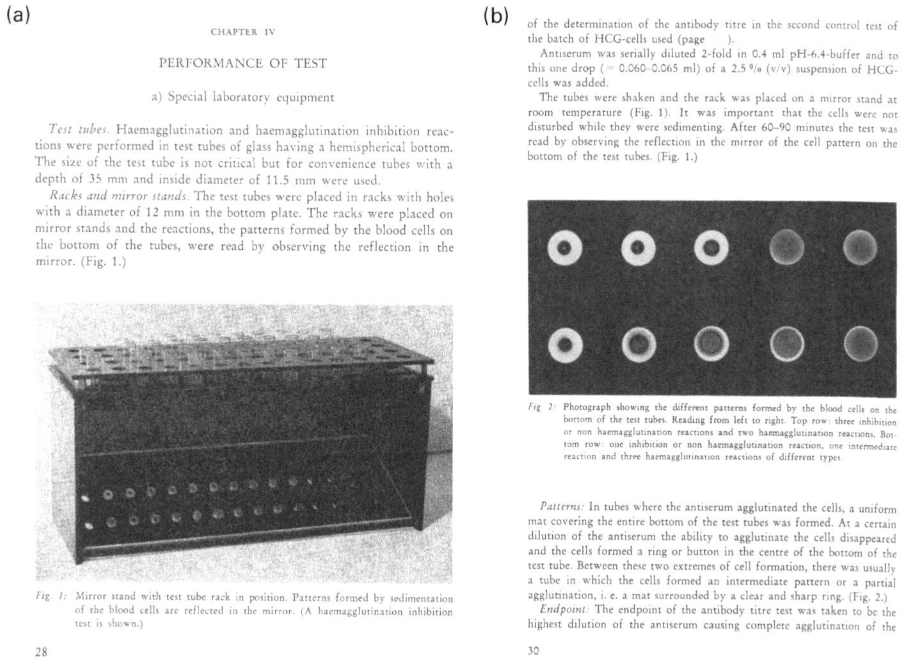

(a)

CHAPTER IV

PERFORMANCE OF TEST

a) Special laboratory equipment

Test tubes. Haemagglutination and haemagglutination inhibition reactions were performed in test tubes of glass having a hemispherical bottom. The size of the test tube is not critical but for convenience tubes with a depth of 35 mm and inside diameter of 11.5 mm were used.

Racks and mirror stands. The test tubes were placed in racks with holes with a diameter of 12 mm in the bottom plate. The racks were placed on mirror stands and the reactions, the patterns formed by the blood cells on the bottom of the tubes, were read by observing the reflection in the mirror. (Fig. 1.)

Fig. 1: Mirror stand with test tube rack in position. Patterns formed by sedimentation of the blood cells are reflected in the mirror. (A haemagglutination inhibition test is shown.)

28

(b)

of the determination of the antibody titre in the second control test of the batch of HCG-cells used (page).

Antiserum was serially diluted 2-fold in 0.4 ml pH-6.4-buffer and to this one drop (= 0.060–0.065 ml) of a 2.5 % (v/v) suspension of HCG-cells was added.

The tubes were shaken and the rack was placed on a mirror stand at room temperature (Fig. 1). It was important that the cells were not disturbed while they were sedimenting. After 60–90 minutes the test was read by observing the reflection in the mirror of the cell pattern on the bottom of the test tubes. (Fig. 1.)

Fig. 2: Photograph showing the different patterns formed by the blood cells on the bottom of the test tubes. Reading from left to right. Top row: three inhibition or non haemagglutination reactions and two haemagglutination reactions. Bottom row: one inhibition or non haemagglutination reaction, one intermediate reaction and three haemagglutination reactions of different types.

Patterns: In tubes where the antiserum agglutinated the cells, a uniform mat covering the entire bottom of the test tubes was formed. At a certain dilution of the antiserum the ability to agglutinate the cells disappeared and the cells formed a ring or button in the centre of the bottom of the test tube. Between these two extremes of cell formation, there was usually a tube in which the cells formed an intermediate pattern or a partial agglutination, i. e. a mat surrounded by a clear and sharp ring. (Fig. 2.)

Endpoint: The endpoint of the antibody titre test was taken to be the highest dilution of the antiserum causing complete agglutination of the

30

8.2 Pages from Wide's doctoral thesis, completed in Uppsala and published in Copenhagen in 1962, showing (a) photographs of a mirror stand with test tube rack and (b) the patterns formed by blood cells on the bottom of test tubes. Wide 1962, 28, 30; courtesy of Wide.

At a meeting in July 1961, several Organon executives argued for abandoning the project to free company resources for more promising ventures. But Jacobus Polderman, an enthusiastic research pharmacist, and other supporters successfully pushed for continuation.[16] Meanwhile, Wide and Gemzell moved to Uppsala University, where the latter had been offered a professorship in obstetrics and gynecology.[17] Wide defended his doctoral thesis in 1962, reporting an accuracy of 99.8 percent in some 2,230 tests.[18] Although clinicians were initially skeptical that such a simple test could actually work, the university hospital there eventually replaced the rabbit test with Wide's.[19] Several other hospitals in Sweden followed suit, and Organon launched "Pregnosticon" in the Netherlands in May 1962.[20] By then, word of Organon's initiative had spread and competitors were already developing rival products.

TEST-TUBE SIMPLICITY

In Britain, Burroughs Wellcome, a prosperous pharmaceutical company with a long tradition of laboratory research, began work on its own immunoassay in 1961.[21] Arthur James Fulthorpe and others at the Biological Division of Wellcome Research Laboratories in Beckenham, Kent, collaborated with a hospital and two group laboratories in nearby Lewisham to compare the new immunoassay to the male toad test in over seven hundred urine specimens, including over two hundred from female employees at the Wellcome laboratory (six turned out to be pregnant).[22] William Barr of the Edinburgh pregnancy diagnosis station worked with Fulthorpe to compare Burroughs Wellcome's "Prepuerin" to *Xenopus* in over 1,500 tests (figure 8.3). Although Prepuerin was "simple and easy to perform," the standard of Hogben testing in Edinburgh was "extremely high": 99.2 percent agreement with clinical

(a)

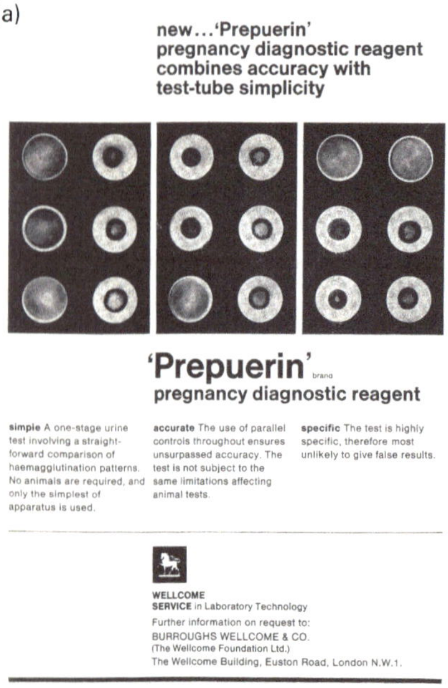

(b)

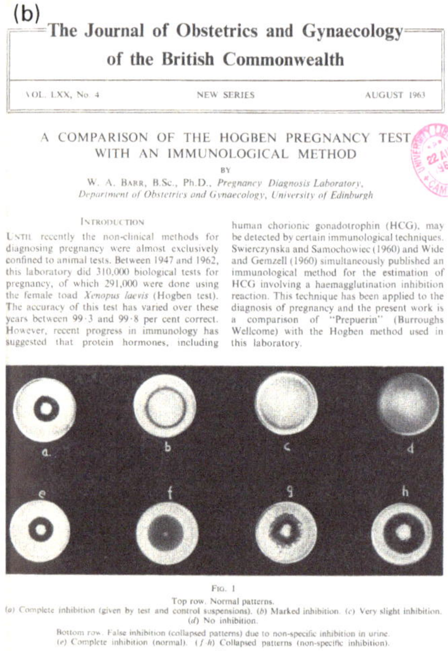

The Journal of Obstetrics and Gynaecology of the British Commonwealth

VOL. LXX, No. 4 NEW SERIES AUGUST 1963

A COMPARISON OF THE HOGBEN PREGNANCY TEST WITH AN IMMUNOLOGICAL METHOD

BY

W. A. BARR, B.Sc., Ph.D., *Pregnancy Diagnosis Laboratory, Department of Obstetrics and Gynaecology, University of Edinburgh*

INTRODUCTION

UNTIL recently the non-clinical methods for diagnosing pregnancy were almost exclusively confined to animal tests. Between 1947 and 1962, this laboratory did 310,000 biological tests for pregnancy, of which 291,000 were done using the female toad *Xenopus laevis* (Hogben test). The accuracy of this test has varied over these years between 99·3 and 99·8 per cent correct. However, recent progress in immunology has suggested that protein hormones, including human chorionic gonadotrophin (HCG), may be detected by certain immunological techniques. Swierczynska and Samochowiec (1960) and Wide and Gemzell (1960) simultaneously published an immunological method for the estimation of HCG involving a haemagglutination inhibition reaction. This technique has been applied to the diagnosis of pregnancy and the present work is a comparison of "Prepuerin" (Burroughs Wellcome) with the Hogben method used in this laboratory.

a b c d e f g h

FIG. 1

Top row. Normal patterns.
(*a*) Complete inhibition (given by test and control suspensions). (*b*) Marked inhibition. (*c*) Very slight inhibition. (*d*) No inhibition.

Bottom row. False inhibition (collapsed patterns) due to non-specific inhibition in urine.
(*e*) Complete inhibition (normal). (*f*–*h*) Collapsed patterns (non-specific inhibition).

551 19

8.3 (a) Advertisement showing photographs of agglutination inhibition reactions obtained with Prepuerin. *Journal of Obstetrics and Gynaecology of the British Commonwealth* 70, Apr. 1963, iii. (b) Lead article in the same journal with similar images: Barr 1963a, 551, Cambridge University Library T323.b.13.88.

diagnosis compared to 97.4 percent for Prepuerin. Most disagreements occurred because of a cross-reaction with the pituitary luteinizing hormone (LH).[23] False positives had not been an issue with *Xenopus*, and pregnancy testers debated their significance regarding the new test kits.

As Prepuerin became commercially available, Alexander Warrack and his colleagues at the Sheffield pregnancy diagnosis center compared it with *Xenopus* in randomly selected urine specimens. By May 1963, they had clinically confirmed the results in 311 of 729 tests. The center maintained a colony of 4,000 toads and performed some 20,000 tests a year for doctors in the Midlands, northern England, and Wales. As we saw in chapter 5, testing ground to a halt in 1954 owing to supply issues and again in 1958, when a bacterial epidemic nearly wiped out the colony. Unsurprisingly given this track record, Warrack was keen to emancipate his laboratory from the "vagaries" of a "creature as temperamental in its outlook as it is difficult to obtain in regular quantities." Despite also obtaining false positives, he praised Prepuerin as "easy to perform and easy to read."[24]

Eileen Shuttleworth, a pathologist at the Cumberland Infirmary, Carlisle, collaborated with Fulthorpe and Warrack to test Prepuerin against *Xenopus*. In October, having clinically confirmed the results in 165 women, she reported in the *Lancet* that Prepuerin was easier, more economical, and quicker than "using living animals."[25] Richard Oliver, a pathologist at the Mayday Hospital in Croydon, preferred Pregnosticon because it was "performed on undiluted urine in the ampoules provided, without the need of additional glassware." He had performed 250 tests in the "few months" since the product had become commercially available.[26] Organon concentrated on marketing in the Netherlands, Belgium, and other nearby countries, but its UK branch in Surrey supplied the FPA with free kits to compare against *Xenopus*.[27]

The company's promotional literature boasted that the "many years of work of a large scientific team" had combined "simplicity" with "great reliability" and that "several investigators" had already reported "good results." Gynecologists at the university women's clinic in Basel, for instance, testified that Pregnosticon was "vastly superior" to all other pregnancy tests. For best results, Organon recommended a "preliminary check" when starting a new pack, each of which contained reagents sufficient for twenty tests and freeze-dried "test samples" of "pregnancy urine" and "normal

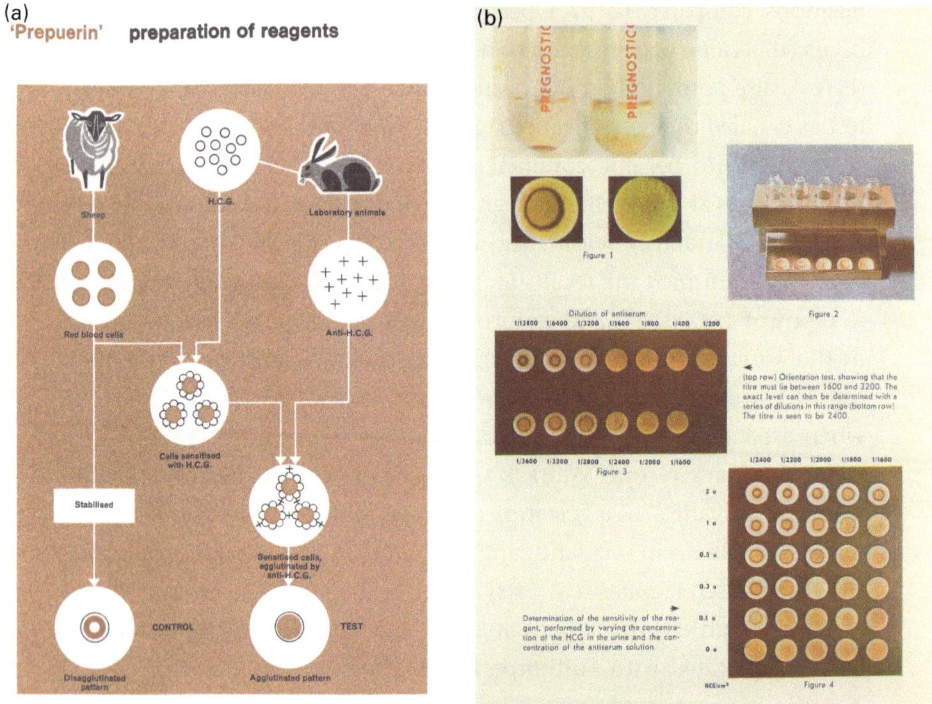

8.4 (a) Detail of promotional brochure showing the production of reagents and reaction patters from raw materials (sheep and rabbit blood). *Prepuerin* (London: Burroughs Wellcome, 1963); (b) Detail of brochure showing the calibration of the reaction with different concentrations of hCG and reagent. *Pregnosticon* (Morden: Organon Laboratories, *c.*1962). Wellcome Collection WF/M/PL/247 and SA/FPA/A7/90.

urine" for checking the pack, a precautionary measure that would later be abandoned as the company gained confidence in its new product.[28]

Burroughs Wellcome boasted that its proprietary test combined "accuracy" with "test-tube simplicity." Prepuerin met the "need" for a "simple," "sensitive," "accurate," "specific," "economical," and "labour-saving" technique that did not require "inconvenient" animals or "special apparatus." Rather, it involved the "straightforward comparison of haemagglutination patterns," long familiar to clinicians and laboratory workers through blood grouping. Any hospital laboratory that stocked Prepuerin would be able to "provide economically a speedy local service" (figure 8.4).[29]

THE ULTIMATE IN PREGNANCY TESTING

In 1961, the same year Burroughs Wellcome began testing Prepuerin in Britain, Ortho Pharmaceutical, a New Jersey–based subsidiary of the American behemoth Johnson & Johnson, launched a two-stage agglutination test using polystyrene latex particles instead of sheep erythrocytes.[30] Electron microscopists in postwar America had enthusiastically used the unusually consistent latex spheres Dow Chemicals produced in Michigan to establish a uniform standard of magnification.[31] The inert particles presented medical workers with an attractive alternative to red blood cells, which deteriorated unless specially treated. In the mid-1950s, doctors at Mount Sinai Hospital, New York, reported the first latex fixation test, using Dow spheres, for the serological diagnosis of rheumatoid arthritis.[32]

Ortho first supplied reagents to doctors in Gainesville, Florida, who reported in *JAMA* that the "simple and rapid procedure" compared favorably with *Xenopus*.[33] Based on their experience, a *BMJ* editorial predicted that latex might soon replace Wide and Gemzell's "temperamental" method.[34] And *New Scientist* called the new latex test "better" than the Hogben test.[35] Bruce Hobson, scientific director of the Edinburgh station, leapt to the defense of *Xenopus*, which was correct, he claimed, "99 times out of 100." Hobson conceded that immunoassays might eventually become the "method of choice," but for accuracy, he preferred the Hogben test.[36]

Undeterred, the UK branch of Ortho in Buckinghamshire provided free latex tests to the Portsmouth and Isle of Wight Area Pathological Service for trial at a local hospital. A decade earlier, a local GP had blamed the locally high demand for pregnancy diagnosis on the presence of a naval port; the service laboratory, which had its own *Xenopus* breeding program, performed some two thousand Hogben tests every year for about one hundred area doctors.[37] But animal breeding and testing allocated "considerable" resources away from the service's "heavy routine commitments," and so the Portsmouth pathologists were easily persuaded of the merits of latex over toads.[38]

Although Hobson had visited Uppsala for six months in 1962 on an Eleanor Roosevelt international cancer fellowship to work with Wide on Pregnosticon, he took every opportunity to defend *Xenopus* (figure 8.5a).[39] In June 1963, he reminded readers of the *BMJ* that only a "few hundred

8.5 Hobson and Wide became close friends and published some twenty collaborative studies in thirty years. (a) Photograph of Hobson in Sweden; courtesy of Wide. (b) Portrait of Sharman from a compendium on Jews and medicine in Scotland: Collins 1988, 93, Cambridge University Library 514:55.c.95.55.

comparisons" had been made, mostly on urine specimens from "normal pregnant and non-pregnant women." In his laboratory, "less than a third of the 24,000 to 25,000 tests done each year [were] for women with normal pregnancies." His staff had performed over three thousand parallel tests, and he was unwilling to replace *Xenopus* with a new method that produced false positives. Toads never ovulated in error, at least not in Edinburgh, and so Hobson's choice was "between an immunological test that will tell nearly 600 women each year that they are pregnant when they are not, and the biological test that will fail to detect the pregnancy of some 170 women."[40] He preferred the latter.

For disenchanted toad testers, however, false positives were not a deal breaker. The Portsmouth group, for instance, argued in the *BMJ* that Hobson and those who agreed with him had simply made a "virtue out of an inherent property of the test" and should reexamine their assumptions.[41] False positives, Hobson insisted in a rejoinder, were to be "deplored" not only because they "mislead doctors" but also because of the "effect of telling a patient who may be unmarried or anxious to have a child that she is pregnant when she is not." Hobson cautioned that immunoassays should be used "with discretion" lest they "fall into disrepute."[42] Siding with Portsmouth, Warrack argued that a false positive was not "any less misleading or unhelpful" than a false negative; they were "equally deplorable."[43] But Hobson countered that his "clinical colleagues" were "unanimous in agreeing that false-positive results would cause more trouble than false negatives," the idea being that nonpregnancy was the default state and easily controlled by retesting in a week's time.[44]

Ortho also provided tests to Albert Sharman, a gynecologist and infertility specialist at Glasgow's Royal Samaritan Hospital for Women (figure 8.5b). Since the late 1930s, Sharman had experimented with injectable pregnancy tests, hoping these might free him from reliance on a laboratory. For this pragmatic reason, coupled with his ambitions as an innovator (he had invented a kymograph for tubal insufflation and a cervical mucus test), he enthusiastically collaborated with Ortho.[45] Sharman compared his results with Ortho's reagents to those of Aschheim–Zondek tests routinely performed at the Royal Maternity Hospital and concluded after six hundred attempts that latex beat mice. He and a pathologist colleague experimented with smaller quantities of antiserum using slides instead of tubes, a modification that significantly reduced the reaction time to two minutes.[46]

Medical News, a recently launched weekly tabloid for GPs, reported that Sharman was "flabbergasted" by the "new two-minute pregnancy test," which involved "mixing together on a slide an anti-serum, a latex precipitin and urine." Having obtained an accuracy of 100 percent in one hundred clinically verified tests, he had "never come across anything so exciting."[47] Sharman sent the Department of Health for Scotland the relevant clipping from *Medical News* and requested an "observer to watch the test," which, he predicted, would "replace *all* biological tests" and save the department "very many thousands of pounds per annum."[48]

As usual, the view from Edinburgh was more cautious. Barr's detailed review in *Medical News* did not dispute the "speed" or "simplicity" of immunoassays but maintained that at least in Edinburgh, where *Xenopus* was "extremely reliable," their accuracy would need to improve before they could "supersede" toads. His most serious criticism regarded false positives: "The social implications of this inaccuracy, particularly in the case of unmarried women or where a pregnancy is unwanted, are felt to be of considerable importance." This might not amount to much where the testing was on a small scale, but Edinburgh performed more than 24,000 tests a year, so 2 or 3 percent false positives meant misdiagnosing as pregnant around 500 women annually. Toads, on the other hand, missed 170 pregnancies annually. Barr advised using the new tests "with discretion" until the issue of false positives had been resolved, and then they would be "extremely valuable."[49]

When Sharman announced Ortho's slide test in the *Lancet*, it was not yet commercially available, but experimental results had convinced him that a "revolution" was under way. Ortho had reduced the sensitivity of its reagents to minimize false positives, and undeterred by more cautious voices, Sharman confidently predicted that immunoassays would "make obsolete all biological tests." The slide test was so simple that a medical graduate could be trained in it "virtually as a side-room method."[50] A Birmingham endocrinologist concurred that a "major breakthrough" was at hand, especially now that Ortho had successfully adjusted the sensitivity of the reaction to exclude false positives.[51] A Durham pathologist, likewise writing in the *Lancet*, saw "no reason why this elegant test should not for most purposes replace the Hogben test," but, sensing the latent threat to his professional jurisdiction, cautioned that "its place should remain in the clinical laboratory," not the clinic.[52]

In an article that anticipated Predictor by nearly a decade, the *Scottish Daily Mail* boasted that "British doctors" had found an "instant and foolproof method of telling a woman whether she is expecting a child." Sharman's test, the *Mail* claimed, was "so simple that families could have their own do-it-yourself kit; so cheap that it will save the National Health Service millions of pounds." It would take only "two minutes to mix the three ingredients on a glass slide and watch for the reaction which determines pregnancy." This "instant" test would cost "only a few coppers each time" and was "nothing like so complicated or long as the old biological tests using either mice or toads."[53]

In March 1964, Ortho commercially launched its much anticipated slide test as Gravindex (figure 8.6). Sharman endorsed it in *Family Doctor* as "by far the simplest, quickest and most accurate method for the diagnosis of

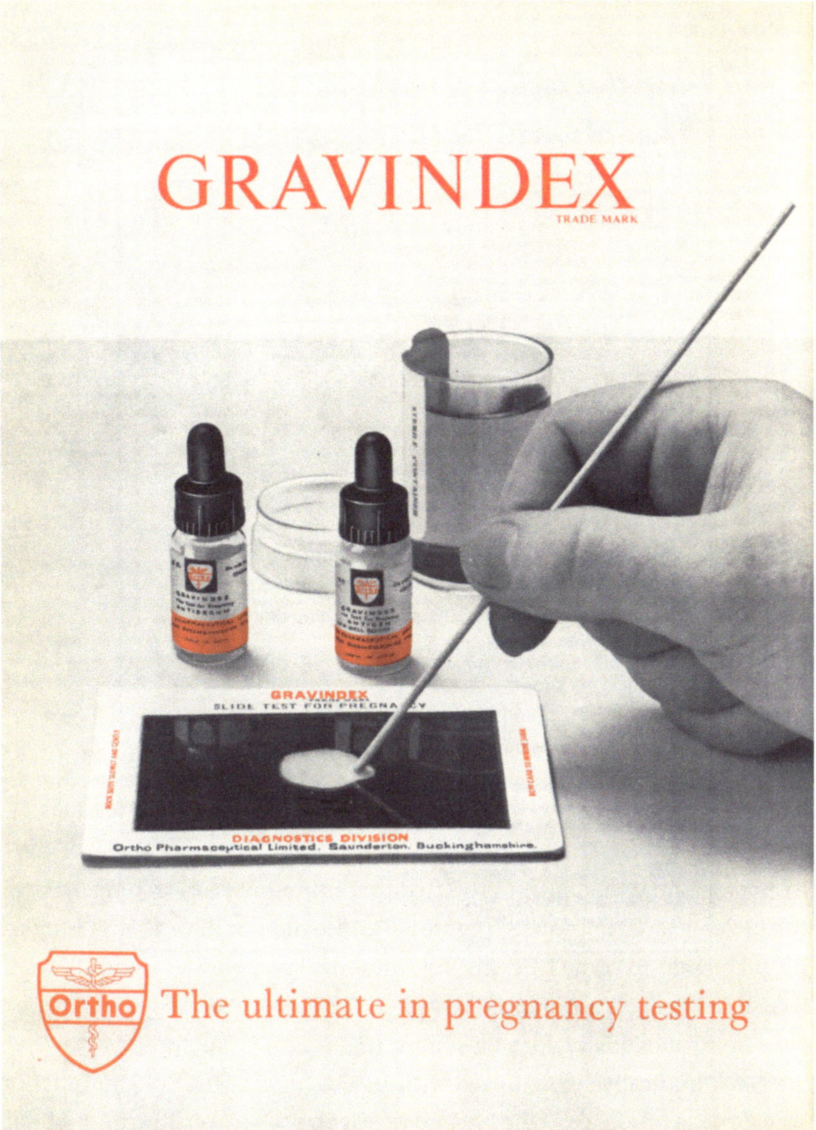

8.6 Slick advertisement for Ortho's Gravindex, marketed as the "ultimate in pregnancy testing." *Journal of Obstetrics and Gynaecology of the British Commonwealth* 72, Jun. 1965, iii–iv, Cambridge University Library T323.b.13.

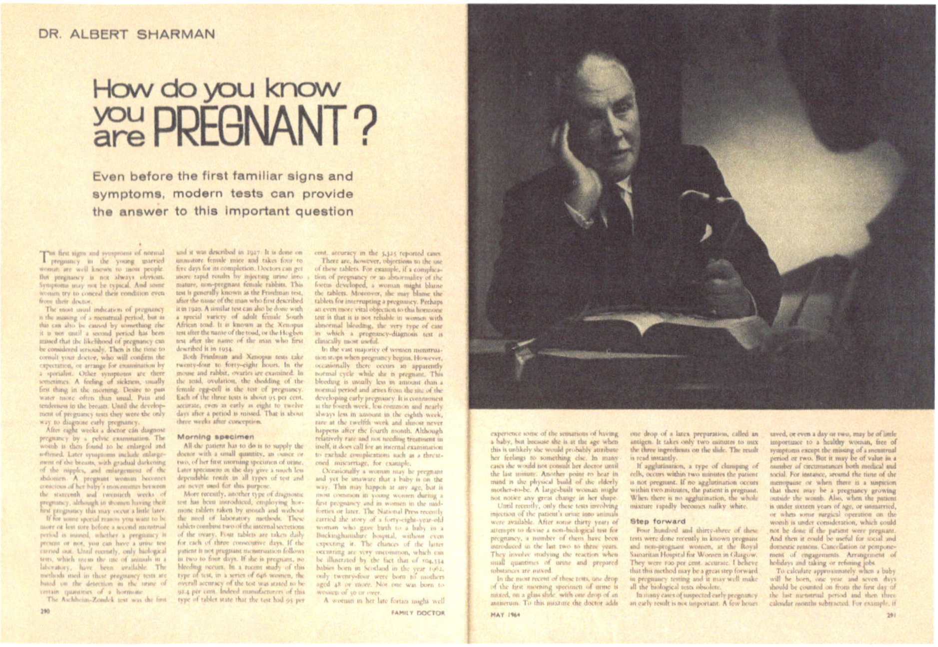
DR. ALBERT SHARMAN

How do you know you are PREGNANT?

Even before the first familiar signs and symptoms, modern tests can provide the answer to this important question

8.7 Sharman's article on pregnancy testing in *Family Doctor* (1964), 290–291, showing a staged photograph of a grandfatherly doctor resting on a large tome and holding a stethoscope viewed from the patient's perspective. The position of the camera just over the patient's shoulder reverses the more typical portrayal of the medical encounter as discussed in chapter 4. Cambridge University Library L300.b.142.14, with permission by the BMA.

early pregnancy" and again predicted that immunoassays would "make all the biological tests obsolete" (figure 8.7).[54] He did not have to wait long for his prediction to come true. Just four months later, boisterous headlines in British newspapers announced that the FPA's 2,000 "redundant" toads were "in a hole" and looking for a new home (figure 8.8).[55] The conservative *Daily Telegraph* ironically sympathized with new victims of the "old automation story" made obsolete by the "march of science."[56] Universities and schools snapped up the toads, but not before the story was picked up by the Associated Press, resulting in scribbled letters from American schoolboys offering to disburden the FPA. Remarkably, the association responded to each of the boys, informing them that "good homes" had already been found.[57]

Laboratory workers at Aberdeen's City Hospital compared tube tests to *Xenopus* before "dispersing" their colony of about five hundred toads.

Bolton Evening News

Wanted—homes for 2,000 rejected toads

THE Family Planning Association said today that it wishes to dispose of its 2,000 South African female toads, following a decision to simplify pregnancy diagnosis tests.

For many years the association has been flying the toads to Britain, but the new tests—claimed to be "ever more reliable"—will be carried out with a chemical manufactured here.

TOUGHEST

The toads are housed in spacious and specially-heated tanks at the association's London headquarters, where they spend a comfortable life feeding on a diet predominantly of liver.

The man faced with the job of finding them new homes is Brig. Rupert Elstone, general secretary of FPA.

He says it is one of the toughest assignments he has ever faced — including his 40 years in the Army!

Eastern Daily Press. 17/7/64

Second-hand toads

PRESUMABLY one can still please an Arab by expressing the hope that his wives will be fruitful and his sons as numerous as the sands of the desert. Leaning back in his Cadillac, however, the modern sheikh may secretly feel something less than joy at the prospect. This we infer from the fact that the Family Planning Association, which does about 100 pregnancy tests a week in its laboratory in the Tottenham Court Road, has done work for clients in Bahrein.

It is not social attitudes alone which are changing. The laboratory's methods are also changing. At one time the diagnosis of pregnancy involved killing mice or rats. Then it became possible to use toads without disturbing their lives unduly—all they suffered was surprise that they were laying more eggs than usual. Now the tests are entirely chemical, and the collaboration of the toad is no longer necessary.

Our artist stepped in just as the twilight of the toad was descending. The last 2000 toads had been declared redundant, and the association was trying to get rid of them. "We had telephone inquiries for two whole days," said an official of the association. Most went to laboratories, some to private doctors. They fetched about £300—not bad for second-hand toads, because at five shillings each they had cost the association about £400.

Southampton Evening Echo 22/7/64

Professor jumps at toads in a hole...

"Echo" Staff Reporter

ONE THOUSAND redundant workers from London are heading for Southampton to work for the university.

They will arrive almost unnoticed, because they change colour to blend with their background—and anyway they are only a few inches long.

It was a buyer's market when Southampton University clinched the deal, for their 1,000 xenopus laevis females—toads if you prefer—will cost them about the quarter of the normal price.

HE JUMPED

These little creatures were formerly employed by the Family Planning Association for pregnancy tests but when a more efficient method was discovered they were put up for sale. Professor K. A. Munday, head of the physiology and biochemistry department, jumped at the chance.

"This is a species of toad that we are very interested in," said the professor. "The Family Planning Association were most concerned with finding them good homes, and we are engaged in the sort of research that is of interest to the Association. Normally we have to import them from Cape Town at about 10s. each"

The next step is to get them to Southampton cheaply. Negotiations with British Railways are going on.

Every luxury has been laid on for the toads; they will live in heated aquaria with circulating filtered water.

Wolverhampton Express 9/7/64

Give a toad a home!

NEWS IN BRIEF

THE Family Planning Association said today that it wishes to dispose of its 2,000 South African Zenopus Laevi female toads, following a decision to simplify pregnancy diagnosis tests. The toads are housed in specially heated tanks. They feed on liver. Man faced with the job of finding them new homes is Brigadier Rupert Elstone, general secretary of the F.P.A. He says it is one of the toughest assignments he has ever faced—including his 40 years in the army!

* * *

A new, larger Renault car of "revolutionary" conception, will go on sale next year, the head of the state-owned Renault company, M. Pierre Dreyfus, said at Lille, France, yesterday.

* * *

Mrs. Sherri Finkbine (31), who went to Sweden in 1962 for an abortion, rather than have a thalidomide baby, is expecting another child.

* * *

Twenty-five Russian young farmers were arriving in England today to attend the Great Yorkshire Show at Harrogate on July 14, 15 and 16.

* * *

President Roosevelt's former summer home on Campobello Island—near the U.S.-Canadian frontier—is to become an international park, jointly financed by the U.S. and Canada.

* * *

A total of 87,621 new motor vehicles—almost a third fewer than last year—were registered in Italy in May. Motorists have been hit by higher taxes and dearer petrol as a result of the Government's auterity measures.

8.8 Tongue-in-cheek newspaper clippings collected by the FPA proclaimed the end of an era. Wellcome Collection SA/FPA/A3/11; on clippings: te Heesen 2014.

When Gravindex became available toward the end of the trial, they decided to test an additional one hundred specimens and reported in the *Practitioner* that *Xenopus* was "replaceable" by commercial test kits "provided the results [were] interpreted along with the clinical findings."[58] In August 1964, even Hobson replaced *Xenopus* with Pregnosticon. Organon had also decreased the sensitivity of its test to minimize false positives, and a further trial in Edinburgh suggested an accuracy of 99 percent.[59] The services of the toad colony established by Crew in the late 1930s and so carefully maintained for some twenty-five years were no longer required.

A PROFITABLE SIDELINE

Immunoassays not only made *Xenopus* obsolete but also heralded new social arrangements. In 1965, Warrack reflected that with the availability of "newer and more rapid laboratory methods," doctors and patients were growing dissatisfied with the old "restrictions," and it was "becoming increasingly difficult to refuse to satisfy the natural desire for a married couple to know as early as possible."[60] But the NHS was disinclined to offer what it regarded as a nonessential service and encouraged GPs to refer patients to a family planning clinic for a £1 test.[61] In June 1965, about a year after the FPA and Edinburgh had switched to immunoassays, one doctor, unable to "get pregnancy testing done" at his local hospital in the low-income northwestern London suburb of Kilburn, took matters into his own hands.[62]

Stanley Solomons, an Oxford-trained GP, called his company Hadley Laboratories Ltd. Surpassing Beric Wright's previous attempt to circumvent medical gatekeeping, Solomons advertised directly to the public. His professional status as a medical practitioner and the fact that the laboratory occupied the same premises as his Kilburn clinic would land him in trouble with the General Medical Council.[63] But he had a successful run of two years, during which time he placed some two thousand advertisements in journals, newspapers, and magazines. Charging two guineas per test, his company showed a credit balance of £1,000 in June 1967. His wife Janice Solomons, not medically qualified, owned forty-nine of the one hundred shares and helped with the clerical work. They offered a same-day service, tested all specimens for albumin (which could mislead the immunoassays),

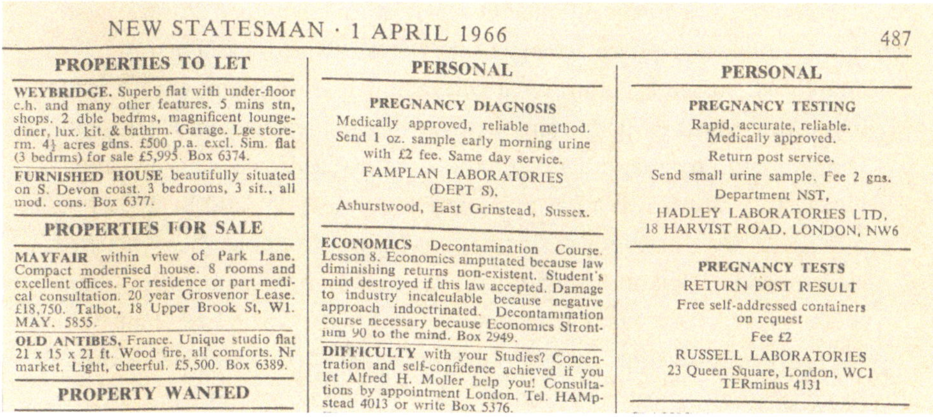

NEW STATESMAN · 1 APRIL 1966 487

PROPERTIES TO LET

WEYBRIDGE. Superb flat with under-floor c.h. and many other features. 5 mins stn, shops. 2 dble bedrms, magnificent lounge-diner, lux. kit. & bathrm. Garage. Lge store-rm. 4½ acres gdns. £500 p.a. excl. Sim. flat (3 bedrms) for sale £5,995. Box 6374.

FURNISHED HOUSE beautifully situated on S. Devon coast. 3 bedrooms, 3 sit., all mod. cons. Box 6377.

PROPERTIES FOR SALE

MAYFAIR within view of Park Lane. Compact modernised house. 8 rooms and excellent offices. For residence or part medical consultation. 20 year Grosvenor Lease. £18,750. Talbot, 18 Upper Brook St, W1. MAY. 5855.

OLD ANTIBES, France. Unique studio flat 21 x 15 x 21 ft. Wood fire, all comforts. Nr market. Light, cheerful. £5,500. Box 6389.

PROPERTY WANTED

PERSONAL

PREGNANCY DIAGNOSIS
Medically approved, reliable method.
Send 1 oz. sample early morning urine with £2 fee. Same day service.
FAMPLAN LABORATORIES
(DEPT S),
Ashurstwood, East Grinstead, Sussex.

ECONOMICS Decontamination Course. Lesson 8. Economics amputated because law diminishing returns non-existent. Student's mind destroyed if this law accepted. Damage to industry incalculable because negative approach indoctrinated. Decontamination course necessary because Economics Strontium 90 to the mind. Box 2949.

DIFFICULTY with your Studies? Concentration and self-confidence achieved if you let Alfred H. Moller help you! Consultations by appointment London. Tel. HAMpstead 4013 or write Box 5376.

PERSONAL

PREGNANCY TESTING
Rapid, accurate, reliable.
Medically approved.
Return post service.
Send small urine sample. Fee 2 gns.
Department NST,
HADLEY LABORATORIES LTD,
18 HARVIST ROAD, LONDON, NW6

PREGNANCY TESTS
RETURN POST RESULT
Free self-addressed containers on request
Fee £2
RUSSELL LABORATORIES
23 Queen Square, London, WC1
TERminus 4131

8.9 Classified ads in *New Statesman*, Apr. 1, 1966, 487, for the three laboratories are among the largest and most expensive on the page, appearing alongside smaller ones for flamenco guitar lessons and German language courses in Trier. Cambridge University Library NPR.C.622.

offered a free retest in the event of a negative result, did not dispense any medical advice or treatment, included with each test a "detailed" information sheet warning of the possibility of false positives, and referred "clients" to their own family doctors in case of health concerns.[64]

It did not take long for Hadley Laboratories to attract the attention of the BMA and the recently formed Advertising Standards Authority (ASA).[65] In line with the BMA's longstanding position that pregnancy testing belonged in the hands of doctors, the ASA ruled in October 1965 that the nonmedical press should refrain from publicizing commercial labs but had "no objection to such advertisements appearing in the medical press, for medical readers, who might wish to make use of the service in connection with their attendance upon patients."[66] The ASA ruling did not put an end to Solomons's campaign. The *New Statesman*, Britain's leading left-wing political weekly and no stranger to controversy, defiantly continued to publish classified ads for Hadley Laboratories.[67] In November 1965, it began taking ads for a second company, Famplan Laboratories, a postal service operating out of Sussex, and in February 1966, for a third, Russell Laboratories of Holborn. These newcomers operated along similar lines as Hadley Laboratories but competitively charged a reduced fee of £2 (figure 8.9).

Solomons was medically qualified, but other laboratory directors were not. Brian Block, a pharmacologist, and Derek Lawford, a biochemist, who established Russell Laboratories at Queen Square in Holborn, were already in business together (performing toxicological analyses) when Block's wife had "wanted a pregnancy test" and they "realized there was a big demand." Russell Laboratories began as a "profitable . . . sideline" and by 1968 had moved to larger premises at Brent Crescent near Wembley.[68] A. H. Lloyd established Bell Jenkins Laboratory in Portsmouth in December 1966 under the supervision of a "qualified laboratory technician" who had "trained in the Army medical corps." Lloyd had approached a "panel of three doctors" to supervise the pregnancy testing, but they had been warned off by the BMA.[69] A registered nurse qualified as a laboratory technician established a laboratory in Lincoln in September 1966.[70] And a physiologist also qualified as a technician performed pregnancy tests in a spare bedroom (previously a darkroom) of his Kilburn flat, the second such service in that down-market postcode.[71]

Despite continued opposition from the BMA, the advertising embargo did not last long. In August 1966, the state-funded Consumer Council, established a few years earlier to promote an individualistic model of "consumer interest," announced the ASA would be allowing advertisements in the "general press at the discretion of publishers, subject to a number of safeguards and a prescribed form of advertisement." Commercial laboratories were acting within their legal rights, and women were free to use them.[72] Their new ruling assumed that pregnancy testing would soon be "amply available" on the NHS at hospital laboratories, rather than discouraged, "except in cases of medical or social need."[73] A year later, the Ministry of Health placed Pregnosticon and Prepuerin on central supply with the expectation that hospital pathology departments would accept all received specimens regardless of the GP's reason for requesting a test.[74] But it was too little, too late. As a *BMJ* editorial noted, demand was "so great" that some labs were "compelled to exercise selection," and it fell to the clinician to "judge the need" and "indicate the reasons for his decision to the laboratory staff."[75]

Regulators tolerated commercial labs in part because of the expectation they would vanish once the health service got a handle on pregnancy testing. Instead of trying to curb publicity, the ASA proposed guidelines: tests should be carried out by qualified technicians and positive-testing clients advised to see a doctor.[76] The ASA also recommended a minimalist form of advertisement limited to the name, address, and telephone number of the laboratory; a request for clients to provide a sample and indicate their age; and the fee. Some journals applied their own standards. *Medical News* required references from six doctors and only accepted advertisements from Bell Jenkins; the *Lancet* accepted a few others. By the end of 1967, more than twenty laboratories around Britain were advertising in the *Daily Telegraph, Guardian, Lancet, London Weekly Advertiser, Medical News, New Statesman, Nursing Mirror,* and *Private Eye* (table 8.1).

When it came to publicity, Block and Lawford innovated more than most. The pair changed the name of their company to Belmont Laboratories and began advertising in London Underground and British Rail stations in 1966. The Code of Advertising Practices Committee granted approval, and the London Transport Board signed a one-year contract for £2,000. But when passengers complained, the commercial advertising manager decided to remove the posters when the contract expired. The British Transport Authority, however, had not received any complaints and continued to display 330 posters in sixty-seven railways around Britain, including in Liverpool and Manchester.[77] Belmont and Famplan also advertised in *Help Yourself to London: A Guide to Services, Facilities and Things to Do,* which noted that in "these liberal days," pregnancy testing services were in "great demand."[78]

INFAMOUS CONDUCT IN A PROFESSIONAL RESPECT

Direct-to-consumer pregnancy testing raised questions for the NHS at a time of changing social attitudes regarding contraception and abortion.[79] Contraceptive pills became available from the NHS for therapeutic purposes in 1961 and by 1964 were being used by around half a million women.[80] Younger members of Harold Wilson's Labour government, which came to power in 1963 with a narrow majority and was reelected with a much larger majority in 1966, supported family planning as part of a broader progressive agenda that included abortion, divorce, and homosexual law reform.[81] Clinics began offering contraceptive advice to unmarried women in 1964, and in 1967, major pieces of legislation decriminalized abortion

Table 8.1 Commercial laboratories advertised in British medical journals, newspapers, and magazines; based on clippings included with Lloyd to Davis, December 30, 1967, The National Archives MA 156/278.

Laboratory	Address	Fee
Hadley Laboratories	18 Harvist Road, Kilburn NW6	£2
Famplan Laboratories	Furnace Wood, East Grinstead, Sussex	£2
Russell Laboratories/ Belmont Laboratories	23 Queen Square, Holborn, WC1 / 188 Brent Crescent NW10	£2
Bell Jenkins Laboratories	4 Charlotte Street, Portsmouth	£2
Wellbeck Laboratories	11 Park Square West NW1	£2 2s
Antigen Medical Laboratories	36 Queen Anne Street W1	£1 1s 6d
Forte Laboratories	1 St. Swithin's Square, Lincoln	?
Abbey Laboratories	19 Waterloo Street, Glasgow	?
Diagnostic Laboratories	Cowhill Lane, Aston-under-Lyne, Lancashire	£2
Cook Laboratories	?	37s 6d
Lanark Laboratories	56 Fortune Green Road, Manchester M4	£2
Analytical Laboratories	26 Corporation St., Manchester M4	£2
Gravida Laboratories	Dunraven House, Riverside, Bridgend, Glamorgan	£2
Boden Lab Service	158 Stanningley Town Street, Pudsey, Yorkshire	£2
Medical Services Lab.	69/71 Monmouth Street, WC2	£2
Lab. Dept. 2.	The Guarde House, Chidcock, Bridport, Dorset	2gns
Tevic Laboratory	34 Grasmere Ave., London W5	£2
Wickham Laboratories	Wickham, Hampshire	£2
Lanco Laboratories	20 26 Briddon Street, Manchester M3	£2
Bristol Laboratories	82 Colston Street, Bristol 1	?
Linhope Laboratories	?	£2

and expanded access to family planning irrespective of medical need, age, or marital status.[82] Negotiations between the BMA and Ministry of Health concluded in 1966 with an agreement that GPs could charge for nonmedical family planning services and established prescription fees for the pill, IUDs, and other devices.[83] Oral contraception became the only medicine for which NHS patients could be charged a fee by their doctors.[84]

As with contraception and abortion, the NHS provision of pregnancy testing on "medical" and "social" grounds was debated in Parliament. In 1964, when Alan Thompson, the Labour Member of Parliament (MP) for Dunfermline Burghs, asked whether Michael Noble (Baron Glenkinglas), the Tory secretary of state for Scotland, would "take steps to ensure" the "widest possible use" of Sharman's slide test, Noble responded that it "would not be proper" for him to influence a "matter of clinical judgement."[85] In March 1967, Nicholas Scott, the liberal Tory MP for Paddington, asked Kenneth Robinson, the Labour minister of health and long-time supporter of birth control and abortion law reform, "why it had been decided that pregnancy tests should in future be available without restriction under the National Health Service."[86] He wanted to know whether it had "always been possible for these tests to be carried out where there have been good clinics" and whether this decision would result in a "great increase in the demand for tests with no clinical justification at all?" Robinson replied that it had not always been possible and that hospital departments were "hard pressed." The new method was far easier and did not "consume so much of the technician's time," so he expected the pathological service would be "able to take this additional load."[87]

The following month, Lena Jeger, the Labour MP for Holborn and St. Pancras who had advocated abortion law reform since the 1950s, asked Robinson if he "would legislate to control the advertising of pregnancy testing laboratories." Robinson replied that, although he had "been advised that it was better for a woman who thought she was pregnant to consult her doctor," he had "no power to control these advertisements" and was "not satisfied that legislation would be justified."[88] And in December, Peter Jackson, the Labour MP for High Peak in Derbyshire, argued that Belmont Laboratories, whose contract had recently been canceled by the London Transport Board, was not a "quack body" but a "perfectly respectable" organization whose "perfectly innocuous" advertisements conformed to the British

Code of Advertising Practice.[89] Jackson suspected that complaints had been made by a modern Mrs. Grundy, that personification of priggishness and propriety, and accused the publicly owned board of a double standard in censoring publicity for family planning and pregnancy testing but not for cigarettes, alcohol, or Sir Oswald Mosley's Union Movement.[90]

The BMA continued to disapprove of commercial laboratories, but the Ministry of Health was powerless to stop them. Unlike therapeutic drugs, no legislation governed diagnostic services, and the ASA, a voluntary organization with no legal power, was the only regulatory body overseeing commercial pregnancy testing. Whereas professional etiquette had stymied Wright's designs, independent commercial laboratories had no reason to appease the BMA or RCOG. Hadley Laboratories, the pioneering company, was a different matter. In contrast to the technicians, physiologists, pharmacologists, biochemists, and registered nurses who followed his lead, Solomons's professional status as a GP exposed him to the General Medical Council (GMC) and the prospect of being struck off the Medical Register.

The Medical Act of 1858 had created the GMC to regulate qualified doctors as distinguished from unqualified competitors.[91] It could remove or threaten to remove from the register a doctor judged guilty of "infamous conduct in a professional respect."[92] In the late nineteenth century, cases of infamous conduct involved unqualified assistants, adultery, indecent publications, breach of confidentiality, fraud, and shopkeeping.[93] As part of this professionalization, the GMC closed ranks against self-advertising as a distasteful form of commercialism and, from 1925, the BMA's ethical committee formally discouraged "indirect advertising," which covered dealings with the nonmedical press.[94]

In May 1967, the disciplinary committee of the GMC, chaired by its president Henry (Lord) Cohen of Birkenhead, accused Solomons of advertising in his own name and charged him with "infamous conduct." Solomons claimed that the appearance of his name in a single advertisement in *Rikerservice*, a free monthly classified advertising service for doctors, in June 1965 had been an "unfortunate mistake by a girl working for him part-time" and confirmed that no other advertisements used his name or address. The usual discrete "professional plate" indicated his premises, and his company, Hadley Laboratories, was not a "screen" for advertising his general practice. His receptionist informed any "patient" who telephoned or visited the clinic because of seeing an advertisement that she would "not be allowed to see him professionally" and referred him elsewhere.[95]

Deploying then current rhetoric of individual rights and the public interest, a Medical Defence Union solicitor argued that a "woman had a right to obtain information about her condition" and that Solomons was acting in the "public interest." If the GMC prevented medical doctors like Solomons from advertising, they would be handing pregnancy testing over to "unqualified people." But the committee remained concerned that Solomons had regularly advertised a company located in the same building as his own practice, the profits of which were only available to himself and to his wife. It insisted that Solomons liquidate Hadley Laboratories and "dissociate himself entirely from direct-access pregnancy testing." Solomons agreed and the GMC postponed judgment.[96] A year later, satisfied that Solomons had held his end of the bargain, the committee found him "not guilty of infamous conduct, thus concluding the case."[97]

AN ELEMENTARY HUMAN FEMALE RIGHT

From March 1966, reports and letters in the *New Statesman* set the terms of a public debate over pregnancy testing that would range widely in newspapers and magazines for months to come. This new form of visibility built on changes in the relationship between medicine and the media that had been accumulating since the late 1940s, when the state had mobilized massive publicity campaigns to inform an increasingly affluent, educated, and vocal public about the NHS. Until then, doctors and reporters had mostly cooperated to consolidate the professional authority of medicine, but journalists had their own agendas, and in the wider canvas of sixties antiauthoritarianism, they increasingly clashed with a paternalistic BMA over public access to medical knowledge.[98] Direct-to-consumer pregnancy testing, at the nexus of debates over birth control, abortion, medical authority, women's rights, consumerism, and the NHS, was a lightning rod.

Journalist Hilary Haywood interviewed fourteen laboratory directors and representatives of the BMA and NHS for her in-depth report in the public health section of the *New Statesman*. She doubted commercial labs were "pinching patients" from the NHS, observed that doctors did "not seem to want to do this service," and speculated that hidden reasons behind the controversy included concerns about blackmail and illegal abortion—risks that persisted under the current ASA ruling. She asked whether regulation would not be better than an apparently ineffective ban. For Haywood, the "most important" angle was that of the "client herself" and the fact that labs referred to clients and the BMA to patients was "vital and pertinent."[99]

With reference to *Born Free*, a just released film about a British couple in Kenya who raise an orphaned lioness named Elsa, she developed an individualist critique of the NHS for disregarding the freedoms of nonpregnant women:

> Now none of us is born a patient, we're individuals; and we don't become patients until we're ill or pregnant. A woman whose result proves negative is *not* a patient. Surely she should be free to regard a non-pregnancy as entirely her own business and not be penalised in this most personal of matters? Born free? Only, it appears, if you're a female lion; not if you happen to be a female human being. Not under the NHS anyway—only thanks to the anonymity of the direct laboratory services.[100]

Haywood's critique hinged on the assumption that a pregnant woman would do the responsible thing and take herself to a doctor.[101] Meanwhile, she argued, it was not "frivolity" that caused laboratories to "flourish" even when they were denied advertising; there were "more amusing ways of spending a couple of pounds." Drawing on the language of individual choice, she concluded that because the "woman herself bears the child," it was "for her to choose how soon she should know that she is pregnant."[102]

Writing in bed because she was "being treated for a threatened abortion," agony aunt Claire Rayner vigorously championed commercial laboratories in the *New Statesman*. Mobilizing newly available resources for justifying the early determination of pregnancy, she characterized the first three months of "intra-uterine life" as the "most vital" because this was when cells differentiated to "form organs" and were "most susceptible to damage," including from the "effects of maternal rubella and thalidomide."[103] "Had I *not* had a test and known for certain I was pregnant," she explained, "I might have assumed I was experiencing delayed menstruation, gone about my normal life, and very probably lost my—to me—precious foetus. I wonder how great a waste of foetal life occurs because of doctors who believe pregnancy testing is 'seldom necessary?'" Articulating a notion of uterine ownership that would define subsequent debates over abortion rights but used here in the service of preventing miscarriage, Rayner argued that pregnancy testing "should be available to *every* woman who has reason to suspect pregnancy. And since it is *her* uterus that is involved, what right has any paternalistic doctor to demand that the ability to find out about such involvement should only be through him?"[104]

Derek Stevenson, secretary of the BMA, countered in his own letter that possibly pregnant women needed "medical advice" to correctly interpret test results and avoid disaster.[105] But Haywood maintained that free retesting safeguarded against false negatives and that a positive-testing woman would probably go a doctor "anyway."[106] Alan Massam, a *Medical News* writer and soon-to-be founding member of the Medical Journalists' Association, echoed Rayner's argument—and those of feminists in the 1930s—that laboratory tests enabled women to "know about their pregnancy early and so take precautions against miscarriage." So long as "NHS rules" left pregnancy "to confirm itself," the "very real need for . . . 'private enterprise' tests" would persist.[107]

Pregnancy testers defended their business in letters of their own. Janice Solomons agreed that in a "perfect world," pregnancy tests would be "available free and on demand," but in reality, the NHS was hampered by a "restricted budget, no space and no staff," and many clients of Hadley Laboratories had been "refused pregnancy tests by their doctors (often because it just wasn't available)." Solomons likewise repeated the 1930s argument that pregnancy testing "saved many women from the danger and misery of an 'abortion' when they were not pregnant." But the "real issue" in the 1960s was (individual) "freedom" (from medicalization): women were "not born into a guild of doctors and a mass of patients, but people, and our bodies are our own." She argued for "regulation," not "suppression," and offered her "willing cooperation" to the ASA and BMA in the "interests of the public."[108] Her husband Stanley put it unambiguously in *Medical News*: "We test samples for clients, not patients. There is no question of our clients being patients in any sense at all."[109]

Of course, the opinions expressed in the *New Statesman*, a leftwing magazine that defiantly carried classified ads for the businesses defended in its public health and letters sections, did not represent a consensus. That the recently founded Sunday edition of *The Daily Telegraph* ran its own investigation by a crime correspondent, Peter Gladstone Smith, signals the contested status of pregnancy testing and conservative leanings of the paper. Gladstone Smith doubted the wisdom of women using commercial laboratories "without the knowledge of their doctors" and sided with the BMA's view that direct-to-consumer pregnancy testing was "unjustifiable, unwise and, unless medically interpreted, unreliable." In the same article, Block, who supported the "public being able to do as they please," told him, "We are not saying we are better, cheaper or quicker than doctors; all we are saying is—we are here." *The Telegraph* medical correspondent likewise insisted that a test result could not be interpreted properly unless "linked with a clinical examination" but also admitted that "shame" might prevent a woman from going to her family doctor. As a (not particularly satisfying) compromise, he suggested that a woman could go to "another doctor for a private consultation."[110]

The *Guardian* reported that Niall MacPherson (Lord Drumalbyn), chairman of the ASA, insisted that advertising "outside the medical press" could "lead to abuse and abortions." And Jeremy Potter, managing director of *New Statesman*, told the *Guardian* that, although "unhappy about rowing with the ASA," the magazine was going to continue publishing the advertisements as a "matter of principle" (although presumably also for the money). Drawing on the usual rhetoric, he argued that it was an "elementary human female right to know whether or not one is pregnant," and so the advertisements fulfilled a "very important end." But Potter also betrayed a degree of elitism in common with the BMA and its supporters when he conceded that publicity for pregnancy testing "might not be suitable in the mass circulation press."[111] Not immune to paternalism, the *Guardian*'s anonymous medical correspondent defended the medical establishment against accusations that it was "being a dog in the manger" and denounced commercial pregnancy testing as a "money-making project" that unfairly exploited "human need" with tragic consequences.[112]

This provoked sharply worded responses from Hadley and Russell laboratories. Janice Solomons countered that the BMA was "not an official body" but "simply the largest medical trade union, and facing revolt from its own members." Its views were "not those of most doctors," and it had become the "maiden aunt of medicine." All medical services, moreover, were "money-making," and even medical correspondents were "paid for writing." If commercial pregnancy testing was a form of exploitation, then so were the Cabinet and BMA. Block and Lawford took issue with the "near-libellous remark" that the "small fee" they charged was unfair and countered that the consequences were not tragic but "merely amazing, sad, frustrating, etc."[113]

In May 1966, the *Daily Sketch*, a struggling rival to the more successful *Daily Mail*, presented the "facts in the row over 'instant' pregnancy tests" under the bold headline, "A Woman's Right to Know" (figure 8.10). *Sketch* reporter Edward Connelly sympathized with what he idiosyncratically called "PTP" (pregnancy testing by post). Firms were "in it for the money," but they were operated by "responsible professional people of integrity." He did not begrudge them "cashing in on the failure of British medicine" to make "freely available" a service required by one million women a year. Connelly interviewed Sharman, who proclaimed that "every woman is entitled to know, as swiftly as science can tell her, whether or not she is expecting a child." He had tested 2,500 women at the Royal Samaritan Hospital free of charge since developing Gravindex three years ago. An immunoassay had "undoubtedly . . . enabled singer Barbra Streisand to announce to a surprised world, almost eight months before the expected birth, that she expects a child in mid-December." But countless women "spent anxious weeks, perhaps months, wondering if they are bearing a child." And, according to Sharman, some GPs did not even know about Gravindex. Going beyond the headline, he suggested that every woman had the right to obtain his test "for free," a position that was in equal measures socialist, feminist, and self-serving. A contract with the NHS would mean significant profits for Ortho with royalties for Sharman.[114]

The Times, a latecomer to the debate, launched a women's page in May 1966 as part of a broader initiative to boost circulation in an increasingly competitive market.[115] Fifteen months later, it reported that GPs were only able to request a test for a patient who was in "poor health," "approaching the menopause," or "separated from her husband." An investigation had confirmed "demand" but also that "however well-run individual laboratories may be, there is scope for less scrupulous operators whose only concern is for commercial profit—backed by the minimum of qualifications and facilities." Despite the "apparent safeguard" of the ASA, it was "perfectly possible to by-pass their conditions." It was "surely not right for an advertising body to be the only watchdog." The "theoretical safeguard,"

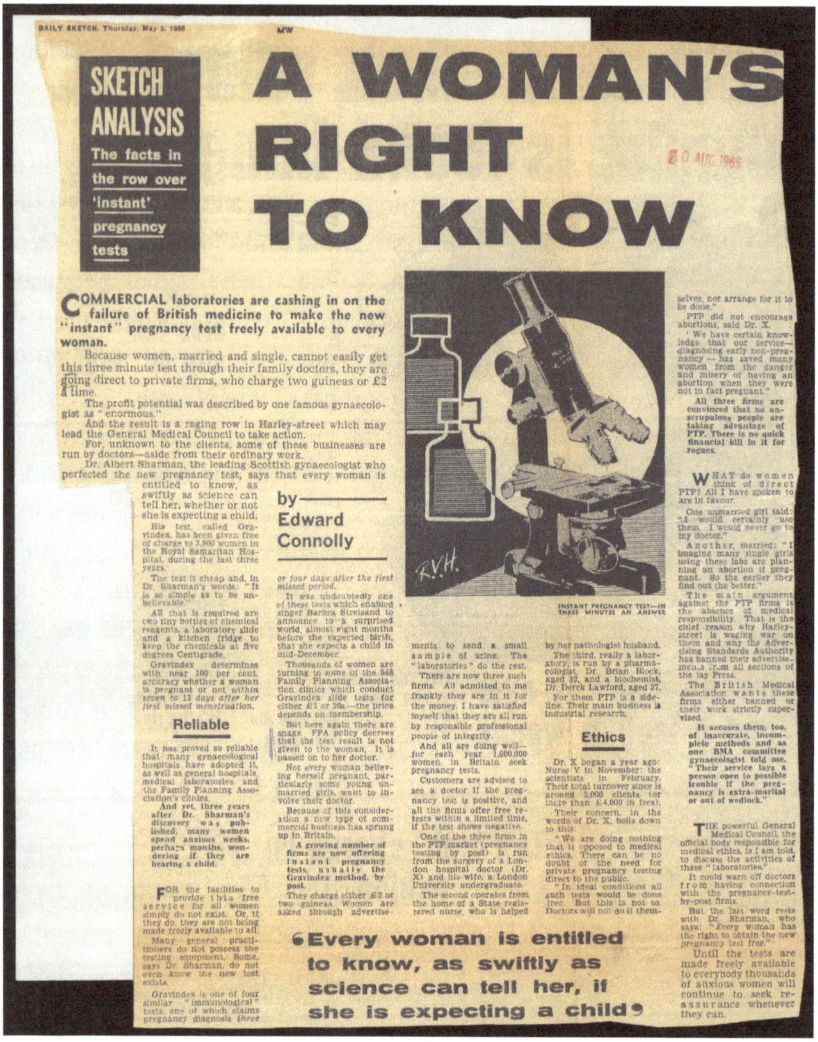

DAILY SKETCH, Thursday, May 5, 1966

MW

SKETCH ANALYSIS

The facts in the row over 'instant' pregnancy tests

A WOMAN'S RIGHT TO KNOW

COMMERCIAL laboratories are cashing in on the failure of British medicine to make the new "instant" pregnancy test freely available to every woman.

Because women, married and single, cannot easily get this three minute test through their family doctors, they are going direct to private firms, who charge two guineas or £2 a time.

The profit potential was described by one famous gynaecologist as "enormous."

And the result is a raging row in Harley-street which may lead the General Medical Council to take action.

For, unknown to the clients, some of these businesses are run by doctors—aside from their ordinary work.

Dr. Albert Sharman, the leading Scottish gynaecologist who perfected the new pregnancy test, says that every woman is entitled to know, as swiftly as science can tell her, whether or not she is expecting a child.

by Edward Connolly

His test, called Gravindex, has been given free of charge to 2,500 women in the Royal Samaritan Hospital, during the last three years.

The test is cheap and, in Dr. Sharman's words: "It is so simple as to be unbelievable."

All that is required are two tiny bottles of chemical reagents, a laboratory slide and a kitchen fridge to keep the chemicals at five degrees Centigrade.

Gravindex determines with near 100 per cent. accuracy whether a woman is pregnant or not *within seven to 13 days after her first missed menstruation.*

Reliable

It has proved so reliable that many gynaecological hospitals have adopted it, as well as general hospitals, medical laboratories and the Family Planning Association's clinics.

And yet, three years after Dr. Sharman's discovery was published, many women spend anxious weeks, perhaps months, wondering if they are bearing a child.

FOR the facilities to provide this free service for all women simply do not exist. Or, if they do, they are not being made freely available to all.

Many general practitioners do not possess the testing equipment. Some, says Dr. Sharman, do not even know the new test exists.

Gravindex is one of four similar "immunological" tests, one of which claims pregnancy diagnosis *three or four days after the first missed period.*

It was undoubtedly one of these tests which enabled singer Barbra Streisand to announce to a surprised world, almost eight months before the expected birth, that she expects a child in mid-December.

Thousands of women are turning to some of the 548 Family Planning Association clinics which conduct Gravindex slide tests for either £1 or 30s.—the price depends on membership.

But here again there are snags. FPA policy decrees that the test result is not given to the woman. It is passed on to her doctor.

Not every woman believing herself pregnant, particularly some young unmarried girls, want to involve their doctor.

Because of this consideration a new type of commercial business has sprung up in Britain.

A growing number of firms are now offering instant pregnancy tests, usually the Gravindex method, by post.

They charge either £2 or two guineas. Women are asked through advertisements to send a small sample of urine. The "laboratories" do the rest.

There are now three such firms. All admitted to me frankly they are in it for the money. I have satisfied myself that they are all run by responsible professional people of integrity.

And all are doing well—for each year 1,000,000 women in Britain seek pregnancy tests.

Customers are advised to see a doctor if the pregnancy test is positive, and all the firms offer free re-tests within a limited time, if the test shows negative.

One of the three firms in the PTP market (pregnancy testing by post) is run from the surgery of a London hospital doctor (Dr. X) and his wife, a London University undergraduate.

The second operates from the home of a State registered nurse, who is helped by her pathologist husband.

The third, really a laboratory, is run by a pharmacologist, Dr. Brian Block, aged 32, and a biochemist, Dr. Derek Lawford, aged 37.

For them PTP is a sideline. Their main business is industrial research.

Ethics

Dr. X began a year ago: Nurse Y in November: the scientists in February. Their total turnover since is around 2,000 clients (or more than £4,000 in fees).

Their concern, in the words of Dr. X, boils down to this:

"We are doing nothing that is opposed to medical ethics. There can be no doubt of the need for private pregnancy testing direct to the public.

"In ideal conditions all such tests would be done free. But this is not so. Doctors will not do it themselves, nor arrange for it to be done."

PTP did not encourage abortions, said Dr. X.

"We have certain knowledge that our service—diagnosing early non-pregnancy — has saved many women from the danger and misery of having an abortion when they were not in fact pregnant."

All three firms are convinced that no unscrupulous people are taking advantage of PTP. There is no quick financial kill in it for rogues.

WHAT do women think of direct PTP? All I have spoken to are in favour.

One unmarried girl said: "I would certainly use them. I would never go to my doctor."

Another, married: "I imagine many single girls using these labs are planning an abortion if pregnant. So the earlier they find out the better."

The main argument against the PTP firms is the absence of medical responsibility. That is the chief reason why Harley-street is waging war on them and why the Advertising Standards Authority has banned their advertisements from all sections of the lay Press.

The British Medical Association wants these firms either banned or their work strictly supervised.

It accuses them, too, of inaccurate, incomplete methods and as one BMA committee gynaecologist told me, "Their service lays a person open to possible trouble if the pregnancy is extra-marital or out of wedlock."

THE powerful General Medical Council, the official body responsible for medical ethics, is I am told, to discuss the activities of these "laboratories."

It could warn off doctors from having connection with the pregnancy-test-by-post firms.

But the last word rests with Dr. Sharman, who says: "*Every* woman has the right to obtain the new pregnancy test free."

Until the tests are made freely available to everybody thousands of anxious women will continue to seek reassurance whenever they can.

INSTANT PREGNANCY TEST—IN THREE MINUTES AN ANSWER

R.V.H.

"Every woman is entitled to know, as swiftly as science can tell her, if she is expecting a child"

8.10 Newspaper clipping with for the FPA with the relevant section marked in blue pen. The illustration by R.V.H. correctly depicts the two bottles of reagents needed for the test but (not for the first time) includes an extraneous microscope to add an air of scientific authority. Edward Connolly, "A woman's right to know," *Daily Sketch*, May 5, 1966, 8, Wellcome Collection SA/FPA/B10/47.

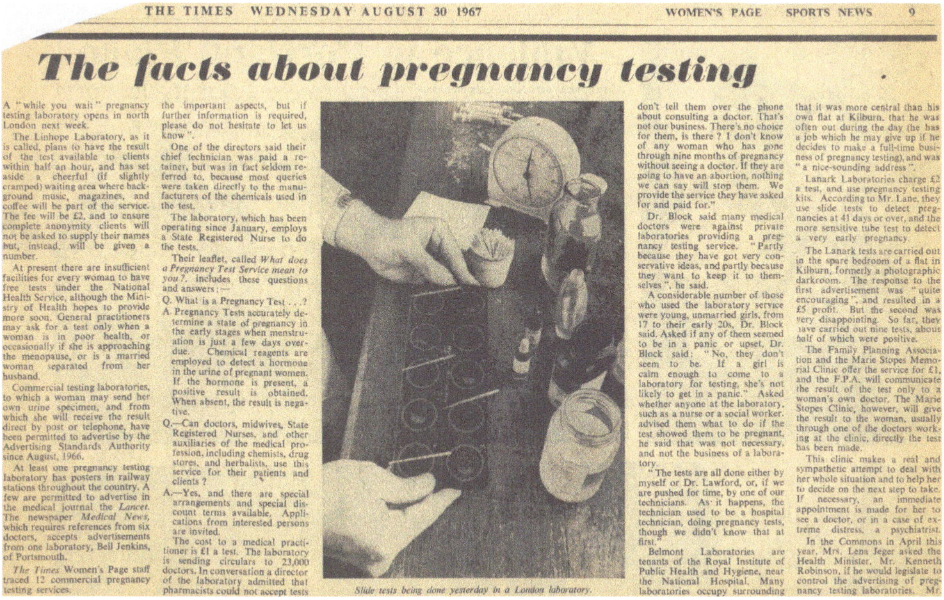

THE TIMES WEDNESDAY AUGUST 30 1967 WOMEN'S PAGE SPORTS NEWS 9

The facts about pregnancy testing

A "while you wait" pregnancy testing laboratory opens in north London next week.

The Linhope Laboratory, as it is called, plans to have the result of the test available to clients within half an hour, and has set aside a cheerful (if slightly cramped) waiting area where background music, magazines, and coffee will be part of the service. The fee will be £2, and to ensure complete anonymity clients will not be asked to supply their names but, instead, will be given a number.

At present there are insufficient facilities for every woman to have free tests under the National Health Service, although the Ministry of Health hopes to provide more soon. General practitioners may ask for a test only when a woman is in poor health, or occasionally if she is approaching the menopause, or is a married woman separated from her husband.

Commercial testing laboratories, to which a woman may send her own urine specimen, and from which she will receive the result direct by post or telephone, have been permitted to advertise by the Advertising Standards Authority since August, 1966.

At least one pregnancy testing laboratory has posters in railway stations throughout the country. A few are permitted to advertise in the medical journal the *Lancet*. The newspaper *Medical News*, which requires references from six doctors, accepts advertisements from one laboratory, Bell Jenkins, of Portsmouth.

The Times Women's Page staff traced 12 commercial pregnancy testing services.

the important aspects, but if further information is required, please do not hesitate to let us know".

One of the directors said their chief technician was paid a retainer, but was in fact seldom referred to, because most queries were taken directly to the manufacturers of the chemicals used in the test.

The laboratory, which has been operating since January, employs a State Registered Nurse to do the tests.

Their leaflet, called *What does a Pregnancy Test Service mean to you?*, includes these questions and answers:—

Q. What is a Pregnancy Test . . .?

A. Pregnancy Tests accurately determine a state of pregnancy in the early stages when menstruation is just a few days overdue. Chemical reagents are employed to detect a hormone in the urine of pregnant women. If the hormone is present, a positive result is obtained. When absent, the result is negative.

Q.—Can doctors, midwives, State Registered Nurses, and other auxiliaries of the medical profession, including chemists, drug stores, and herbalists, use this service for their patients and clients?

A.—Yes, and there are special arrangements and special discount terms available. Applications from interested persons are invited.

The cost to a medical practitioner is £1 a test. The laboratory is sending circulars to 23,000 doctors. In conversation a director of the laboratory admitted that pharmacists could not accept tests

Slide tests being done yesterday in a London laboratory.

don't tell them over the phone about consulting a doctor. That's not our business. There's no choice for them, is there? I don't know of any woman who has gone through nine months of pregnancy without seeing a doctor. If they are going to have an abortion, nothing we can say will stop them. We provide the service they have asked for and paid for."

Dr. Block said many medical doctors were against private laboratories providing a pregnancy testing service. "Partly because they have got very conservative ideas, and partly because they want to keep it to themselves", he said.

A considerable number of those who used the laboratory service were young, unmarried girls, from 17 to their early 20s, Dr. Block said. Asked if any of them seemed to be in a panic or upset, Dr. Block said: "No, they don't seem to be. If a girl is calm enough to come to a laboratory for testing, she's not likely to get in a panic." Asked whether anyone at the laboratory, such as a nurse or a social worker, advised them what to do if the test showed them to be pregnant, he said that was not necessary, and not the business of a laboratory.

"The tests are all done either by myself or Dr. Lawford, or, if we are pushed for time, by one of our technicians. As it happens, the technician used to be a hospital technician, doing pregnancy tests, though we didn't know that at first."

Belmont Laboratories are tenants of the Royal Institute of Public Health and Hygiene, near the National Hospital. Many laboratories occupy surrounding

that it was more central than his own flat at Kilburn, that he was often out during the day (he has a job which he may give up if he decides to make a full-time business of pregnancy testing), and was "a nice-sounding address".

Lanark Laboratories charge £2 a test, and use pregnancy testing kits. According to Mr. Lane, they use slide tests to detect pregnancies at 41 days or over, and the more sensitive tube test to detect a very early pregnancy.

The Lanark tests are carried out in the spare bedroom of a flat in Kilburn, formerly a photographic darkroom. The response to the first advertisement was "quite encouraging", and resulted in a £5 profit. But the second was very disappointing. So far, they have carried out nine tests, about half of which were positive.

The Family Planning Association and the Marie Stopes Memorial Clinic offer the service for £1, and the F.P.A. will communicate the result of the test only to a woman's own doctor. The Marie Stopes Clinic, however, will give the result to the woman, usually through one of the doctors working at the clinic, directly the test has been made.

This clinic makes a real and sympathetic attempt to deal with her whole situation and to help her to decide on the next step to take. If necessary, an immediate appointment is made for her to see a doctor, or in a case of extreme distress, a psychiatrist.

In the Commons in April this year, Mrs. Lena Jeger asked the Health Minister, Mr. Kenneth Robinson, if he would legislate to control the advertising of pregnancy testing laboratories. Mr.

8.11 Detail of women's page devoted to "the facts about pregnancy testing" in the newspaper of record featuring a photograph of yesterday's slide tests with accoutrements: the obligatory two bottles of reagents, a test-test plate, mixing sticks, syringes, a timer, and what appears to be an assortment of bottles and jars repurposed by clients as urine containers. *The Times*, Aug. 30, 1967, 9, Wellcome Collection SA/FPA/B10/47, with permission by The Times/News Licensing.

it continued, that publications and other media would not accept advertisements made matters worse "by driving it underground." If the Ministry of Health was "unable to put its own scheme into operation quickly," the article concluded, it should "at least devise a method to ensure that commercial companies providing such a service are properly supervised—by legislation if necessary" (figure 8.11).[116]

HOGBEN TEST'S LAST CROAK

Beyond advertising and journalism, taboo-breaking novels and films embedded pregnancy tests in risqué dramas and satires as part of broader trend that made sex and reproduction even more visible across media and in shops.[117] A middle-class, unmarried mother in Lynne Reid Banks's *The*

L-shaped Room (1960) spends a "surprising amount of money" on a "special test" that her doctor claims is "unnecessary."[118] A gynecologist in Helen Lourie's *A Question of Abortion* (1962) offers to "have a test made" for a young unmarried patient, possibly an agent provocateur, whose normally "clockwork" cycle is three weeks late.[119] Barbara, in David Lodge's *The British Museum Is Falling Down* (1965), tells her husband that her doctor "wouldn't prescribe any more tests—not on the National Health, anyway. Besides, by the time the result came through, I'd know anyway."[120] The trapped narrator of Andrea Newman's *The Cage* (1966) vaguely remembers having read something in a women's magazine and tells her boyfriend, "I think you can have some kind of test with animals when you're a fortnight late."[121] And a middle-class "boss's daughter" in Barry Hines's *The Blinder* (1966) knows she is pregnant because she "went to the doctor's" and he "gave [her] a test."[122]

Two pounds was a lot of money in the 1960s, so working-class characters in kitchen sink cinema continued to rely on the canonical signs. A pregnant teenager in *A Taste of Honey* (1961) is exhilarated when the "baby" "kick[s] her, confirming "it's alive." The eponymous antihero in *Alfie* (1966) is alerted by a calendar to his girlfriend's condition (figure 8.12a). And a factory girl discloses her condition when she runs to the lavatory to be sick in *Up the Junction* (1968). The condition of a middle-class student in *A Touch of Love* (1969), however, is established in the opening scene, which shows her restlessly circling a positive test result in the British Museum Reading Room. Although the full name of the lab falls outside the frame, you can just make out the address and postcode of Belmont Laboratories (figure 8.12b). Now a Hollywood cliché, this may have been the first cinematic pregnancy test. It stands as a testament to the unprecedented visibility of commercial pregnancy testing following four years of public controversy and to class as a determinant of access to an expensive service.[123]

The 1968 edition of William Johnstone's long-running *Textbook of Midwifery* finally replaced the image of the dissected Aschheim–Zondek mouse with one of an agglutination inhibition reaction. Johnstone's successor, Robert James Kellar, rewrote the section on pregnancy diagnosis at a time when animals were "being replaced by a variety of commercially available immunological pregnancy tests."[124] By the end of the decade, Bruce Hobson reported "absolutely no demand from anywhere in Scotland for a

8.12 Continuity and change, and the relevance of class, are made apparent in two diagnostic moments of 1960s British cinema. (a) Alfie Elkins (Michael Caine) is alerted by a calendar to his girlfriend's condition in *Alfie* (Lewis Gilbert, 1966): produced by Lewis Gilbert and Sheldrake Films, distributed by the BFI. (b) Rosamund Stacey (Sandy Dennis) circles her positive test result in *A Touch of Love* (Waris Hussein, 1969): produced by Amicus and Palomar, courtesy of STUDIOCANAL Films Ltd.

confirmatory Hogben test" and had "disbanded" his residual *Xenopus* colony.[125]

In early 1970, a *Guardian* headline belatedly proclaimed the Hogben test's "last croak" and predicted that "life for the female immigrant [*Xenopus*], famed for its cooperation in the old Hogben test, will be easier hereafter." The recently formed Department of Health and Social Security had decided to close the Sheffield center, which performed only two Hogben tests in 1968 and none in 1969. Instead of the "unfashionable" toad, the NHS favored the "much quicker, cheaper, and simpler agglutination tests," which most hospital laboratories performed.[126] After nearly four decades, the era of bioassays and large, centralized services had drawn to a close. In its place came a new economy of pregnancy testing built on the mass production of immunological test kits, direct-to-consumer advertising, same-day laboratory services, and the willingness of enough women—not as patients but as consumers with newly articulated rights, freedoms, and "public opinion" on their side—to pay out of pocket.

9

OVER THE COUNTER

The Ministry of Health implemented a more liberal approach to "social" pregnancy testing in August 1967, just a few months before the Abortion Act was passed. The new policy placed the reagents for Pregnosticon and Prepuerin—but not Gravindex or any of the other more rapid but somewhat less accurate slide tests—on central supply. The Ministry did not approve of commercial laboratories, but neither did it want to divert resources toward regulating them. The new approach was intended as a policy of containment. Policymakers hoped that a free on-demand service would curb the business of pregnancy testing. They hoped this change in policy would suffice to remove the need for commercial laboratories and that no further action would be required: the NHS would lure women back to their GPs, the state would regain control, and the commercial market would correspondingly shrink.

But the new policy left GPs in much the same position as before; they remained dependent on the cooperation of pathologists, many of whom were overworked, unsympathetic to "social" pregnancy testing, or both. Even an entirely cooperative hospital service could take a week or more to return a test result. This was much longer than the twenty-four hours offered by commercial laboratories, not to mention the minutes it would take a GP using Gravindex. Patients knew this and increasingly pressured doctors to offer not only free pregnancy testing on demand but also a

Table 9.1 Tests available in Britain, of which the most prevalent were centrally supplied Pregnosticon and Prepuerin, followed by homegrown Gravindex and imported Planotest; based on data in Saroja Ramaswamy, "Critical Review of Pregnancy Tests," *Medical Newsletter* (London: FPA, 1972), FD 23/4486, TNA. For tests available in the United States and more internationally: "Pregnancy Tests: The Current Status," *Population Reports*, Series J, No. 7, November 1975, J109–J124.

Test	Company	Place	Type	Date
Pregnosticon	Organon	Oss, NL	Tube	1962
Ortho (Latex)	Ortho	Raritan, NJ	Tube	1962
Prepuerin	Burroughs Wellcome	London, UK	Tube	1963
UCG	Wampole	Stanford, CT	Tube	1963
Gravindex	Ortho	Raritan, NJ	Slide	1964
Hyland	Hyland	Los Angeles, CA	Slide	1964
Pregslide	Wampole	Stanford, CT	Slide	1967
Planotest	Organon	Oss, NL	Slide	1967
DAP	Denver	Toronto, ON	Slide	1968
Pregnosticon All-In	Organon	Oss, NL	Tube	1969

service that was as quick as those available through Belmont and a handful of other private services. To make matters worse, the government did little to publicize its new policy, and the press continued to perpetuate the old situation. As late as March 1969, for instance, *The Sunday Times* reported that "at present a woman has no right to demand an early pregnancy test under the NHS."[1]

By then, rival slide tests, including Organon's Planotest, were encroaching on the market for Gravindex, and before long, the BMA was petitioning central government to make one or more of these directly available to GPs (table 9.1). The two main obstacles were the lack of an obvious mechanism for paying GPs to perform pregnancy tests on the NHS and the perception, especially reinforced by Bruce Hobson's input, that they would do the tests badly. Mounting pressure from patients and technological change ensured the BMA raised the issue at intervals. Pharmacists, unconstrained by government policy and enlisted by the entrepreneurial directors of Belmont

Laboratories, began offering over-the-counter pregnancy testing services in the summer of 1969. This chapter is about the professional rivalries that intensified as commercial demand and government inaction drove a wedge between the BMA and the Pharmaceutical Society and shifted power from GPs to pharmacists.

ON PRESCRIPTION?

When it came to pregnancy testing, GPs were in a double bind. Slide tests promised to place pregnancy diagnosis in their hands, and they were free to pay out of pocket for the reagents. But they were not allowed to charge NHS patients or claim expenses back on the NHS. So in November 1967, the BMA asked the Ministry of Health to consider approving one or more slide tests reimbursable on the NHS, just like prescription drugs. By then, Planotest cost £2.10s for ten tests (about a tenth of what private laboratories charged), and local medical committees around Britain were pressing the BMA for access on the NHS.[2] A few months later, in January 1968, the *Guardian* reported that negotiations between the BMA and the Ministry might "result in the issue of pregnancy testing kits to family doctors."[3] But the Ministry maintained that slide tests were less reliable than the tube tests on central supply and so were unlikely to be made available to GPs on prescription; it would consider the issue of uncooperative hospital laboratories on a case-by-case basis.[4]

As slide tests became more generally available and cost-competitive, some doctors even proposed allowing pharmacists to use them. Alex Franklin, a London practitioner, suggested to the Ministry of Health that pharmacists be permitted to perform pregnancy tests at an "agreed sum of ten shillings." He worried that an "increasing number of firms" were making a "substantial profit out of the fear of women" and enclosed a price list allegedly showing the "very large profit" that "pregnancy test racketeers" made by charging the apparently "nationally agreed price" of £2 a test (figure 9.1).[5] As per usual, the Ministry of Health politely but firmly rejected his suggestion.

When the NHS was created, certain diagnostic reagents, including litmus paper, were listed as prescribable on form "EC10," a thin piece of white paper mostly used for prescribing drugs and a ubiquitous hallmark

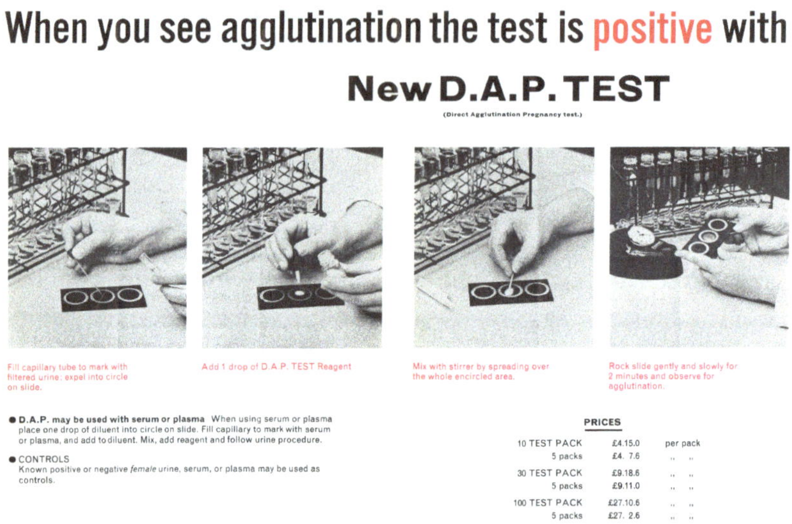

9.1 Price list accompanying a promotional flyer for the new DAP test, manufactured by Denver Laboratories, a Canadian company, and distributed in Britain by George Gurr of London. In contrast to most tests, which inhibited agglutination, DAP, as the name indicates, functioned through direct agglutination. The bulk cost of £27 10s 6d for a pack of one hundred works out to 66p a test, so a commercial service charging £2 a test (more than seven times the cost to the lab) would clear 1£ 8s 14d per test, not counting overheads, labor, and so on. The National Archives MH 159/78.

of the health service.[6] In addition to the list of prescribable reagents, special arrangements had been made in the early years of the health service for the direct reimbursement to GPs of the cost of a small number of reagents used in the "public interest" for testing immunity to diphtheria and tuberculosis, as well as in the diagnosis of tuberculosis, scarlet fever, and protein allergies. These were listed in Appendix V to the *Handbook for General Medical Practitioners*, first published by the Ministry of Health in 1950. By the late 1960s, however, the proportion of doctors making use of this arrangement, which had never been large, had diminished to the point that the newly created Department of Health and Social Security (DHSS) viewed it as vestigial. This was the situation when, in February 1969, the BMA's General Medical Services Committee asked the department to reconsider making reagents for pregnancy testing available to GPs.[7]

The department resisted this sort of request, as had the Ministry, on the grounds that there was "no strong reason of public policy" to promote the use of reagents by GPs, "rather the reverse," as pregnancy tests were "more reliably . . . carried out in hospital laboratories, where they will be done at the doctor's request." Moreover, the department was disinclined to "add to a list which has outlived its original purpose, which in any case it served indifferently."[8]

Whenever the department sought expert advice on the use of slide tests or their reagents by GPs, it turned to Bruce Hobson, director of the Edinburgh pregnancy diagnosis center. Hobson, who held to exceedingly high standards and was friends with Leif Wide, endorsed Organon's Planotest for the "best results" but was critical of its "readability." Reactions could be "quite easily read in good daylight" but only with experience; reading them in artificial light was "not easy." Hobson maintained that the "only way anyone can work a test (any laboratory test not just pregnancy tests) up to a high degree of accuracy is by constant use of the material." The problem with GPs was that most were "unlikely to do more than one test per week," which was "just playing at it." The medical profession, Hobson warned, would "lose a great deal more than they [would] gain" by doing pregnancy tests "badly." Underperforming NHS laboratory services could be improved, but Hobson objected to GPs "being given test kits" not because they were incapable of learning how to use them properly but because most would not get enough practice to become "proficient." He did not "subscribe to the policy that a test badly done is better than no test at all."[9]

The DHSS took Hobson's advice to heart and informed the BMA, in August 1969, that to maintain results at an acceptably high standard, GPs would need constantly to perform tests under controlled temperature and lighting, conditions that were "very unlikely to be met by the GP working in his own surgery." The department's policy remained unchanged. GPs were still welcome to commercially source "the materials to carry out tests for pregnancy in the same way as they provide other diagnostic materials and equipment," but the department saw no reason to make "special arrangements for direct supply of pregnancy diagnosis kits to general practitioners through National Health Service channels."[10]

In February 1970, the department once more turned to Hobson for the latest information about the relative accuracy of the slide tests compared with Pregnosticon, his favorite tube test. Had anything changed since his review in the *Practitioner* one year ago? The department was "anxious to have the very latest information" as well as Hobson's opinion on the "suitability of Pregnosticon-Planotest and other slide tests as consulting room tests for General Practitioners." This was, by now, an "old story," and the department's line had not wavered. Nevertheless, technological change reopened the discussion. Had Hobson changed his mind about the potential "accuracy of Planotest in the hands of General Practitioners and others in non-specialist laboratories carrying out occasional tests, possibly without adequate controls?"[11]

Of course not. For Hobson, the only good way to do pregnancy testing was his way, in a centralized laboratory with "properly trained" staff. He even had doubts about the quality of the NHS service under the new policy. As for GPs, the "busy" doctor would simply "not have enough time to do the tests properly." Slide tests seemed "simple," but they were less sensitive and gave "equivocal results" more often than the tube tests on central supply. Hobson had "no first hand information" about the accuracy of side tests in inexpert hands and was dismissive of a recent article in the *Practitioner* because the reported 99.4 percent accuracy had been achieved by a certified laboratory technician with help from a company doctor: "So much for that report." An earlier paper, Hobson claimed, gave the "results which one might expect from someone not trained in laboratory techniques." In the hands of a medical registrar, Pregnosticon had an error of 4 percent. In Hobson's laboratory, the error was less than 1 percent. Ultimately, however, he did "not suppose that anything [he had] written or said now or in the past [would] make a blind bit of difference." Hobson expected to see GPs "with their tests eventually."[12]

Following his advice, the department held fast to its official line that slide tests, "even when performed under laboratory conditions," were "markedly less accurate" than centrally supplied tube tests; that good results with slide tests "could only be achieved under controlled conditions"; that "casual users [could] very easily misread negative results"; and that accuracy depended on the "very careful mixing of reagents." It followed that adding slide tests to the list of reimbursable diagnostic materials did "not seem to be justified."[13] But some pregnancy testers, including pathologists and laboratory directors, disagreed.

A consultant pathologist in the Liverpool region, for instance, reported good results with slide tests but had no evidence from clinical trials to counter Hobson's weighty advice.[14] And John Murray, a laboratory director at Queen Charlotte's Maternity Hospital, attacked the department's line as "based on wrong information." "Whoever advised the Department that slide tests are markedly less accurate than tube tests," he wrote in June 1970, was "way behind the times." As far as he was concerned, any "competent" GP or nurse with "common sense" was "capable of using modern slide test kits with a high degree of accuracy after a little instruction." Moreover, a "great deal of expense and waste of time and labour could be avoided" if only practitioners were "able to do these tests at their surgeries while the patient waits." "Highly skilled laboratory technicians" were simply "not necessary for these very simple tests." Murray's "only reservation"—one he unknowingly shared with Hobson—was that some GPs "may not do them often enough to maintain accuracy of technique or potency of materials, but the same could be said of almost any test." He suggested making slide tests generally available to health centers and group practices "where the throughput of tests is great enough to warrant them."[15] But the department, using his own reservation against him, rejected his proposal.

So, despite repeated appeals by the BMA, individual doctors, and even a few pathologists and laboratory directors, most GPs were still not able to do their own pregnancy testing in 1971, when Predictor debuted in Britain (see chapter 10). The idea of while-you-wait pregnancy testing in the consulting room foundered. Meanwhile, despite the state's effort at containment, the market for private pregnancy testing continued to expand. Beginning in the summer of 1969, pharmacists finally caved in and began offering services of their own. The rest of this chapter recovers the intra- and interprofessional debates in 1969, two years before Predictor, over a controversial practice that cemented the role of pharmacists as intermediaries in pregnancy testing.

700 DRUG STORES

At first, the BMA and the Pharmaceutical Society saw eye to eye on pregnancy testing and worked together to persuade the Department of Health to reign in commercial laboratories. To this end, they jointly petitioned the department for support in making "tripartite" representations to the ASA.[16] By then, the department was used to receiving letters, mainly from private individuals but also from MPs offended by public advertisements for pregnancy testing. The department's official line remained that a woman's decision to "spend 2 guineas for a private test" was a "matter of personal discretion" and while the government might reasonably be expected to protect consumers from harm, there was no evidence that commercial laboratories were providing a "seriously misleading service."[17] Although the ASA's measures had been implemented as a stopgap until central government introduced legislation, the department was satisfied with the oversight it provided. The BMA and Pharmaceutical Society did not get very far with the department. They were "free to make their own approach" to the ASA, but the government would not be joining them.[18]

Few pharmacists performed pregnancy tests before 1969, and those who did, much like laboratories before 1965, refrained from public advertising and dealt exclusively with doctors. Paragraph 22 of the 1964 version of the Statement upon Matters of Professional Conduct could not have been clearer: "Specimens for pregnancy diagnosis should only be accepted through a medical practitioner to whom the report will be sent by the pharmacist or independently. Such facilities should not be advertised." As historian Stuart Anderson has noted, the position of the Pharmaceutical Society "mirrored the prevailing medical view of pregnancy testing, but also represented the deference of pharmacists to medical authority."[19]

The situation began to change when Brian Block and Derek Lawford, the directors of Belmont Laboratories, the largest and most enterprising of the commercial services, made a bid to turn "drug stores" into nodes for urine deposit and result collection in their expanding mail-order empire. This forced the hand of the Pharmaceutical Society, which responded to the encroachment by staking a claim to the dispensing pharmacy, independent of medical oversight, as an appropriate site for pregnancy testing. The result further eroded medical authority, bypassed the GP, and firmly established the pharmacy as a place where women could go to find out whether they were pregnant.

The term *drug store*, which in the US long referred to a registered pharmacy dealing in prescription drugs, gained currency in Britain in the 1960s with the emergence of a new kind of retail outlet that did not include a pharmacy or employ a registered pharmacist. Drug stores (or shops), in the British sense, sold a range of over-the-counter medicines but did not dispense prescriptions. They undercut and ate into the traditional business of local chemists, particularly with the rise of Superdrug. Founded in 1964, Superdrug had three stores by 1968 and ambitious plans for expansion.[20] The opening of Le Drugstore in 1967 in two locations in Paris, widely reported in the British press, further raised the profile to the term, as did the Chelsea Drugstore, "with its boutiques and chemist's shop" in the King's Road, London.[21] Drug stores, then, were trendily American in the late 1960s and a commercial threat to pharmacists. Belmont Laboratories' much publicized campaign to enlist them in pregnancy testing posed pharmacists with an ethical, professional, and commercial dilemma, not least regarding their already somewhat frayed relationship with GPs.

On July 26, 1969, the *Pharmaceutical Journal*, the official organ of the Royal Pharmaceutical Society of Great Britain, published a letter by J. Langer, an East London pharmacist who had heard on the radio that Belmont Laboratories was "distributing free sample containers to 700 drug stores for collecting urine samples for pregnancy testing." Pharmacists, Langer pointed out, were prohibited from taking specimens "direct from the public" because it was "not considered ethical." Pregnancy testing was "surely a professional side of [the pharmacist's] work that should not be handed over to unqualified persons." Langer had been offering a limited service for four years but did not publicize this either to doctors or the public. The Council of the Pharmaceutical Society was, "after years of deliberation," finally reconsidering its policy on pregnancy testing, but a decision would not be made until the next annual general meeting in May 1970. He advised acting "now, not in a year's time."[22]

A leading article in the same issue of the journal concurred that Belmont Laboratories was forcing the issue. If the public was to have "direct access" to pregnancy testing, then the pharmacy, not the drug store, was surely the "proper intermediary." Pharmacists could no longer postpone "facing up to the new climate of public opinion" lest "other less desirable arrangements" came to pass. The article raised the possibility that pharmacists could be authorized to offer a pregnancy testing service either directly or as agents, "provided that certain safeguards" were adopted and discreet advertising confined to the pharmacy. Proposed safeguards included the usual one urging the recipient of a positive result to consult her doctor without delay.[23]

Langer's letter and the leading article provoked several pharmacists to weigh in. R. Benz of Farnham, Surrey, suspected all registered pharmacists were "dismayed at the thought of 700 'drug-stores' offering a service of this nature, while our hands are effectively tied." He strongly recommended a referendum to "ascertain quickly" the membership's views. M. J. Morris of Oxford could not understand why Langer was not free to inform doctors of his services. The Pharmaceutical Society had been urging its members "for years to emulate our Continental colleagues and carry out this type of work" but at the same time preventing "from making this known to the medical profession." Morris had no wish to fill his dispensary with the "smell of stale urine" but felt that Langer should be "allowed to carry on the ethical side of his business while the rest of us continue to merchandise toothpaste." A. Norman of South Benfleet, Essex, hoped the Society would "not be hurried into any wrong decisions on standards of conduct . . . because of cool relations with the medical profession." Most women went to their doctors "to have a pregnancy confirmed (very wisely in my opinion)," whereas those who patronized Belmont Laboratories were "no doubt hoping for a negative result." Peter J. Wildblood of northwest London asked, "What 'bad' reasons can women have had for having an interest in quick and accurate diagnosis of pregnancy?"[24]

Langer reaffirmed his position, warning that if the Society did not change its policy, "all" pregnancy testing would be done by "every Tom, Dick and Harry" bar pharmacists. Doctors, he suggested, would be "far happier to see the tests performed by pharmacists, with the results always reported back to them, rather than by the so-called laboratories." Most of his tests were for "women of menopausal age," the vast majority of whom were not pregnant: "They have been worrying themselves sick for nothing and are most grateful for a speedy and accurate result that one gets from a urine test." Moreover, Langer's fee was "much lower" than the standard £2 charged by laboratories. If only the Society would give its members a "much freer hand," the pharmacy and not the commercial laboratory or drug store could become the "natural place a woman will think of going to for a pregnancy test." This, Langer concluded, would "improve our professional image to the general public, and also improve doctor-pharmacist relations." If the Society deliberated too long, drug stores would become the "appointed agents" with pharmacists left to "merchandise toothpaste," as Morris had put it.[25]

The general press, having regularly intervened in the controversy over public advertising for the past four years, was quick to report on the possibility of over-the-counter pregnancy testing. *The Times* presented the pharmacist's professional code of conduct as outdated and quoted an official of the Pharmaceutical Society as saying that "change in morals and in social customs" had to be factored in.[26] The *Daily Sketch*, which had previously supported commercial pregnancy testing as a "woman's right to know," now reported that women would "soon be able to ask their chemists if they are pregnant," while *The Daily Telegraph* quoted a Pharmaceutical Society spokesman as promising that individual chemists who decided to participate in the "pregnancy testing scheme" would not be faced with any disciplinary action.[27] Some local papers took a less rosy view, even stoking fears of disreputable laboratories with links to an underground abortion racket. *The Scarborough Evening News*, for instance, noted the "boom" in pregnancy testing followed "close on the heels" of a "recent wave of alarm" that London was becoming the "abortion capital of the world, with women coming from overseas to seek abortions under Britain's new laws."[28] "Most dangerous of all," *The Chingford Express* speculated, there was "nothing to stop an individual carrying out a pregnancy test and then telling a woman she is pregnant when she is not. If he (or she) knows the child is not wanted, there is then plenty of opportunity to con the woman into paying for an abortion. All they have to do is anaesthetise her for a short period. And no one will ever be the wiser. Could it happen?" The same article further warned that commercial laboratories threatened to enlist hairdressers, newsagents, or the "man who has the shop on the corner" as intermediaries. Newsagents and corner shops often had small advert cards in the window that notoriously in some cases covertly advertised sexual services, a sleazy subtext that (along with South Asian settlement) may have been in play.[29]

Despite mixed reviews in the general and trade press, the Council declared in the *Pharmaceutical Journal* that at least until May 1970, when the matter would be decided at the annual general meeting, it would not discipline pharmacists for participating in "pregnancy testing arrangements."[30]

Block and Lawford took this as their cue to begin recruiting pharmacists to their expanding diagnostic empire. Having recently set up the limited company, Pharmacy & Professional Services (PPS), for just such a purpose, the pair began mailing "details of its scheme to all general practice pharmacies" along with a reprint of the *Pharmaceutical Journal*'s leading article of July 26. Around one in five responded, mostly in favor of the scheme.[31] *Medical News* predicted the "smouldering" issue of over-the-counter pregnancy testing would soon "ignite."[32]

Boots, which owned with subsidiary Timothy Whites around 2,500 of the 13,500 pharmacies in Britain, rejected the PPS scheme officially because it doubted demand would be sufficiently large.[33] Some Co-Operative Chemists followed suit. Block and Lawford began mailing sample pregnancy test kits to all other general practice pharmacists. With PPS, they intended to "make pregnancy testing 'respectable,' an ordinary matter where a woman could go to a pharmacy to obtain the service." In an interview with the *Pharmaceutical Journal*, Block estimated that commercial laboratories performed around one thousand tests a week (Belmont Laboratory accounted for a fifth of these), a rate that, "with pharmacists as an intermediary," could be increased tenfold.[34]

The sample kits consisted of three boxes addressed to PPS, each containing a plastic urine container, explanatory letter, request and report forms, information leaflets for customers, and two display stickers. Pharmacists, for whom there would be no capital expenses, charged patients the usual two pounds and would be invoiced for thirty shillings, leaving them with a margin of ten shillings. Pharmacists might be able to perform the tests themselves, but Block argued—as Hobson had long insisted of the Edinburgh center—that centralization was more economical. In addition to space, pregnancy testing required reagents (around £5 for twenty tests), as well as urine containers, droppers, slides, filter paper, printed forms and so on.[35] Gone were the days of large, expensive animal houses, but entrepreneurs could still plausibly claim that commercial pregnancy testing benefited from economies of scale.

The BMA was quick to oppose the scheme. On September 10, 1969, an anonymous consultant gynecologist defended the BMA's position on BBC1's investigative news program, *24 Hours*, on the grounds that a pregnancy test was "just one of the investigations that may be done to prove that a woman is pregnant or not" and that a positive result did "not tell you that she has a normal pregnancy." Block, who was invited to defend his scheme on television, countered that no woman who believed herself to be pregnant "crawled away into a bush" to give birth. A pregnant woman needed to see her doctor at an "early stage," and Belmont Laboratories had always advised the positive-testing woman to see her doctor "at once." The gynecologist cautioned that she might seek a "back street abortionist" or one of the "'less scrupulous' members of his own profession," or even "attempt to terminate her own pregnancy," and that testing by a doctor "minimised" the "dangers." But these old rationales may have lost their bite since the Abortion Act had come into effect, and Block maintained that it was the "woman's prerogative to know if she was pregnant as quickly as possible and under the new scheme she would be able to go into a pharmacy for the kit and get a result within 24 hours of sending the sample."[36]

Siding with the BMA, *Medical News* strongly condemned the Pharmaceutical Society's attitude: "The dangers involved in relying entirely on an imperfect test for confirmation of pregnancy, without a clinical history of examination, are so great that it seems utterly irresponsible that anyone who stands in a professional advisory capacity to the public should contemplate sanctioning such a move."[37] The Sheffield *Morning Telegraph* reported the BMA's warning that women who went to the chemist instead of their doctor "may be risking their lives" (figure 9.2). But the paper also quoted a local representative of the Pharmaceutical Society as saying, "We would like whenever possible to co-operate with doctors, but we realise perhaps a lady does not want to."[38] *The Times* quoted a defiant Block: "We sent out packs for the service to nearly 11,000 chemists last week. Only one has refused to join. The pharmacists are united as never before."[39] And the *Bellshill Speaker*, a Lanarkshire newspaper, combined Scottish thrift with liberal feminism when it portrayed PPS as a "needless . . . waste of money" and a "major breakthrough" that irked a "dictatorial," "indifferent," and "insensitive" medical profession by extending "individual freedom," "choice," and "independence" to women.[40]

Block and Lawford took out classified ads in major newspapers, and even bad press—in the papers, on radio, and on television—helped spread the word.[41] On September 20, 1969, a leading article in the *BMJ* maintained that pregnancy testing was a doctor's responsibility (for all the usual

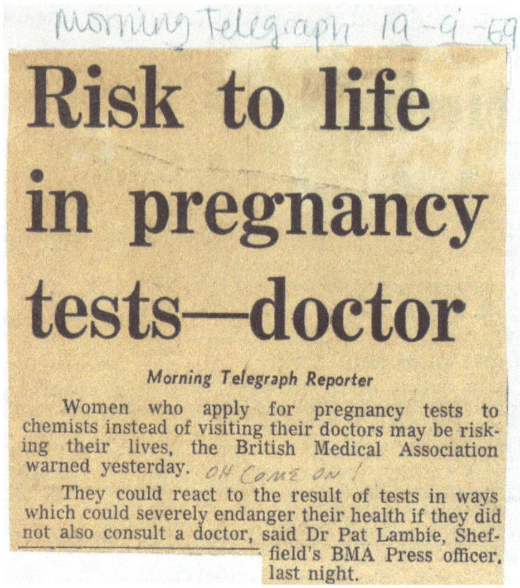

Morning Telegraph 19-9-69

Risk to life in pregnancy tests—doctor

Morning Telegraph Reporter

Women who apply for pregnancy tests to chemists instead of visiting their doctors may be risking their lives, the British Medical Association warned yesterday. OH COME ON!

They could react to the result of tests in ways which could severely endanger their health if they did not also consult a doctor, said Dr Pat Lambie, Sheffield's BMA Press officer, last night.

9.2 Surely one of the more extreme headlines generated by pregnancy testing. A previous reader, probably someone affiliated with the FPA, for whom the clippings were collected, penciled in the margins: "Oh come on!" *Morning Telegraph*, Sept. 19, 1969, Wellcome Collection SA/FPA/B10/47.

reasons) but also engaged in some soul searching. Why, the article asked, "do women use a postal service instead of going to their doctor?" This had become a troubling question for the medical profession. Some women might be reluctant to "discuss their state with any person at all," but "too often, regrettably, the woman believes that she may get an unsympathetic hearing from her own doctor, particularly if she is not married; while in London and other large cities many girls are not registered with a doctor at all." If the profession was serious about opposing what it called the "lay diagnosis of pregnancy," then it needed to "accommodate women who don't want to 'wait a few weeks' before knowing the worst—or the best; and it should urge the Health Department to make regulations to prevent doctors from being out of pocket if they help their patients in this way." So long as some women perceived their doctors as "unsympathetic," the article concluded, then "laboratories offering a service by post or over the counter will do a thriving trade."[42]

For the *Economist*, a weekly liberal newspaper, the evident demand from women made a "nonsense" of the BMA's reaction that doctors "should always be involved in pregnancy testing." Many doctors, it claimed, "knowing that a clinical examination can tell nothing certain in the very early weeks of pregnancy, will often advise patients who want a result in a hurry to apply to one of the commercial firms instead of waiting for the same test to be done at a hospital." Busy doctors, it concluded, "should welcome this new service as a relief and recognise the Pharmaceutical Society's old role for what it was—a pretty piece of inter-professional restrictionism."[43] But opposition, as letters in the *Pharmaceutical Journal* show, came not only from the BMA but also from within the pharmaceutical profession, especially from pharmacists outside London.

An Essex pharmacist "didn't quite get it": "Surely a patient who suggests being pregnant should always be referred to her doctor, yet apparently it will become quite ethical for a pharmacist to act as distributor for the 'mail-order, do-it-yourself' pregnancy testing service."[44] A. P. Wilcox of County Durham lamented that "hippies" had "taken over Bloomsbury Square," where the Pharmaceutical Society was headquartered. Having received a "neat package containing pregnancy test packs, a supply of leaflets on the new service and two small window stickers, one bearing the legend 'Now available here—pregnancy testing service—confidential advice given—free leaflets,'" he rhetorically asked what advice he was "qualified to give; is it how to become pregnant; how to not become pregnant; or, possibly, how to obtain an easy termination of pregnancy." Members of the public "could be forgiven for assuming that there is no difference between a 'drug store' and a pharmacy," and citing a document he had recently received on contraceptives, he testily advised the Council to "urge the free supply of baby foods for those infants who have managed to arrive despite the pill and the pregnancy test."[45] Gladys Millington in Devon saluted Wilcox for "admirably" expressing her own view: "Times may be hard, but surely the line must be drawn somewhere. I am seriously considering 'shaking the dust of pharmacy' off my feet."[46]

Others, meanwhile, openly criticized the BMA's position. C. S. Hunter accused the medical profession of the "usual stuffy self-righteousness and self-importance." Pregnancy was a "highly personal business" and "many patients" came to "serious psychological harm by not being sure and not having the courage to find out in time." "Why," he asked, "must they confide their inner secrets to doctors if it should turn out to be unnecessary?"

Hunter, whose letter in *Medical News* was subsequently reported in the *Pharmaceutical Journal*, could "think of no unpleasant repercussion to the patient" and foresaw only that "one or two of our few careless colleagues may be in for a surprise or two, but is this a bad thing? It might return these gentlemen to a position of their feet instead of their glutei." The claimed accuracy of 98 percent was "'pretty good' and comparable with the best that doctors' professional mystique has to offer."[47]

PARAGRAPH 22

Albert Howells, president of the Pharmaceutical Society, "strongly defended the Council's decision to omit the paragraph on pregnancy testing from the Statement upon Matters of Professional Conduct" in his address to the Society's first Scottish regional conference at the Dunblane Hydro Hotel on October 5, 1969 (figure 9.3). The NHS was evidently not meeting the demand for "immediate" pregnancy testing and "public opinion," an ascendant keyword and resource in the battle of words, "or a very large section of it, had undergone dramatic change." Women now wanted an "immediate answer to a question which overhung their entire future," and they were "getting the answer through a heavily advertised postal service, which at no time had brought them into personal contact with a member of the nation's health team." This "totally impersonal service was surely more open to criticism than one which took the patient into the pharmacy, where advice was available."[48]

The profession might yet decide to "supply not only the kit but also the result of the laboratory diagnosis." In the case of a positive result, pharmacists "could advise the patient to consult her doctor and also inform her of the social security available to her." Alternately, they might "carry out the tests" themselves, a task for which they were "fully qualified," not at the "dispensing bench" but "under proper laboratory conditions." Until May, when the Council would decide the matter, pharmacists had a free hand. Howells did not claim that all or even most pharmacists would fully support "participation in a service which the pharmacist did no more than supply the patient with a kit," but he asked those present to accept the Council's decision as "entirely justified" and "no more" than a reflection of a "social development which could not be ignored."[49]

REGIONAL CONFERENCES

President defends pregnancy testing

WHEN the President of the Society (Mr. Albert Howells) spoke at the first Scottish regional conference of the Society, he strongly defended the Council's decision to omit the paragraph on pregnancy testing from the Statement upon Matters of Professional Conduct.

The conference was arranged by the Scottish Department of the Society and was held at Dunblane Hydro on October 5. It attracted an attendance of over 100 representing members in general practice, hospital and academic pharmacy. Other speakers were the Vice-President of the Society (Mr. W. M. Darling), Mr. J. P. Bannerman (member of Council) and Mr. W. Lund (of the Society's Department of Pharmaceutical Sciences). The Chairman of the Scottish Executive (Mr. A. Roxburgh) presided.

The PRESIDENT said: "There is one pleasant duty that I intend to discharge at once. And that is to thank you and your colleagues in Scotland for assailing the Secretary of State for Health, not with your claymores but with your pens, on the subject of rural dispensing in England and Wales. It was through your good offices, Mr. Chairman, that Scottish pharmacy demonstrated its support for the profession south of the border, and I cannot let this occasion pass without expressing our gratitude." There was no doubt, said the President, that a very large number of Scottish pharmacists had despatched their own personal protests to the Secretary of State, and that the latter had been impressed not only by the weight of protest itself but by the fact that the authors of it, far from seeking anything for themselves, had been roused to action by an injustice elsewhere in Britain.

The President went on to speak of planned distribution. "I firmly believe", he said, "that planning is vital both to the economic future of the profession and to the status of pharmacy within the community." In Scotland, over the past four years, more than 150 pharmacies had closed; and a growing number of communities had been left with no pharmacy whatever. It was not surprising, therefore, that the annual conference of the Scottish Association of Executive Councils should express its grave concern. "We, too," said the President, "are concerned."

At the meeting of the British Pharmaceutical Conference in Belfast, he (the President) had expressed his concern—the Society's concern—at a trend which was evident throughout Britain and which had produced a net loss of 267 pharmacies in the first six months of 1969. He had said: "To sacrifice the absolute independence of decision and action that we enjoy today may be to pay but a small price for a professional design which, while guaranteeing the public's access to a vital service, will also guarantee pharmacy a new and much needed economic stability".

The President (Mr. Albert Howells)

The President continued: "In that sentence there is a proposition which I believe will commend itself to an increasing number of our members. I believe that our profession is entitled to plan its future; I believe that our members are entitled to reasonable security in the practice of their profession; and I believe that the public are entitled to expect from us a plan which is designed to serve the public interest. What cannot be denied is that what many of our colleagues are facing now is not economic stability but economic collapse, not security but insecurity. The result to date: a net loss in Britain of 2000 pharmacies since 1954. For more and more communities, not a pharmaceutical service but a pharmaceutical vacuum. Obviously, the legitimate interests of pharmacy and of the public are being simultaneously destroyed."

The Committee on a Planned Pharmaceutical Service had lost little time in producing its first report, which proposed a solution in areas of low population density. The plan proposed mobile pharmacies, part-time pharmacies, and collecting points for prescriptions; it was a plan which had special relevance for Aberdeenshire, Inverness-shire, Sutherland, and Argyllshire—counties which had been named in the Press as containing communities which had been left with no pharmacy.

The President emphasised that there was no question whatever of mobile pharmacies being allowed to intrude into the area of established pharmacies; before the Society agreed to the introduction of mobile pharmacies, very special safeguards against any form of intrusion would need to be incorporated in the regulations.

The plan had been submitted to the Scottish Home and Health Department, to the Department of Health and Social Security, and to the Welsh Office, and he had every reason to think that it had been received with keen interest. A preliminary meeting between representatives of the Health Departments and of the Committee to discuss the plan was to be held very soon.

The Committee, of course, had no intention of sitting back on its laurels. It had produced no more than a fraction of the grand plan that was envisaged, and it must now examine other areas.

The advent of health centres gave added urgency to the task of planning, the President continued. Only recently the Scottish Under-Secretary had announced that six health centres were under construction in Scotland, plans for another 14 had been approved, and 33 more were at the planning stage. Glasgow alone was to have 20 health centres. "From that one stark fact," he said, "I do not have to proceed far to explain the Society's reluctance to be joined in the development of health centres. Twenty pharmacists operating as chemist-contractors in 20 health centres, and monopolising medical prescriptions in the city, could wreak economic havoc among their colleagues in general practice pharmacy and thereby destroy the present pattern of the pharmaceutical service, to the undoubted detriment of the public."

The President then referred to the subject of pregnancy diagnosis. "It is certainly a controversial topic," he said. Those present would all be aware that the Council's decision to omit Paragraph 22 from the Statement upon Matters of Professional Conduct, and the subsequent

9.3 The president's defense of pregnancy testing dominated coverage of the Pharmaceutical Society's first regional conference in Dunblane, Scotland. *Pharmaceutical Journal*, Oct. 11, 1969, 442, Cambridge University Library P328.b.11.200, with permission by the Pharmaceutical Journal.

W. M. Darling, vice president of the Society, supported Howells's position. New factors—the "availability of simple tests," the ASA's decision to permit public advertising, the "changing attitude of the population," and the Abortion Act—had all to be considered. Because it was impossible to amend Paragraph 22 until May, the Council had been forced to deal with the "question of whether pharmacies should act as agents for a company conducting pregnancy tests . . . on an *ad hoc* basis." If such a service was going to exist anyway, then "in the public interest," the pharmacy was the "best place for it." Darling hoped that whatever decision was finally reached in May, it would not "bring the profession into disrepute." But not everyone in the room agreed.

An Edinburgh pharmacist was "extremely critical" of the Council's "short sighted" decision on pregnancy testing. Individual members could choose to retain Paragraph 22, but the "horse" had "bolted." What would happen, he asked, to "those who had taken up this money making business" if in May the "membership indicated that it wanted Paragraph 22 retained"; would they be controlled or "allowed to continue while other members were controlled?" A Colchester pharmacist favored pregnancy testing, but only by pharmacists who were "properly qualified"; he doubted the "right" of "elderly" pharmacists "to do such work." And a third, from Glasgow, who had "never been asked to undertake a pregnancy test," felt the Society was "escalating the problem out of all proportion." The audience of over one hundred general, hospital, and academic pharmacists applauded when he objected to "being used as a professional postman."[50]

Despite such objections, Block and Lawford were soon receiving around one hundred specimens a week from pharmacists who had joined the scheme. Doctors, meanwhile, continued to voice their opposition. T. M. Winstanley of Ruthin, North Wales, accused the Royal College of Obstetricians and Gynaecologists in the *Lancet* of "masterly inactivity over a matter which smacks of commercialism at the possible expense of clinical well being." He was "amazed" the medical profession had relegated pregnancy testing, with all its implications, "to the shop counter." The chief pharmacist at Warneford General Hospital in Leamington Spa agreed that results "should always be reported to the doctor for his interpretation, rather than to the patient" but added that she now had "direct access to such tests, whether she goes to the pharmacist or not."[51] The status quo of direct, doctorless pregnancy testing through minimally regulated commercial laboratories and possibly even drug stores (or worse) made it easier for pharmacists to defend their new role as the lesser evil.

BLATANT TRESPASS

Some pharmacists embraced their new role as mediators of a woman's right to know. Others grudgingly accepted it with an air of resignation. One of the latter anonymously mused on the new circumstances in the *Pharmaceutical Journal*. His pharmacy door now displayed a notice advertising a "willingness to accept samples for pregnancy diagnosis" but not because he personally endorsed the scheme or was "urged by any missionary inclination to provide such a service." Rather, his employer believed the notice would "marginally improve trade and the 'image' of the premises, and since [he had] no ethical objections to it, it is displayed." This pharmacist could not muster "any enthusiasm for the controversy." On the one hand, he did not want to be involved in pregnancy testing but only because it was "not really our job." On the other, since mail-order services were "already widely advertised," pharmacists did "no harm by acting as extra post-boxes." He disputed Howells's implication that pharmacists were "eminently suited to offer confidential advice on social security benefits or elementary health precautions in pregnancy." He knew little about benefits and was "sure that few pharmacists have space to afford the sort of privacy which the giving of such advice requires." Few women were keen on discussing "advice on their procreative function or financial circumstances within easy earshot of shop staff and customers." Of those who required "advice or encouragement after positive diagnosis," few would "wish to air their uncertainties 'over the counter.'"[52]

Over-the-counter pregnancy testing was, for this pharmacist, harmless, just not particularly useful. What of its "practical and 'professional' implications"? He had "no desire to act as a collecting depot for samples for diagnosis" and saw this as "yet another chore added to the many which already [shortened his] time spent practicing pharmacy." He was mildly more attracted by Organon's recent proposal to sell Planotest directly to pharmacists, but "neither the duties of technician nor of postmaster really appeal[ed] as a diversionary activity." He acknowledged that rapid pregnancy testing was an "undoubtedly useful social service since legal abortion (and often peace of mind) requires early diagnosis" and accepted that most women would either consult their doctors or use a mail-order company before going to the pharmacy as a "last resort." For individual pharmacists, the decision to participate in the PPS scheme was above all an "economic one: the profit margin is high, the outlay small." He would have preferred "not to take part," but his employer "thinks we ought to. So we do."[53]

Laboratory technicians had less clout than doctors, but they too registered their displeasure at the encroachment of pharmacists. Twelve technicians signed a letter in *The Gazette of the Institute of Medical Laboratory Technology* condemning the Pharmaceutical Society's "rather disturbing" decision as "undoubtedly motivated by profit." Pregnancy diagnosis, they insisted, was a "medical laboratory examination" that "should be undertaken only by registered technicians." Over-the-counter testing was "blatant trespass" comparable to technicians declaring themselves "willing and able to dispense prescriptions of certain types." Pregnancy testing was "not difficult," but neither was "counting tablets." The signatories lamented the "unfortunate but unavoidable" proliferation of private laboratories "under present law" and explained that registered technicians were "barred from performing pregnancy tests except at the request of a doctor and from disclosing the result to a patient" lest they be found "guilty of infamous conduct." They wanted pharmacists to back down or, failing that, technicians to be "granted the same freedom of action."[54]

But not all laboratory technicians agreed, and many wanted hospital laboratories to do less, not more, pregnancy testing. Some were happy for the private sector to disburden them—and the taxpayer—of an activity they regarded as low priority. A fellow of the Institute of Medical Laboratory Technology argued in *The Gazette* that pregnancy testing was "one of the few 'medical' laboratory procedures that does not determine whether a person is sick or well." He was sure that "many busy pathology departments would be only too pleased to see the back of these tests." Women were "entitled" to a prompter and more discreet service than that provided by the NHS, even if that meant paying out of pocket. Another technician did not think hospital laboratories should be responsible for the "enormous number of tests performed for domestic rather than medical reasons." There was "no good reason for the community to bear the cost of informing a perfectly healthy woman that she is pregnant, a little earlier than she would otherwise know." Yet another deplored NHS pregnancy testing as a "form of state collectivism of which we already have more than enough."[55] Others, however, defended the professional territory staked out by the signatories.

George Lunay, an Australian pathologist, wrote in the *BMJ* that "neither the pharmacy nor the surgery of the general practitioner is a suitable place for pregnancy testing." The procedure "should be performed solely by experienced personnel in the clinical laboratory, where it may be carried out under properly controlled conditions and correctly interpreted in the light of the relevant clinical history." He doubted whether all pharmacists were qualified to perform the slide test "accurately and safely." In his experience, 5 percent of reactions were "not easily read" and required comparison with "positive and negative control tests . . . before finally reporting on them." Aspirin and other drugs could produce false results, and the old fears of dangerously misdiagnosing a hydatidiform mole or placental tumor lingered. So did concerns that a misdiagnosed nonpregnant woman would tragically "seek the services of an abortionist when in fact there is no pregnancy to terminate, or when there is an incomplete evacuation of a 'molar' pregnancy or choriocarcinoma."[56] But the decades-old arguments had less power in a new era of legal abortion and public opinion.

The *Pharmaceutical Journal* countered that Lunay's letter did "not acknowledge the fact that a high proportion of the pregnancy tests undertaken through pharmacies are done in specialised laboratories." Nor did it accord with findings announced in the *Practitioner* (the same ones Hobson dismissed) that GPs were capable of reliably using one of the commercially available slide tests. A more recent review agreed with the manufacturers that slide tests "could be carried out quickly and easily by untrained, intelligent personnel with the minimum of technical instruction." Detailed instructions included with Organon's Planotest or Wellcome's Prepurex (figure 9.4) discussed the interpretation of results and pointed out that certain pathological conditions could produce false-positive results. For the *Pharmaceutical Journal*, this was a good enough safeguard against possible mistakes and their potentially harmful consequences.[57]

Block, who had a vested financial interest in relegating pharmacists to the role of "post-boxes," was critical of Lunay's "less enlightened" views

9.4 Advertising leaflet for Wellcome's "improved" Prepurex, *c.*1970: (a) front unusually opted for the "wise" owl motif; (b) back shows the components of 22-test and 110-test kits as well as contrasting positive and negative endpoints that also evoke the owl's distinctive eyes; and (c) inside pages show the many steps involved. Wellcome Collection WF/M/PL/248.

but conceded in the *Pharmaceutical Journal* that consistently accurate testing took practice. He required a novice technician to perform "nearly 200 tests under close observation before being allowed to carry out a test alone." The pharmacist, in contrast, would "never have the opportunity of this degree of practice and cannot hope to achieve the same degree of accuracy as trained personnel doing tens of thousands of tests a year." Making suppositories was after all a "relatively simple" procedure but, asked Block, "how many intelligent people would make them well after reading well-written instructions?" On the other hand, GPs increasingly prioritized the "convenience of the woman" and "perhaps even the BMA will soon give patients the same consideration. Can they really object to women obtaining in 24 hours through the pharmacy an item of information that takes 7 to 10 days to receive from a doctor?"[58]

Dismayed by the BMA's "arrogance," a Chelmsford doctor argued in the *BMJ* that there were "many, many, many good reasons why chemists should not only be allowed, but should be encouraged, to do pregnancy testing," including the "patient's convenience," which must be taken seriously.[59] The *Which? Supplement on Contraceptives* maintained that a "sympathetic and helpful doctor is always the best person to go to if you think you may be pregnant" but admitted that "as long as doctors or hospitals cannot provide a quick and impersonal pregnancy test service, the commercial laboratories are supplying a useful need." By January 1970, the weekly increasing number of tests performed by PPS had "well overtaken what Belmont was doing." This indicated to Lawford that the "public liked the new service through pharmacies."[60]

A DISCREET NOTICE

In May 1970, the Pharmaceutical Society published its new Statement Upon Matters of Professional Conduct, the first substantial revision since 1964. As once more Stuart Anderson has noted, the revised statement "indicated that although the dispensing of medicinal products, or the professional services of a pharmacist, should not be advertised directly or indirectly, an exception could be made for 'a discreet notice, relating to Pregnancy Testing Services, [which] may be exhibited at any pharmacy.'"[61] This retroactively sanctioned the notice on the door of the

9.5 Some pharmacies, like this one on Nicholson Street in Edinburgh, c.1970s, prominently advertised over-the-counter pregnancy testing alongside prescriptions and other basic services. When I visited again in 2016 the sign had been painted over. Photo: Jana Goldsmith, 2011.

unenthusiastic pharmacist who had contemplated its implications in the *Pharmaceutical Journal*. Other pharmacists would paint "Pregnancy Testing" on the paneling over window displays (figure 9.5). By June, more than three thousand pharmacies had used PPS, which did about one thousand tests a month. More than half the clients asked for results to be telephoned not directly to them but to the pharmacy.

In London, the service was used mainly by "younger pharmacists and more widely diversified pharmacies," but chemists all over Britain participated.[62] Quantitative information about the women who used the service is scant, but a survey carried out for Belmont by Survey Computing Ltd. toward the end of 1970 disclosed that 75 percent of clients used the service because they wanted quicker results and 16 percent because they did not want to see a doctor. About half of all clients (55 percent) were unmarried. Seventy-two percent wished to terminate their pregnancy if positive, a ratio that dropped to 45 percent of married women. Only 32 percent of all women wanted to keep the child if they turned out to be pregnant.[63] Moral critics of pregnancy testing were not entirely wrong about women's motives, but they had fewer allies than before and insufficient clout to prevent women from paying for a commercial test and then deciding whether to legally terminate the pregnancy. Patterns of consumption, if not "public opinion," had shifted, and with it so had professional practices and ethical codes of conduct. Pharmacies had become established as places women went for a pregnancy test. The next step—retailing self-testing kits directly to women—was not long in coming.

10

DO IT YOURSELF

On Tuesday, June 16, 2015, at 1 p.m., the original prototype of Predictor, the "first home pregnancy test," went up for auction at Bonhams in New York. Prior to auction, the test and its inventor, Margaret (Meg) Crane, were little known. Wikipedia listed her as an "American scientist of dubious mortality" until she corrected the entry to a "still very much alive graphic designer."[1] Archivist and historian Lesley Hall had drawn my attention to a press release in the Wellcome Collection (figure 10.1), but I knew nothing of Crane until she emailed me out of the blue a month before the auction: "I happen to have invented the first home pregnancy test in 1967 and have 2 patents for it. Yes, it resembled a small chemistry set, but it was the first chance a woman had to test herself at home. My prototype and one of the first marketed products is up for auction on June 16 at Bonhams Auction House in New York."[2]

Predictor, as I later learned from Crane, came to auction because of a *New York Times Magazine* article on the invention of the home pregnancy test that failed to mention her in 2012.[3] Her brother and his daughter, a journalist, decided to take matters into their own hands and so got in touch with a curator at the Smithsonian Institute who established contact with Crane and eventually asked her to donate the prototype.[4] Procedure required her to have it valuated, and the specialist at Bonhams who performed the valuation convinced Crane to put it up for auction.[5] The Smithsonian's National

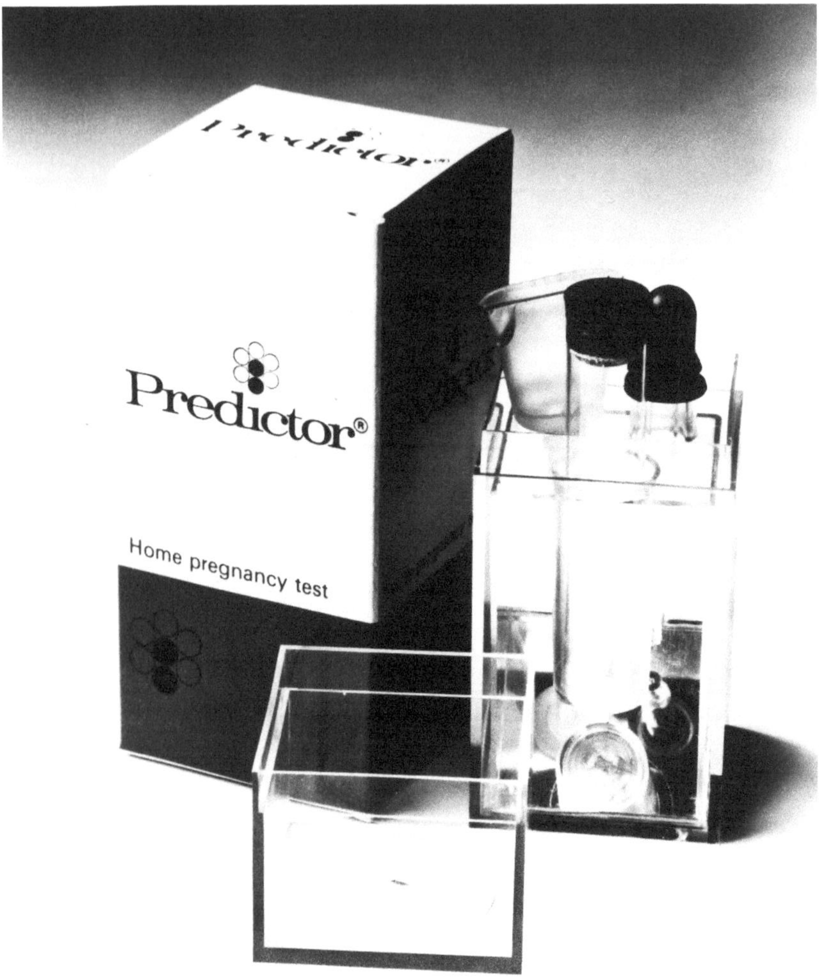

10.1 Publicity photograph of Predictor, launched in Britain by Chefaro UK in November 1971. Wellcome Collection PP/GRA/B.4.

10.2 "Lot 37" of *Voices of the 20th Century* exceeded the estimated value of $6,000 to $9,000. The hammer price was $9,500 and with the buyer's premium, the prototype and finished product garnered $11,875. Crane's prototype (2.9 × 1.7 × 1.7 in.) next to the retail product for the Canadian test market (3.6 × 1.5 × 1.5 in.). *Voices of the 20th Century* (New York: Bonhams, 2017), 35, with permission by Bonhams.

Museum of American History in Washington, D.C., acquired "Lot 37" of *Voices of the 20th Century* for $11,875 (figure 10.2).[6]

Following the auction, fanfare around Predictor contributed to a growing appreciation of Crane's rightful place in the design history. Her invention debuted in Britain in 1971, a full seven years before it would be marketed in the United States, so it makes sense to investigate what happened there first. Predictor, as we shall see, was not an overnight success. Twentisec, an unreliable pregnancy test, beat it to the market and damaged the reputation of the sector. Initially restrictive rules governing where and how Predictor could be advertised didn't help. Central government, unsurprisingly, kept self-testing at arm's length, while stakeholders drew on newly available phrases, including "women's liberation," "permissive society," and "sexual freedom," to interpret rising demand.[7] This chapter is about, among other things, the medical debate over "social" pregnancy testing that exposed worsening tensions within the NHS between general practice and hospital services. It begins in the late 1960s in New York, with graphic designer Meg Crane and her invention story.

THE INVISIBLE DESIGNER

Margaret M. Crane was born in 1939 in Jersey City and studied fashion illustration and graphic design at Parsons School of Design in Manhattan (figure 10.3). In March 1967, she was hired to work on cosmetics by Organon Pharmaceuticals, the New Jersey–based subsidiary of the eponymous Dutch company that in 1962 had successfully launched Pregnosticon, Leif Wide's innovation. Crane first encountered pregnancy testing in the

10.3 Crane from around the time she started working at Organon, in 1967. Photo: Anna Kaufman Moon, courtesy of Crane.

cosmetics laboratory, where Pregnosticon underwent quality control testing and development. There she saw "rows of coded test tubes positioned over a long reflective, angled surface" and learned that Organon performed around half of all pregnancy tests in the United States, for doctors who posted urine samples (figure 10.4).[8] In the account of her invention prepared for the auction, she recalls having been "impressed" by the "simplicity" of the tests: "They remained in the back of my mind and I began to think that a woman could perform one herself; they were easy to read and required little equipment."[9]

In the same narrative, Crane attributes her interest in pregnancy testing to abortion politics, which were heating up in the late 1960s: "pregnancy could be a serious factor in a woman's life" and so the "possibility of making the test available to the public became an increasingly important idea to me." Crane made sketches on paper and used cardboard and Styrofoam to construct models. Next, she approached Joseph Ruvane, Organon's executive vice president, with the "idea of a consumer test," but he worried the potentially controversial product would cost the company its lucrative medical business. Undeterred, Crane continued to press the matter with Ruvane, who agreed to raise her idea at a meeting with AKZO, the successor company to Organon Netherlands. Ruvane was granted a "small budget to investigate market possibilities," but company managers and laboratory staff opposed the project.[10]

Some feared, with Ruvane, that marketing a pregnancy test directly to members of the public would threaten Organon's medical market and possibly even the company's other products. Still others objected on "moral grounds" that women had "no right to be testing themselves for pregnancy," that pregnancy testing was "linked only to abortion," and that the project would "bring the wrath of church hierarchies upon them." But the project moved forward, and when "sales projections began to look good," marketing "warmed" to the idea of a consumer-oriented pregnancy test.[11]

Organon hired Frank Ennis as marketing director of Intec Laboratories, a new division specially created to manage the project, and an "obsessed" Crane got to work on the design. Like all good invention stories, Crane's has a Eureka moment: "One day, I was staring at a plastic container on my desk. It came from the Azuma store in New York [a gift shop that started out selling Japanese handcrafts]. I used it to hold paper clips. But it was perfect!

10.4 Annotated advert (*c.*1964) for the Pregnosticon TestRac™ showing a positive "doughnut" in four of the ten angled mirrors. Rockefeller Archive Center, The Medical Letter 163, Box 88, Folder 2.

It would need a shelf in it to hold a test tube, an eyedropper, and an angled mirror at the base. The cap could be used to collect urine." Crane presented Ennis with a minimalist design she "knew to be absolutely right" but that Ennis and Ruvane rejected as "too expensive." Ruvane kidded Crane that it would be for wealthy women only, and Ennis asked her to "contact product specialists to get started on ideas for the 'real' product."[12]

Ennis held a meeting for professional product designers, who lined up their frillier prototypes on a conference table. Crane, who had asked to be at the meeting, placed her prototype after theirs. Foote, Cone & Belding, an advertising agency, was assigned to the product and sent their representative, Ira Sturtevant, to the meeting. Sturtevant, as Crane later recalled, "walked down the row of designs, picked mine, and said 'This is the only one you can use.' Ennis said 'Oh, that's just something Meg did for talking purposes.'" Ennis maintained that Crane's prototype was too expensive to produce, so Crane took a few days off to look in and around New York for an affordable plastics company. She found two, including one that quoted her a cost that was "one third less than the least expensive of the 'professional' offerings."[13]

Even so, management continued to object to the still nameless product on moral grounds, so Crane and Ennis together met with Reverend Howard Moody of the Judson Memorial Church in Greenwich Village. Following the defeat of a bill to loosen New York State abortion law, Moody, a former marine sergeant, had recently established a citywide network of Protestant ministers and rabbis to help women obtain "legal therapeutic abortions" and counseling.[14] Moody, unsurprisingly, responded favorably to Crane's idea. So too did Ennis's own minister in suburban Rye. A satisfied Ennis reported back to management, and that seemed to end the objections. The product was christened "Predictor," and clinical testing began at Mount Sinai Hospital, with a group of "low-income women," to determine if it could be used easily and accurately even by the uneducated. Progress stalled, however, when the license for Pregnosticon came up for renewal and other products took priority.

Some months later, Organon's legal chief informed Crane that the company was applying for a patent for Predictor and asked her to sign it over to Intec for the standard, symbolic fee of $1 (figure 10.5). She never received the fee, but she met her soulmate, the adman who picked her prototype

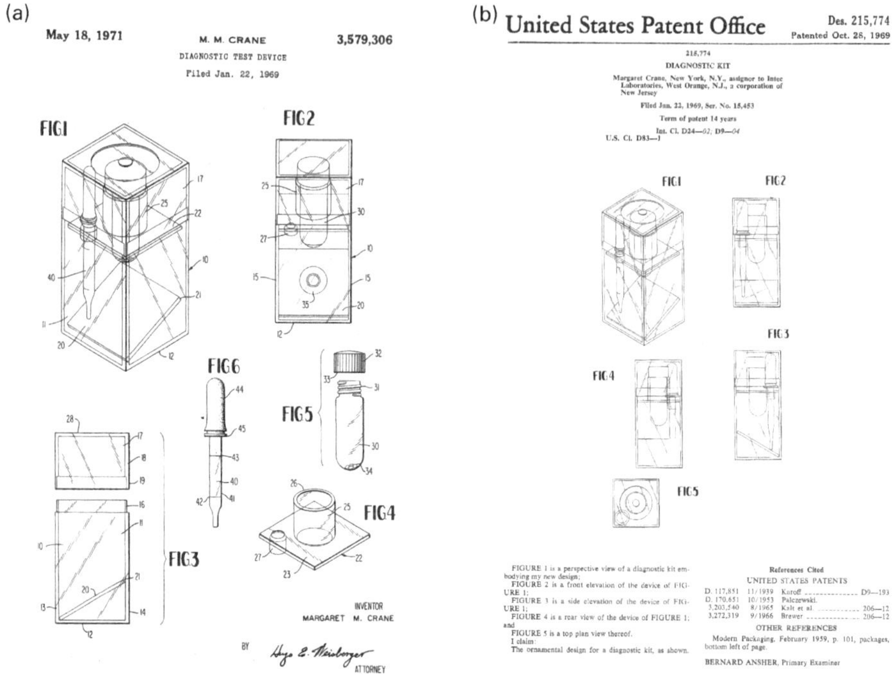

10.5 (a) Product patent (3,579,306) and (b) design patent (215,774) for Predictor filed in Margaret Crane's name in January and October 1969.

out of the line up (figure 10.6). As she told me in an email, "I fell in love with him the minute I saw him and told my roommate that evening that I had just met the man I was going to spend the rest of my life with." He was her "compensation and more." The pair began working together on Predictor and after a few months had moved in together. Crane would later tell people they "met over a pregnancy test, and it wasn't mine and it certainly wasn't his." Together they started their own advertising company, Ponzi & Weill, which continued until Sturtevant's death in 2008.[15]

In October 1970, Ruvane offered the couple the account to introduce Predictor in Canada as the test market for the United States. This came as a surprise because they had assumed the product was "dead." In fact, as Ruvane informed them, Predictor was now market ready and slated for launch in the Netherlands and Britain, with other European countries to follow. Crane and Sturtevant would be part of any US plans that

10.6 Photograph by Shinichiro Tora of Sturtevant and Crane at a Japanese restaurant in the early 1970s; courtesy of Crane.

materialized, depending on how the Canadian test market developed. The approval process for Canada was "uncomplicated and swift": Sturtevant traveled to Ottawa to obtain permission to sell Predictor north of the border and was only in the capital "for a day."[16] Chefaro, a Montreal-based subsidiary of AKZO, introduced a bilingual version of Predictor to Canadians in August 1971 (figure 10.7).[17]

Organon, a "business-to-business pharmaceutical company," hoped to enter the consumer products sector, and the Canadian trial was partly intended to "test their ability to expand into consumer goods," including the newly acquired cosmetics line that Meg had been hired for. But the Canadian experiment was short-lived. Organon was unable, on the back of a single product, to marshal a national sales force with the required expertise in pharmacy sales. Despite increasing sales of Predictor in Canada, Organon withdrew altogether from the consumer market and canceled US plans, thus ending Crane and Sturtevant's involvement. Instead of directly marketing Predictor, AKZO and Organon licensed it to large consumer-oriented companies that in 1978 launched four equivalent products under the tradenames of Acutest, Answer, e.p.t., and Predictor.[18]

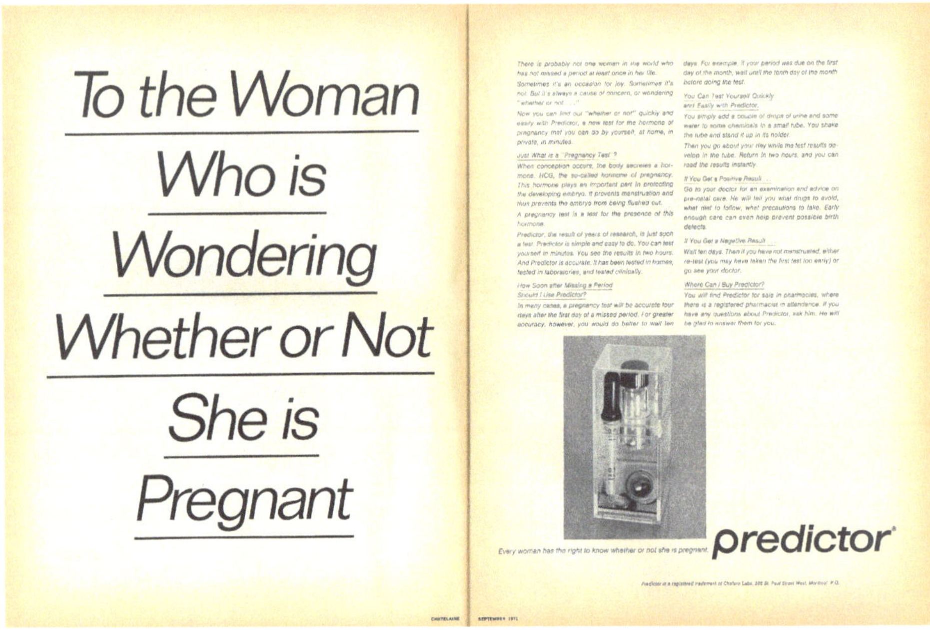

10.7 This two-page spread advertising Predictor in the English version of Canada's preeminent women's magazine, *Chatelaine*, Sept. 1971, 98–99, appropriated the "right to know" language, turning it to commercial ends.

All four tests derived from Pregnosticon and so were of a high, laboratory standard. Prior to the initial 1971 launch, however, smaller companies had sought to enter the Canadian, US, and British markets by quickly bringing out their own proprietary tests that were not based on Crane's design. Denver Laboratories (Toronto) launched Confidelle, an apparently reliable immunoassay developed by a small New Jersey company, in Canada in November 1970.[19] It lacked the commercial heft needed to make a splash, but good press and an absence of significant controversy north of the border may have motivated AKZO to move forward with Predictor. Other companies, however, launched patent "chemical tests" that simply didn't work (figure 10.8). In 1972, the US Food and Drug Administration recalled Ova II, a dubious "non-immunologic" color change test for pregnancy that performed no better than "tossing a coin."[20] In Britain, Twentisec, a dud test based on "estimating the pH of the urine," beat Predictor to market by several months.[21] Like Ova II, it was recalled in 1972, but not before damaging the reputation of the industry just as it was getting off the ground.

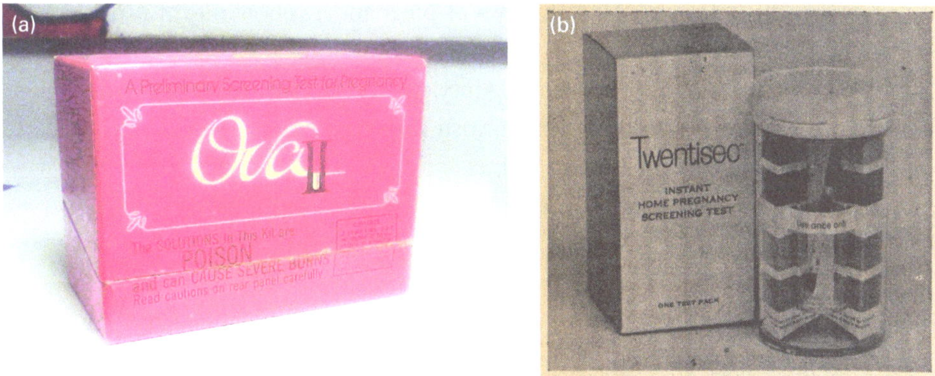

10.8 (a) Box of Ova II, sufficient for two tests, marketed for $4.95 by Faraday Laboratories of Hillside, New Jersey. In addition to the conspicuous "POISON" warning on the box, an insert included detailed instructions for first aid and the disclaimer: "The existence or non-existence of pregnancy must be confirmed by your physician." Photo: Margaret Crane. (b) Twentisec test pack and contents, including a "calibrated disposable urine sample tube, a colour indicator chart and a reagent." "Over-the-Counter Pregnancy Test Kit," *Pharmaceutical Journal*, June 26, 1971, 486, Cambridge University Library P328.b.11.200, with permission by the Pharmaceutical Journal.

TWENTY SECONDS

In Britain in 1971, some hundred laboratories, a thousand FPA clinics, and twice as many pharmacies performed an estimated 1.5 million pregnancy tests, and hormone pregnancy tests (mostly Primodos and Amenoron Forte) peaked at 100,000 prescriptions.[22] Most women who patronized commercial services planned to abort if positive, especially if they were unmarried; only 32 percent of clients "would want to keep child if pregnant."[23] Fees ranged between £1.50 and £3 in currency that from "D-Day," February 15, 1971, was decimalized.[24] Pharmacists also sold "test kits" for urine collection, which seem to have been rumored to be a new "do it yourself" method.[25] So the idea of self-testing was in the air when Twentisec, the first product marketed in Britain as a home pregnancy test, debuted on June 22, 1971.

Made in Britain by Global Laboratories and distributed by Barclay & Sons, Twentisec was not an immunoassay but a simpler color change test. Recommended from two weeks after a missed period, it consisted of a specimen tube, measuring test tube, phial of reagent, and color chart. It worked, at least in theory, by adding the reagent, a halogenated cresol derivative, to an early morning sample of urine, which turned purple—as matched against the color chart—in the case of pregnancy; yellow meant no pregnancy.[26] Prior to its British launch as an "instant home pregnancy screening test," Twentisec had been sold to American doctors for nearly two years.[27] The makers, who purchased the British sales rights from the leading condom manufacturer Julius Schmid, claimed an overall accuracy of around 90 percent. This was significantly lower than the immunoassays, but they expected demand to be high because the result, correct or not, would be known only to the woman herself. At a retail cost of £1.25, or two for £2.20, Twentisec undercut commercial laboratory analysis by about half.[28] Simplicity was a selling point. *The Times* would soon compare Twentisec to "those first science experiments at school where one got the litmus paper to change to blue."[29] But that it required "significantly less sophistication for interpretation than previous methods" should have been a warning.[30]

Daily Mirror science editor Ronald Bedford championed Twentisec as a major "advance in Women's Lib" that gave "every girl with £1.25 in her purse . . . a chance to find out in the privacy of her bathroom the answer to the question that can keep her awake half the night." The test was so simple that it could be used by any woman who could "cope with the simplest recipe or instruction leaflet." Preempting criticism, the managing director of Global Laboratories had told Bedford that Twentisec was for not the "flighty girl" but the "married woman . . . entitled to find out this important fact *for herself*."[31]

The Glasgow *Daily Express*, on the other hand, reported the BMA's objection that Twentisec could give misleading results.[32] Similar reservations were voiced in the trade press, including by the main competition. Derek Lawford, codirector of PPS, warned in *Chemist & Druggist* that women on the pill were likely to obtain false-positive results. This was problematic, he argued, because 60 percent of the tests his laboratory performed were from "women who were taking, or had recently stopped taking, oral contraceptives," a source of anxiety-related demand that had skyrocketed in the 1960s. A company spokesman told the journal that the packaging and insert would warn buyers that Twentisec was "unsuitable" for women on the pill.[33]

Alongside the pill, *Guardian* journalist John Ezard chimed in, Twentisec completed a "chain of sexual privacy which can now stretch from puberty to menopause, unless of course your 'Twentisec' is positive." But the "publicity-drenched" press conference that launched Twentisec was all a bit much for the Cambridge graduate and former *Oxford Mail* reporter, who derided the "morally torturous" interest taken in the "first do-it-yourself home pregnancy test kit" by the "teenybopper trendsetting magazines, 'Honey,' 'Rave' and 'Fabulous,' the heavy pharmaceutical journals, the agony columnist of 'Woman,' the medical press and Rupert Murdoch's 'Sun.'"[34]

Mostly attended by women, the press conference was presided over by Michael Winstanley, a prominent Manchester GP, broadcaster, columnist, and former Liberal MP who headed a panel of doctors that backed Twentisec as a "valuable addition to screening measures which enable people to learn more about themselves." The manufacturers, Ezard mused, envisioned the ideal user as "intelligent, practical, deft, with a good colour sense, capable of decanting their pre-breakfast urine into a small tube, adding two drops (no more) of reagent and comparing the result against a 'Positive/Negative' chart of eight colours enclosed in the kit. And all of this possible in 20 seconds." "But what," he asked "of another type of girl?"

> What about a dim teenager, living at home, with unstable hormone levels, who tests herself secretly in a back bedroom, becomes one of the 10 in 100 to get a false negative, "takes this as gospel," and later finds herself needing an abortion? Had Dr. Winstanley considered figures on the incidence of suicide in teenage pregnancy? Had he considered that private testing removed a girl from potential contact with any source of adult advice?

Winstanley countered that the "last point was already true of a girl who used laboratories or consulted the British Medical Association's booklets on pregnancy symptoms," but the confrontational tone of the conference convinced Global's public relations consultant to advise pharmacists against selling to "girls shelling out their pocket money after school."[35]

Exceptions like Ezard aside, press coverage was generally favorable. *Penthouse* lauded a "useful new addition to the bathroom cabinet equipment of any considerate Casanova" that could tell a "girl in 20 seconds whether a friendly gesture has developed into an unwanted gestation."[36] But most reports were for and by women. *Observer* medical correspondent Christine Doyle reported on a "booming side effect of the permissive society." She bemoaned the lack of "consumer guidance in choosing a test" in the absence of regulatory oversight, on the one hand, or "instantly recognised brands with a time-tested stamp of authority," on the other. The BMA, unhelpfully, remained "indiscriminately disapproving of the lot," and the "only control seems to come from guidance by the Adverting Association's Investigations Department."[37]

"It seems incredible, even in the 'permissive' society," remarked *Times* women's editor Moira Keenan, that there are "enough pregnant or even possibly pregnant women to support this rapidly growing industry; even more surprising that they should be prepared to pay for a service that any National Health doctor should be able to carry out free."[38] Even for a "nation of do-it-yourself addicts," Keenen felt home pregnancy testing was "carrying things too far"—until a conversation with a "sensible woman doctor" allegedly changed her mind. Home pregnancy testing, she concluded, was "no bad thing." On the contrary, it was

> especially a good thing for all those women who go through mental agony every month because they do desperately want a baby or they desperately don't want one. They tend to think themselves pregnant, symptoms and all, but feel they cannot be continually tagging along to join the queue in the doctor's waiting room. Most doctors would soon lose sympathy and patience, and the doctor I talked to said that anyway most male GPs are "pretty chary of having too much to do with women below the waste." Apart from this it takes up to 12 days to get the result—it could be days of anguish and nailbiting. "You feel an awful fool telling even your husband you might be pregnant and then finding you are not," she says. "At least pregnancy screening provides you with some kind of clue before you face the doctor or telephone all your friends and relations."[39]

At first things seemed to be going "extremely well" for Twentisec.[40] Chemists sold more than 75,000 kits in the first month. By December, however, cracks were beginning to show. The Consumers' Association, an organization dedicated to empowering shoppers through comparative product testing, panned Twentisec as "too unreliable and too difficult . . . to be recommended."[41] And the ASA did not grant approval. The coup de grace came when Winstanley advised taking it off the market on the grounds of "too many false negative results.'" First *World Medicine* and then *Chemist & Druggist* reported that Winstanley had "achieved disappointing figures in tests among his own patients." His advice came in "advance of the results from a London teaching hospital involving 'a large series' which seriously conflicted with the American and earlier British results." Global issued a statement to acknowledge Winstanley's findings

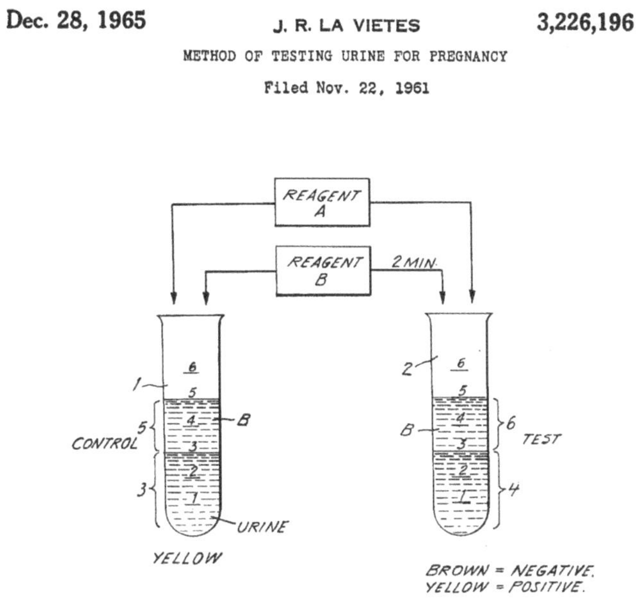

10.9 Detail of a US patent (3,226,196) filed in 1961 for a "chemical" (pH) test of the kind marketed in the US as Ova II and in Britain as Twentisec. Several such patents were filed in the 1960s and 1970s.

and say that it had informed Julius Schmid and was taking "immediate steps . . . to determine the cause of the discrepancies."[42]

Doctors at University College Hospital who tested Twentisec for Global reported 35 percent false negatives and 32 percent false positives; 48 percent of menopausal women also tested positive. On a hunch, they then estimated the pH of all samples from pregnant patients and found that the Twentisec gave (true) positive results for high pH (6.83) and (false) negative results for low pH (6.01). Global informed them that the "main ingredient" in Twentisec was bromocresol purple, a pH indicator. Further testing revealed that positive results could be "made negative by acidifying the urine and conversely any urine can be made to give a positive result by making it alkaline." Twentisec was a test not for hCG but for pH (figure 10.9). Even under "ideal laboratory conditions" Twentisec was "too inaccurate to be used . . . even as a screening test." By the time the damning study was published in the *BMJ*, Winstanley had already advised Global to withdraw Twentisec.[43] It did not remain on pharmacy shelves for long after that.[44] Meanwhile, Moira

Keenan's endorsement of home kits in *The Times* touched off a debate in the *BMJ* that exposed simmering tensions between GPs and hospital laboratories over the rise of "social" pregnancy testing.

SOCIAL REASONS

Provoked by Keenan's article, Anthony Thursz, a West Cumberland consultant obstetrician, warned in the *BMJ* that the NHS could ill afford a "luxury" to become a "routine beginning to a normal pregnancy, for after all, time usually tells." Some NHS laboratories, he complained, were already straining; pregnancy testing at one hospital accounted for an estimated 16 percent of the "entire expenditure on bacteriology."[45] Amplifying Thursz's critique, F. W. Winton, a Dumbarton hospital bacteriologist, insisted that laboratories like his were no place for "routine tests merely to determine whether a normal state of pregnancy exists—solely for 'social' reasons." Such tests, he suggested, belonged in general practice "with suitable reimbursement to the doctor" or in a private laboratory. Winton's laboratory had performed 4,840 tests in the past twelve months, including 228 pregnancy tests in the past two weeks. Of these, 163 (71.5 percent) were for "apparently social reasons," and 65 (28.5 percent) were for "medically acceptable diagnostic purposes." Only 25 requests came from hospitals; the remaining 203 came from GPs, of which only 50 (19.7 percent) were for "justifiable patient-care reasons." The expenditure, 11.4 percent of his lab's budget, represented a "considerable wastage" of time, cost, and materials. Hospital labs, Winton concluded, had more pressing duties and could do without the interruption of some forty telephone calls a week for results.[46]

But times were changing, and in contrast to hospital workers, GPs increasingly sympathized with their anxious patients. To them, Winton seemed out of touch. W. G. Benson, an Exeter GP, conceded that "excessive demand" confirmed controversial Conservative MP "Enoch Powell's thesis that a free Health Service will inevitably lead to an unlimited demand on that service" but nevertheless maintained that "social" reasons could also be "valid" reasons. Much medical work, "including pathology," was, after all, "not strictly necessary" but done "willingly for some such reason as reassurance of our patients." Most GPs probably preferred patients to "consult at a stage when pregnancy can be confirmed clinically, but nowadays they seem to present only a few days after they have missed a period—and they expect an answer." To Benson, this was "understandable," and refusal risked damaging his "rapport" with patients.[47]

Even with a modest patient list, George Grant of Jarrow received "frequent requests" for pregnancy testing. To avoid incommoding the hospital laboratory, he performed the tests himself "while the patient waits" and accepted the cost (25p a test or around £18 a year) as a "normal practice expense."[48] But not all could afford to underwrite pregnancy testing. Jeffrey Segall would have liked to carry out pregnancy testing at his North London surgery, "if the small cost were reimbursed or could be met by the patient," but did not have the means. Macpherson Knowles of Worcester objected that Winton simply had no idea of the "strong emotion felt by a woman awaiting the result of a pregnancy test." He questioned the distinction between the "medical" and the "social," categories that echoed those invoked in a related, parallel debate over the limits of legal abortion under the "social clause" of the 1967 act.[49] Pregnancy testing was always about patient care. GPs were concerned with not only the "bacteriological and pathological states" observed in the laboratory but also the "hopes and fears" of patients. Knowles took "great pride" in "efficient, prompt, and courteous pregnancy diagnosis." This was not an "interruption of . . . real work" but an "important" part of the NHS and of the "greatest value" to patients.[50]

A Huddersfield practitioner agreed with Winton that "many pregnancy tests pander to the impatience of the modern woman to know the worst before nature in the shape of a second missed period gives the answer" but insisted on the "extreme urgency" and humaneness of early diagnosis on companionate and pragmatic grounds if there was any question of abortion:

> Without this test how can the general practitioner set in train the elaborate machinery (referral to a psychiatrist or organization of a trip to London or Birmingham) necessary to remedy the unfortunate patient's condition? Then there are the harassed mothers on the pill who miss a period. Social reasons these may be for this test (in my experience the most used of any laboratory procedure) but as valid surely as any if the patient's good is the primary consideration.[51]

Laboratory personnel, meanwhile, rallied in support of Winton. Increasing demand for pregnancy testing, and laboratory analysis more generally, "disturbed" pathologists at City Hospital, Aberdeen. Their laboratory covered bacteriology, hematology, clinical biochemistry, and toxicology.

In 1960, they analyzed a total of 67,018 specimens, including 1,978 (2.9 percent) for pregnancy testing. By 1970, the total had doubled to 134,433, including 6,221 (4.6 percent) for pregnancy testing. They expected pregnancy tests to exceed regional births by 1971 and by 1990 to equal the "total for women (married and otherwise) between the ages of 15 and 44." What drove this increasing demand, they asked. Was it the pill? The Abortion Act? The "greater freedom in premarital sex relationships"? Or was it "just another result of the 'free' facilities of the Welfare State"? In 1970, the laboratory had spent on pregnancy testing some £1,200 in reagents and £500 in postal charges, not to mention containers, technicians' hours, office time, stationery, and so on. There was "no abatement in demand"; projecting forward from the first ten months of 1971, they predicted an annual total in excess of seven thousand tests.[52]

Such exasperation was widespread. Family doctors' requests for all kinds of laboratory investigations, including pregnancy tests, had in fact doubled from around 22 million in 1961 to 45 million in 1971, straining a health system that would be facing a major financial crisis by the mid-1970s.[53] Overworked laboratory staff who did not come face-to-face with patients but were at the receiving end of urine specimens could be forgiven for dismissing pregnancy testing as a medically inessential "social" service that pandered to "impatient" women. They perceived the abuse of "free" laboratory analysis as a general problem, but pregnancy testing more than any other diagnostic procedure drove a wedge between GPs and hospital laboratories.

The debate in the *BMJ* over "social" pregnancy testing culminated in a lead article that attributed the "sharp increase" in demand to "greater sexual freedom." Delays in obtaining a test result fed a thriving commercial industry that since 1966 had been formally permitted to directly advertise to women. Despite these changes, however, the BMA's position had not wavered: directors of commercial labs should be properly, although not necessarily medically, qualified; the "clinical interpretations of a laboratory finding should be given only by a registered medical practitioner"; and a "positive result should be communicated immediately to the patient's own doctor."[54]

These three conditions reinforced the BMA's understanding of pregnancy testing—and "any action that may follow it"—as fundamentally involving "medical considerations." As with abortion, the demarcation between "medical" and "social" reasons for pregnancy testing was blurring. But the "doctor's job" was to advise patients on "every aspect of pregnancy—its avoidance, whether it is present and the action to be taken if it is—on the basis of all the information he needs, pathological tests included." It was in this paradigm that the BMA considered the launch of Predictor in November 1971:

> This is claimed to be a simple, accurate, and early test which women can perform on themselves before visiting their doctor. But the fact that a woman knows she might be pregnant implies that she should have medical advice whether the result is positive or negative. She may need advice on contraception if this is not being used or has failed; a decision on possible abortion if the right indications are present; support by the social agencies and possible adoption if an unmarried girl decides to proceed with the pregnancy. And in the relatively rare event of a high titre of human chorionic gonadotrophin being found, specialist help may be needed to eliminate the presence of a hydatidiform mole or a chorionepithelioma.[55]

Provoked by Predictor, as well as the fight between GPs and hospital labs, the BMA doubled down in a final bid to regain control. Pregnancy testing should be performed not in the laboratory, pharmacy, or at home but in the clinic, "where they should be available on request." This was at once a conservative and progressive stance: conservative because it continued to insist on the medical control of pregnancy testing and progressive because it implied free, "social" pregnancy testing on demand. The only stumbling block, as per usual, was the state. "Unfortunately," the article lamented, family doctors had "little encouragement to undertake this additional work." They awaited a "clear-sighted decision by the Department of Health to reimburse the extra expense involved [and] restore pregnancy testing to its rightful context—an essential part of general medical care."[56] The department, however, would not agree to reimburse doctors. Instead, it would adopt a laissez-faire policy that allowed women to continue paying out of pocket for pregnancy tests, including, from November 1971, Predictor.

LAUNCHING PREDICTOR

AKZO created Chefaro to launch Predictor in Amsterdam in May 1971. Other European countries soon followed, and some pharmacies reportedly sold out within hours. Prior to launch, the product had been successfully tested on nearly one thousand women, according to a company spokesman. "Any woman who can follow a cooking recipe, can carry out this test," he added, invoking an analogy with a gendered household skill that would become a commonplace in debates over women's competencies in self-testing.[57] A few months later, Chefaro (UK) would launch the product in Britain, where additional premarket testing and market research were first carried out.

Chefaro asked 188 women to test Predictor "in their own homes," not with urine but with water spiked with hCG and plain water in the control group. The women were to follow the written instructions included with every kit, but to make things more challenging, the thumbnail sketches were omitted (figure 10.10). Nearly all (97 percent) "achieved the test successfully and interpreted the result correctly." Next, the company distributed structured questionnaires to a "representative sample" of 549 women between the ages of fifteen and forty-four, "stratified by region and town." Around two-thirds had heard of laboratory tests and, after being shown a brief description of Predictor, a similar proportion agreed with the statements: "It is an excellent method for checking up before going to the doctor," and "It seems like the answer to the times when I am late and want to check whether I am pregnant or not." Finally, Chefaro hired Cooper Research and Marketing, whose psychologist conducted a series of in-depth interviews to establish "women's attitudes towards pregnancy in general and their feelings towards Predictor." Most interviewees, particularly if they were married, "regarded the doctor as a logical and necessary step in the pregnancy cycle." But they also "knew and felt that he would be unwilling to confirm pregnancy at a less than two week stage of development" and expected to be told, "Come back when you have missed the next period," an "often disappointing" and in "some situations" distressing response. Some women avoided the doctor because they didn't want to seem "neurotic or anxious." Others preferred to keep their sexual lives to themselves.

Interviewees agreed that Predictor was a "logical development" but expressed "clear doubts . . . based on newspaper reports on such products to date," an indirect reference to Twentisec. Married women did not see themselves as the intended consumers, but the interview process disclosed various reasons why they too might purchase a home kit. For one,

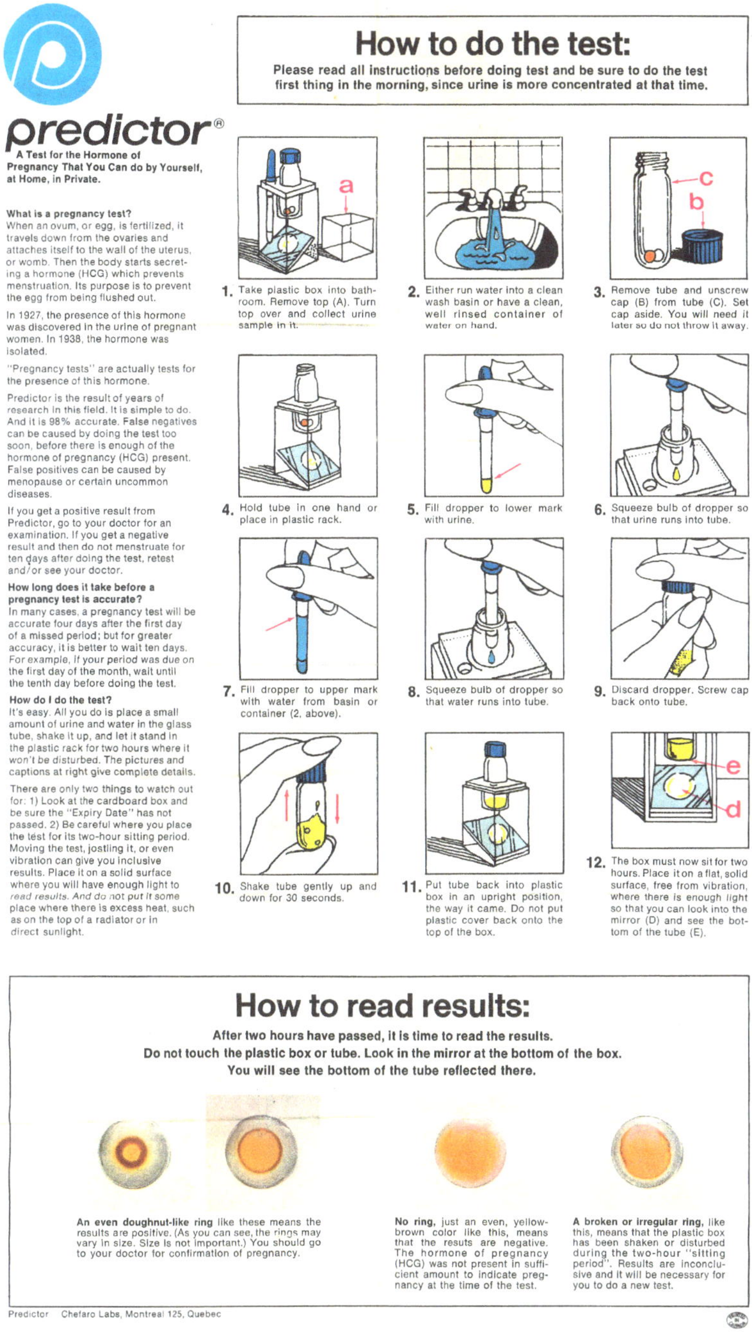

predictor®

A Test for the Hormone of Pregnancy That You Can do by Yourself, at Home, in Private.

What is a pregnancy test?
When an ovum, or egg, is fertilized, it travels down from the ovaries and attaches itself to the wall of the uterus, or womb. Then the body starts secreting a hormone (HCG) which prevents menstruation. Its purpose is to prevent the egg from being flushed out.

In 1927, the presence of this hormone was discovered in the urine of pregnant women. In 1938, the hormone was isolated.

"Pregnancy tests" are actually tests for the presence of this hormone.

Predictor is the result of years of research in this field. It is simple to do. And it is 98% accurate. False negatives can be caused by doing the test too soon, before there is enough of the hormone of pregnancy (HCG) present. False positives can be caused by menopause or certain uncommon diseases.

If you get a positive result from Predictor, go to your doctor for an examination. If you get a negative result and then do not menstruate for ten days after doing the test, retest and/or see your doctor.

How long does it take before a pregnancy test is accurate?
In many cases, a pregnancy test will be accurate four days after the first day of a missed period; but for greater accuracy, it is better to wait ten days. For example, if your period was due on the first day of the month, wait until the tenth day before doing the test.

How do I do the test?
It's easy. All you do is place a small amount of urine and water in the glass tube, shake it up, and let it stand in the plastic rack for two hours where it won't be disturbed. The pictures and captions at right give complete details.

There are only two things to watch out for: 1) Look at the cardboard box and be sure the "Expiry Date" has not passed. 2) Be careful where you place the test for its two-hour sitting period. Moving the test, jostling it, or even vibration can give you inclusive results. Place it on a solid surface where you will have enough light to *read results. And do not put it* some place where there is excess heat, such as on the top of a radiator or in direct sunlight.

How to do the test:

Please read all instructions before doing test and be sure to do the test first thing in the morning, since urine is more concentrated at that time.

a

1. Take plastic box into bathroom. Remove top (A). Turn top over and collect urine sample in it.
2. Either run water into a clean wash basin or have a clean, well rinsed container of water on hand.

c
b

3. Remove tube and unscrew cap (B) from tube (C). Set cap aside. You will need it later so do not throw it away.
4. Hold tube in one hand or place in plastic rack.
5. Fill dropper to lower mark with urine.
6. Squeeze bulb of dropper so that urine runs into tube.
7. Fill dropper to upper mark with water from basin or container (2, above).
8. Squeeze bulb of dropper so that water runs into tube.
9. Discard dropper. Screw cap back onto tube.
10. Shake tube gently up and down for 30 seconds.
11. Put tube back into plastic box in an upright position, the way it came. Do not put plastic cover back onto the top of the box.

e
d

12. The box must now sit for two hours. Place it on a flat, solid surface, free from vibration, where there is enough light so that you can look into the mirror (D) and see the bottom of the tube (E).

How to read results:

After two hours have passed, it is time to read the results.
Do not touch the plastic box or tube. Look in the mirror at the bottom of the box.
You will see the bottom of the tube reflected there.

An even doughnut-like ring like these means the results are positive. (As you can see, the rings may vary in size. Size is not important.) You should go to your doctor for confirmation of pregnancy.

No ring, just an even, yellow-brown color like this, means that the resuts are negative. The hormone of pregnancy (HCG) was not present in sufficient amount to indicate pregnancy at the time of the test.

A broken or irregular ring, like this, means that the plastic box has been shaken or disturbed during the two-hour "sitting period". Results are inconclusive and it will be necessary for you to do a new test.

Predictor Chefaro Labs, Montreal 125, Quebec

10.10 Instructional insert for Predictor, from a Canadian test, clearly showing the thumbnail sketches omitted from Chefaro UK's premarket field testing; courtesy of Crane.

self-testing could alleviate the "embarrassment of false alarms," particularly in menopause, when women were "expected to have doubts over possible pregnancies," but also for "those who wanted a child badly but were having difficulties." It could also fulfill a "desire to share [the experience] with loved ones" and help with planning around work, vacation, housing, or "other family activities."[58]

Predictor went on sale in Britain in November 1971 for £1.75 a kit, just under the going rate for a laboratory test. For the pharmacist, this meant a 57p profit on every item sold. From December, it was supported by a £60,000 advertising campaign aimed at pharmacists and consumers.[59] It was not, however, an overnight success. A retrospective account described the first five years of marketing as "very tough."[60] Initially, only one in five British pharmacists stocked Predictor. In contrast to Dutch sales of 100,000 kits a year, British pharmacists sold a disappointing 30,000 Predictors.[61] The buoyant Dutch market may have enabled AKZO to absorb losses elsewhere until sales picked up. In Britain, Twentisec was partly to blame for poor sales. Predictor launched just around the time Twentisec came under attack, undermining public and professional confidence. To overcome stigma, Chefaro pitched Predictor to pharmacists not as the first home test (it wasn't, if you count Twentisec) but as the "first nationally advertised home pregnancy test you can safely recommend."[62]

Establishing Predictor as a trustworthy product meant persuading pharmacists and consumers of its reliability (figure 10.11). Advertisements in the trade press claimed the "same 99% accuracy that only the best laboratory tests can guarantee."[63] Similarly laid-out ads in the general press addressed women directly: "Are you pregnant?" Comparing the dropper to the "one you get in eye drops," it emphasized the simplicity of the procedure: "all you have to do is put a sample into the test tube, add the fluid from the plastic tube, and let it stand for two hours." And it advised the positive-testing woman to "go and see your doctor. He is the one best able to advise you."[64] This point was further emphasized in the instructional insert, which also "strongly advised" the woman whose repeat test was negative to consult.[65] Alan Giles, the marketing manager for Chefaro, hoped such precautions would win over the BMA. But doctors were not the only critics of Predictor. Boots would not stock home kits until the late 1970s, and some independents objected too.[66]

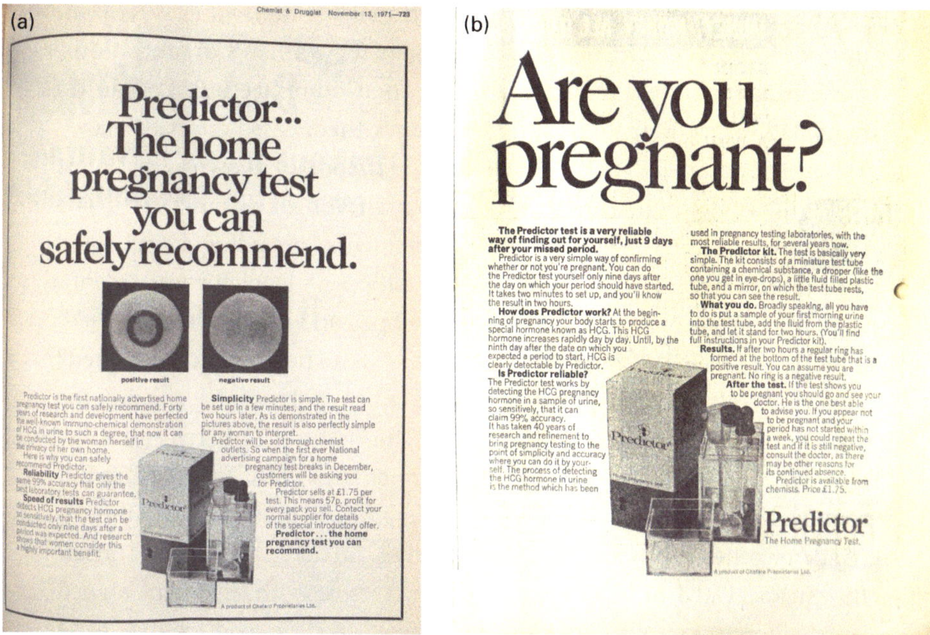

10.11 Marketing to pharmacists and consumers used identical product photography, but differed in other ways: (a) full-page ad for Predictor in *Chemist & Druggist*, Nov. 13, 1971, 723, British Library LOU.658–660; (b) smaller ad in the *Daily Express*, Dec. 7, 1971, 14, here pictured as the same size.

A visit by a Chefaro sales representative provoked a Hertfordshire chemist to reject Predictor in *Chemist & Druggist* as not "in the best national interest." Drawing on still current ideas about "subnormality" within social work and intelligence testing, he foresaw "grave errors occurring by women of sub-normal intelligence or low educational standard due to their inability to understand and follow the instructions" and favored keeping test kits "in the hands of those who have the necessary training and knowledge." Instead of selling Predictor to customers, he proposed charging them the "usual £2" for a test performed by the pharmacist with "no doubt about the result."[67] A more prominent attack came in connection to the Lane Committee, which Sir Keith Joseph, the Conservative secretary of state known for his contentiously Malthusian views, set up in 1971 to investigate the provision of abortion under the 1967 act.[68]

In February 1972, the Women's National Commission (WNC), an umbrella body created by the government for women's organizations, made headlines when its report to the Lane Committee urged a ban on home kits, citing the refusal of Boots to stock them based on inaccuracy.[69] This provoked Chefaro to defend Predictor in a report of its own as the "most widely advertised and distributed Do-it-yourself Home Pregnancy Test." The report explained that apart from "accoutrements" including a "measured dose of distilled water and a convenient dropper," Predictor was "identical" to Pregnosticon All-In, a recently developed and (as the name implied) more self-contained version of Organon's "well-known" and "accepted" tube test. Both were "99 percent reliable," a better "level of accuracy," the report claimed, than any diagnostic test. The "purpose" of Predictor, Chefaro insisted, was "not to disturb the normal DR/Patient relationship but to provide, for those women who want it, an early indication . . . of pregnancy or otherwise which can alleviate their anxiety before they consult their medical advisor." Citing its own market research, the company argued that for "social reasons," many women "welcomed" a "confidential, early indication" and that Predictor could disburden the health service of dealing with "false alarms." Accurate self-testing, the report concluded, would not lead to "unwise action (or abortion)" any more than other available methods and was "*less* likely to lead to the delays in confirmation which concern the WNC."[70]

Accuracy under ideal conditions was of course not the same as accuracy in practice, which, as we have seen with bioassays, could be highly variable. Self-testing added to the equation domestic conditions that were even less controlled than those of the lab, clinic, or pharmacy. So Chefaro enlisted a panel of GPs in Surrey, a Guy's Hospital laboratory technician, and, indirectly, eighty-six possibly pregnant patients to field-test Predictor. The GPs distributed test kits to eighty-six patients and instructed them to return with the result and the tested urine sample for retesting by Pregnosticon and by a "trained nurse" using Predictor. Eighty-three of the women (96 percent) obtained the "correct result," as measured against Pregnosticon. There were "no false positives," and of the three mistaken results, two could be explained by the patient disturbing the kit too soon. Championing self-testing (and Chefaro) in the *BMJ* in 1973, the GPs dismissed "stress factors" (fertility issues, unmarried pregnancy, or contemplated abortion) as irrelevant to accuracy and unambiguously concluded that Predictor, in contrast to the discredited Twentisec, provided "reliable results in the hands of the general public."[71]

Predictor was still the only home kit on the market when the Consumers' Association (CA) surveyed the field in 1974 for *Sex with Health: The Which? Guide to Contraception, Abortion and Sex-Related Diseases* (figure 10.12). After obtaining a laboratory-confirmed perfect score with twenty-two urine samples from pregnant and nonpregnant women, the association proclaimed Predictor accurate. Although "reasonably easy" to use, too much urine or agitation could render the result "useless" and "you would have wasted almost £2" (the cost had gone up to £1.96½). Mindful of not only the consumer's budget but also the BMA's position, the guide endorsed "your GP" as the first port of call "to find out for you whether or not you're pregnant." Failing that, it recommended the nearest branch of the Pregnancy Advisory Service, an abortion provider, a commercial laboratory, or a family planning clinic and presented self-testing as a last resort that could waste the "clumsy" user's money.[72] Nevertheless, in the context of negligible government oversight, the fact that *Which?* gave Predictor a (qualified) pass was a seized on—not only by Chefaro and other champions of home kits but also by government officials and agencies reluctant to allocate resources to reigning in self-testing.

POWERS TO TAKE POWERS

Hindered by the bad press over Twentisec and the reluctance of many pharmacies, including Boots, to stock Predictor, the ill-timed product's breakthrough came belatedly, in 1976, following the endorsements in the *BMJ* and *Which?* These crucially reassured potential advertisers that Predictor was not another dud, thus removing barriers to a more confident, larger-scale marketing campaign that would not be opposed by the ASA.[73] Chefaro pulled out all the stops. Six months of advertising in the capital—in underground trains, on buses, and in the *Evening Standard*, *Girl about Town*, and *Time Out*—delivered a "phenomenal increase" in sales, especially in and around London (figure 10.13).

The following year, Chefaro took out full-page ads in *Miss London Weekly*, *Time Out*, and *Cosmopolitan*. Outside London, it distributed window stickers,

Pregnancy testing 54

Conclusions

Ideally, your GP should be the person to find out for you whether or not you're pregnant. If for any reason he's not, the free test provided by the Pregnancy Advisory Services should be satisfactory, if you're near one of their branches.

However, if there isn't a branch near you, or you don't want to find out whether you're pregnant in an abortion-oriented atmosphere, you could use a commercial laboratory or a family planning clinic (but phone the clinic first, to check they do pregnancy tests). There's nothing much to choose between them for price, so go to whichever is handiest.

If you don't want to go to either of these for any reason, you could use the do-it-yourself Predictor. But you'd have to use it very carefully – if you're at all clumsy with it, you could get a wrong result, and waste almost £2.

Whatever method you use, remember they're not 100 per cent reliable. And there are a few rules to follow:

- wait until your period is at least two weeks late
- collect a small specimen of urine (about two tablespoonfuls) first thing in the morning, in a completely clean bottle (especially not soapy)
- send the sample immediately, enclosing the fee, a note of your age (particularly important for the over-40s), and the date of your last period
- if the result is negative, and your next period still doesn't come, go and see your doctor
- if the result is positive, go to see your doctor. If you want an abortion, read the next section for the best ways to go about it.

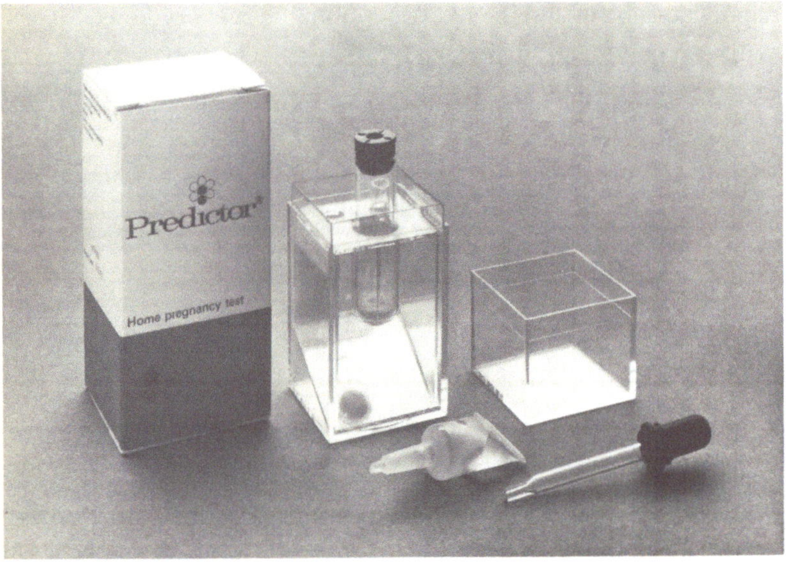

10.12 Product photograph of Predictor disassembled, probably supplied to the Consumers' Association by Chefaro. *Sex with Health: The Which? Guide to Contraception, Abortion and Sex-Related Diseases* (London: CA, 1974), 54, Cambridge University Library L448.c.23.17, with permission by Which? Ltd., May 10, 2022.

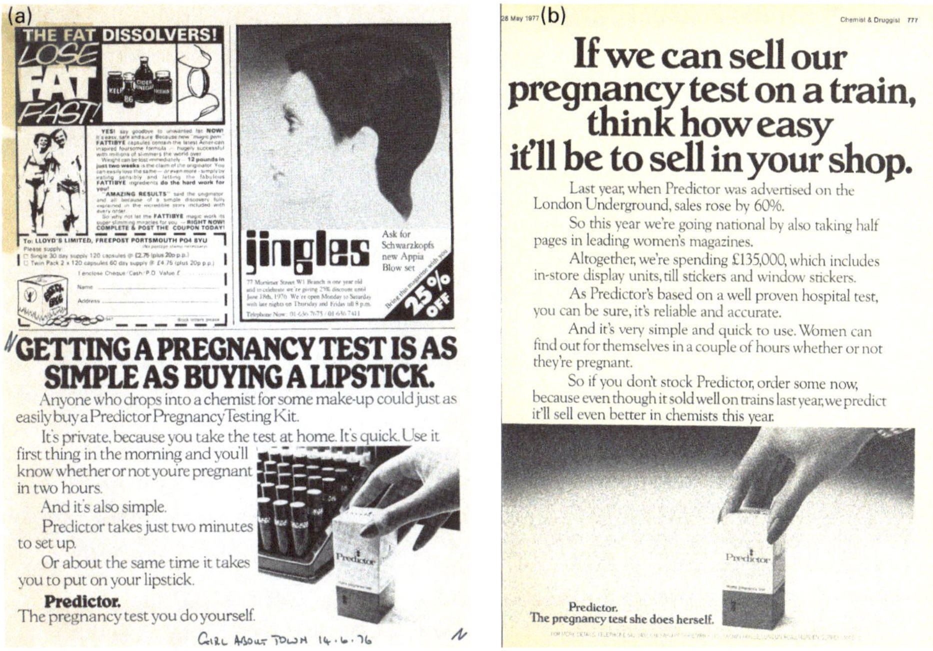

10.13 (a) Half-page advertisement comparing the ease of buying a Predictor to that of a lipstick in the trendy London magazine *Girl About Town*, Jun. 16, 1976, The National Archives MH 156/633. (b) Full-page ad in *Chemist & Druggist*, May 28, 1977, encouraging chemists to stock Predictor based on the success of the previous year's marketing campaign and sales projections, British Library LOU.697.

till stickers, and display units emblazoned with the words "private, reliable, simple, early—the pregnancy test you do yourself" and "Available here" to circumvent the provincial consumers' supposed "natural embarrassment."[74] Nationwide marketing was expensive. Chefaro shelled out £135,000 on its biggest campaign yet, more than double what it had spent launching Predictor. Bullish copy in *Chemist & Druggist* encouraged retailers who did not already stock Predictor, "The pregnancy test she does herself," to "order some now, because even though it sold well on [underground] trains last year, we predict it'll sell even better in chemists this year."[75]

Although commercially successful, Chefaro's use of the London Underground for marketing purposes made enemies in high places. Behind the scenes, Walter Holland, a professor of clinical epidemiology and social medicine at St. Thomas's Hospital, had enough clout to set government deliberations in motion when he complained to Sir Anthony Royle, his MP in leafy Richmond, about an advertisement for Predictor he had seen in a train. A "false answer," he feared, "could cause great distress."[76] Holland's letter of July 1976 reverberated in the corridors of power, where the Lane Committee's recommendations on pregnancy testing (a ban on home kits and licensing for commercial laboratories) were still being debated by the Select Committee set up in 1975 to discuss James White's Amendment Bill.[77]

British law permitted the sale of home test kits even without disclosure of premarket testing. Legislation would not cover medical devices until the 1990s and then only in compliance with European Union law; diagnostic test kits eluded control until 2000.[78] Post-thalidomide drug efficacy and safety regulation belatedly came into effect in 1971, with the implementation of the Medicines Act of 1968. This created the Medicines Division, a national regulatory agency within the Department of Health authorized to confirm or deny approval to new drugs.[79] Predictor, as an apparatus, was not covered by the act, and ministers regarded pregnancy testing as a "relatively low priority" for the "overloaded" agency. So, while powers existed in the act to "extend its controls," they agreed in January 1976 that the Medicines Division "should take on no fresh commitments." To evade responsibility, the agency cited the Consumers' Association's (qualified) endorsement of Predictor to (plausibly) claim "insufficient hard evidence of any adverse effects," not to mention "other pressing problems" that took "priority of our time and resources."[80]

William Roberston, a medical researcher at Leeds General Infirmary who advised the Medicines Division on pregnancy testing, failed to see the problem. A positive result merely indicated to a woman when to "take medical advice" and did not cause "great harm" if mistaken. He summed up the situation in August, following Holland's complaint. He did not rule out the possibility of "exploitation" by disreputable abortion providers, but this was "not relevant for Medicines Act purposes." So, in the absence of "hard evidence" that Predictor posed a "health hazard," Robertson found it "difficult to advise" control.[81]

To act, the Medicines Division would need evidence that Predictor was less accurate than "ordinary path tests." But, Robertson reasoned, "path could only be an element on which a clinician [exercised] judgment," and "even the most rigorous hospital laboratories could never guarantee 100% accuracy." If Predictor fared no worse than Pregnosticon, then the "case would be that much weaker." Pregnancy testing, moreover, distracted from the more pressing task of "clearing applications for new medicines, a commitment sought by Industry and agreed by Ministers."[82] Concerns about "distress" possibly caused by false results did not justify diverting resources away from evaluating "new medications which might constitute a physical health risk." Easily persuaded by Robertson's expedient case "against exercising the powers to take powers," ministers delegated the matter; if Holland felt the offending advertisement was in any way "misleading," he was free to take it up with the ASA.[83]

Holland's complaint ultimately came to nothing, but not before reaching all the way up to the recently appointed Labour minister of state for health, David Owen. Holland's MP, for his trouble, received a letter signed if not actually written by Owen, explaining that because Predictor was "not subject to official controls," the Department of Health did not know by which criteria Chefaro had judged it to be market ready. Owen conceded that false results "could cause distress" but expected consumers to have "other grounds for assessing reliability" and to "seek medical advice" following a "positive result or doubtful negative." The CA had assessed Predictor as giving "reasonably accurate results if instructions are strictly followed." This was good enough for Owen, who had seen no evidence that self-testing constituted a "hazard to physical health." The department would not consider taking steps to control home test kits. "At present," the letter concluded, "we are giving priority, within our limited resources, to the assessment of new medical products."[84] Predictor now had the state's assent.

In a final, futile effort to set legislative gears in motion, Holland wrote to his old friend John Butterfield, the Regius professor of physics (medicine) at the University of Cambridge who also happened to be on the Medicines Commission, an advisory body created by the Medicines Act.[85] Although unsure whether the act covered pregnancy testing, Butterfield affably agreed to assist Holland in his "crusade."[86] About a month later, he reported back that the overburdened commissioners had not "taken any policy decision," nor would they unless some health hazard was involved. A statutory ban would be "less demanding of resources" but also had drawbacks. Namely, it would affect all test kits, "whether efficient or otherwise, and logically would also involve the banning of advertising of pregnancy testing laboratories." Because there was "no power" under the act to bring such laboratories "under control," a ban on advertising would create the unsatisfactory situation wherein only those already in the know would "have recourse to them."[87] Holland gave up after that.[88]

WOMEN'S LIBERATION

There can be no doubt that Predictor was a significant landmark. However, it is easy to ascribe too much revolutionary significance to the advent of self-testing. As we have seen, women were already obtaining test results without the consent of doctors when Predictor debuted in 1971. And many women failed to perceive the domestic privacy offered by Predictor as a clear advantage over alternatives or worth the expenditure. Some refused all forms of mediated knowledge or medical advice. For example, Mandy Green, a hairdresser interviewed by Ann Oakley in the 1970s, refrained from going to the doctor even when she "half felt" pregnant because "if I get to know too soon it's going to be an awfully long nine months. So I didn't want to go any earlier."[89] Early detection was not for everyone. Perhaps more surprisingly, activists in the women's liberation movement viewed Predictor as a form of commodification and instead promoted their own feminist alternative.

As I have shown elsewhere, women's groups around Britain organized free, at-cost, or by-donation drop-in sessions that combined pregnancy testing with sympathetic counseling, sex education, contraceptive advice, and referrals to providers of abortion or prenatal care.[90] They used slide tests (provided in some instances by a sympathetic doctor) to demystify medicine and empower women through supportive, nonmedical, and noncommercial services (figure 10.14). The local women's center in Bristol, for example, performed around five hundred tests (in a one-room basement kitchen) in 1976, on par with the local hospital's eight hundred tests on samples sent by family planning clinics. The volunteer testers tried to cultivate a "friendly informal atmosphere" and saw themselves as filling a "real gap in the N.H.S."[91] They disagreed over whether the aim should be to pressure the health service to emulate the model they'd developed or to continue as a more radical alternative. They did not embrace self-testing.

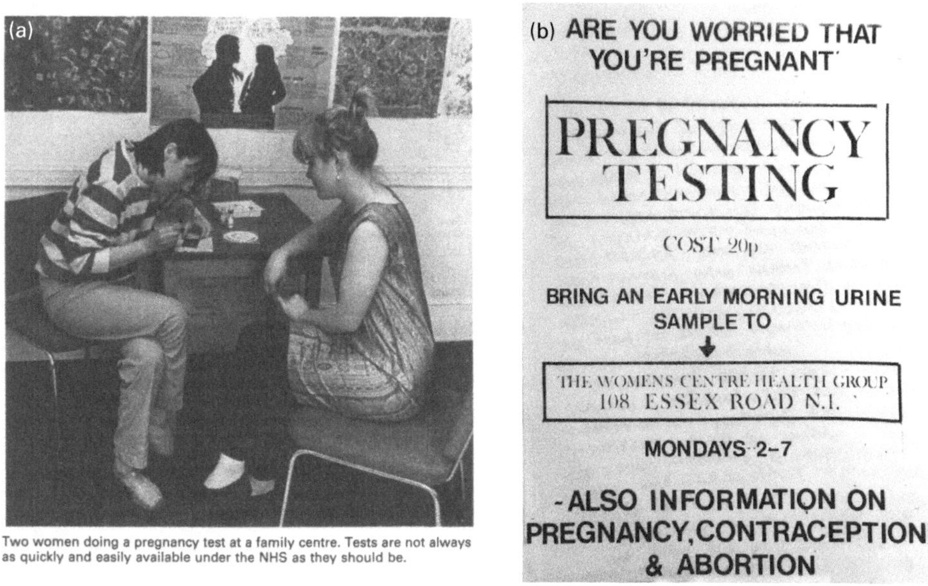

10.14 (a) Pregnancy testing as women helping women; the test kit appears to be Pregnosticon. Photo: Gina Glover (Photo Co-op). Davidson and Rakusen 1982, 60, Cambridge University Library 9323.c.897. (b) Leaflet produced by the Essex Road Women's Centre, *c.*1973, Women's Library, LSE 5ERC/5/1.

Unloved by the women's liberation movement, Predictor should be seen as part of a larger process of commercialization that diminished medical authority, transformed patients into consumers, and consolidated pharmacies as sites of reproductive choice. Its history throws into relief dimensions of the "permissive society" that are less apparent in histories of contraception or abortion. More than the pill or abortion law reform, home pregnancy tests point to retailers and customers as agents of change. But key social and technical innovations prepared the way, especially the small-time entrepreneurs that, from 1965, did so much to commercially sustain a demand that made self-testing thinkable. Crucially, despite the availability of Predictor from 1971, self-testing did not become the norm until much later. It is to the market that Predictor made and the drawn-out process of normalization that I next turn.

11

BLUE LINES

The politically tumultuous 1980s saw an intensification of public interest in issues around human procreation.[1] The rise of fetal monitoring, including through amniocentesis (to detect Down syndrome) and routine ultrasound, refashioned the fetus as a visible, supervised, and public entity and every pregnancy as potentially risky.[2] From the highly publicized birth in 1978 of Louise Brown, the first "test-tube baby," to the Human Fertilisation and Embryology Act of 1990, the tabloid press, parliamentary debate, and lobbying focused scrutiny on in vitro fertilization (IVF).[3] Margaret Thatcher's Conservative Party, elected in 1979, emphasized "family values," and a series of private members' bills and the fledgling "prolife" movement challenged aspects of legal abortion.[4] Medical concerns about risks to the fetus increasingly emphasized maternal "lifestyle," with advice correspondingly prescribing vitamin supplements and proscribing cigarettes and alcohol.[5] Through this period, a rapidly expanding market for home pregnancy tests and, from 1986, home ovulation tests defined a new category of consumer: the thirtysomething career woman whose "biological clock" was ticking.[6]

The previous chapter reconstructed the British reception of Predictor. Although initially sluggish, sales improved following a nationwide marketing campaign after restrictions were relaxed in 1976. Two years later, Predictor and equivalent products were finally allowed on the US market. In Britain, Predictor reigned unchallenged until 1979, when Carter Wallace launched Discover 2. These were joined in 1985 by Clearblue, the brand that did the most to reinvent and expand the market that Predictor made. Covering the four-and-a-half decades since 1979, this chapter reconstructs the decline of Predictor and testing by pharmacists, on the one hand, and the rise of Clearblue and own-label (store-brand) test kits, on the other. It thematizes the limits and ironies of technological progress as well as the commercial process that cultivated new kinds of demand, especially from women hoping to conceive, and consolidated self-testing as the new normal.

TWO FOR ONE

Carter Wallace, a US company with UK headquarters in Kent, backed its November 1979 launch of Discover 2 with a £250,000 promotional spend in women's magazines. By then, the product was already a brand leader in Italy, France, and Germany and had a 37 percent share of the more competitive US market. As the name indicates, Discover 2 innovatively retailed in packs of two. The instructions encouraged testing as early as five days after a missed period, compared to eight days for Predictor, but with the caveat that earlier tests were correct only three times in four—hence the second test, for confirmation.[7] Initially priced at £4.95 for a two-pack, Discover 2 would have seemed like a bargain compared to a single Predictor, which retailed for £4.50 in 1980.[8]

Despite a steady drip of advertisements in women's magazines, the market, which centered on London and the affluent southeast of England, was far from saturated in 1980. A letter in the *Lancet*, for instance, reported that only four in one hundred consecutive patients booking prenatal classes in a Sheffield women's hospital had used Predictor. In comparison, thirty-seven had been tested by a hospital lab, thirty-three by a pharmacist, twelve by a GP, and fourteen had no test.[9]

Advice books, meanwhile, were ambivalent about self-testing. *Woman's Own Birth Control* (1980) rated Predictor as just as "accurate as a chemist's test" if the instructions were followed "to the letter" but also warned that a positive result could be "very upsetting"; it directed readers to a family planning clinic, where testing was accompanied by "helpful and professional advice."[10] Natural childbirth activist Sheila Kitzinger's *Pregnancy and*

Childbirth (1980) advised readers to get tested before the conventional two periods had been missed, either by taking a urine sample to the GP or pharmacist or by using a home kit, in which case it was "important" to follow the instructions "very carefully" and "best" to wait two weeks after a missed period.[11] Somewhat unusually, Margaret Thom's *Having a Baby* (1981) favorably contrasted the domestic privacy of self-testing with mediated alternatives that could leave a woman feeling as though her possible pregnancy had "been announced on BBC News."[12]

In addition to privacy, home tests could be used a few days earlier than the two weeks after a missed period typically advised by laboratories. Interviews conducted by Ann Oakley in the 1980s suggest that some women perceived this as an advantage, especially if they had a particular reason to know as early as possible. For example, Dorothy Alexander, a young married woman who conceived soon after coming off the pill in 1980 and worried about its residual effects on her pregnancy, waited no more than a week before doing a home test, "which was positive."[13]

A review of pregnancy testing in the February 1982 issue of *Which?* conceded that a woman might "want to find out as early as possible" for her own "peace of mind" or to be "as sure as possible . . . before deciding who to go to for advice" but cautioned against placing "*absolute* faith" in "fiddly" home tests. The twenty women who tried Predictor (now £4.75) and Discover 2 (£5.25) for *Which?* "slightly preferred" the former, but the review concluded, there wasn't "much to choose" between them.[14] Professional pregnancy testing services were still going strong, and the Pharmaceutical Society expected NHS cutbacks to further consolidate pharmacies as the preferred site.[15]

Although confidentiality was of paramount importance, pharmacists were required to keep written records for one year before disposing of them, and in May 1983, the *Sun* reported that files giving names, addresses, and pregnancy test results were found dumped outside a pharmacy in Old Broad Street, City of London.[16] But such transgressions were rare, and women were increasingly comfortable taking their urine to the pharmacist in a variety of containers, including prescription vials, Lucozade bottles, and Pan Yan pickle jars. This "appalled" Edwin Evens, an Essex pharmacist, who complained in the *Pharmaceutical Journal* that pharmacists, by accepting "any old bottle," would damage the "professional image" of pregnancy testing. Still more concerning was the "astonishment not only of the patient but of the staff in many cases" when he refused to accept a substandard receptacle. Evens would only accept a sample if tendered in a new container, preferably a 120-mL glass ointment jar, which he had personally labeled with the client's name and the date.[17]

Meanwhile, intensifying competition between Predictor and Discover 2 resulted in reformulations of both products and the launch of Confirm, a two-test version of Predictor. In 1984, when the *Times* announced the "birth of a baby book boom" (figure 11.1), hospitals performed 1.4 million tests; pharmacists, clinics, and pregnancy advisory centers completed a further 850,000; and women purchased 400,000 home tests, bringing the total up to 2,650,000. Boots, which had started selling home tests in the late 1970s, overtook independents in sales, in part because many women perceived the large chain as more anonymous. Market research indicated that consumers appreciated and were willing to pay for the "speed" and "convenience" of a home test that saved them from making a doctor's appointment or waiting two weeks for the result of a hospital test. Moreover, the improved sensitivity of home tests had widened the gap to between three days, for the most sensitive home test, and eight or twelve days, for a standard hospital test.[18]

Chefaro and Carter Wallace continued to spend around £200,000 a year, mostly on ads in women's magazines—advertising rules proscribed marketing home pregnancy tests in broadcast media—but national surveys suggested that only around 60 percent of "women of child-bearing age" even knew they existed (fewer than had heard of lab tests in 1971), and only around 12 percent had used one.[19] Just one of thirty young Afro-Caribbean women interviewed in 1983 for the Training in Health and Race project used a home test; the majority (twenty-three) relied on a GP.[20] As reflected in and to some extent produced through advertising campaigns that targeted some groups more than others, the market skewed toward white, middle-class metropolitans.[21]

Companies sold more than a test result; they also brokered the idea of a healthy pregnancy that began with early detection. The marketing apparatus for Predictor included *Guide to a Healthy Pregnancy*, a free booklet that came with the product; an informational leaflet inserted in *Pulse*, a newspaper for doctors; and a help line for customers that received about two

THE TIMES MONDAY JULY 30 1984 11

MONDAY PAGE

Birth of a baby book boom

Guides for expectant mothers are flooding the market – but how good are they?

Rachel Cullen reads between the lines

'Some pregnant women have an insatiable desire for knowledge'

'Sheila Kitzinger presents a sense of glory in having a baby'

Daunting decision ... just some of the books on offer to the expectant mother

FIRST PERSON

I've lost my mantra but found how to keep cool

Clare Colvin

11.1 The pregnant woman in profile is reading Sheila Kitzinger's *Birth Book* and Lennart Nilsson's *A Child Is Born* can be seen in the stack. Rachel Cullen relegated the latter to a "class of books bought by expectant fathers and hated by expectant mothers" at a time when Janet Balaskas's "active birth" was the "latest rage" and Sheila Kitzinger, who launched a movement in 1962 with *The Experience of Childbirth*, remained "in a class by herself." See Al-Gailani 2018, 564; on Nilsson's bestseller: Jülich 2015; Kitzinger: Raphael 2010. *Times*, Jul. 30, 1984, 11, Cambridge University Library NPR.B.462, with permission by The Times/News Licensing.

dozen calls a week. Carter Wallace maintained its own phone service and introduced permanent display stands to encourage pharmacists to keep for Discover 2 in the front of the shop. Women, reported *Chemist & Druggist* in 1984, were becoming "less embarrassed about revealing that they might be pregnant," but it was still important to keep home tests on display, so that customers could buy them "without having to ask."[22] Newly commercialized biotechnologies, meanwhile, were poised to transform the market.

GROWING THE MARKET

Alongside recombinant DNA technology, the hybridoma technique of producing monoclonal antibodies (MABs) played a central role in the development of the biotechnology industry on both sides of the Atlantic in the 1980s and 1990s. Developed in Cambridge, England, in 1975 and controversially patented a few years later by American scientists, MABs failed to transform cancer therapeutics as initially hoped. But they did have a major impact on the faster-paced and more lucrative diagnostics sector, which did not require lengthy clinical trials to bring products to market. Not least because their "polyclonal" predecessors had cleared the way, MAB-based pregnancy tests were among the most commercially successful products of the biotech boom (figure 11.2).[23]

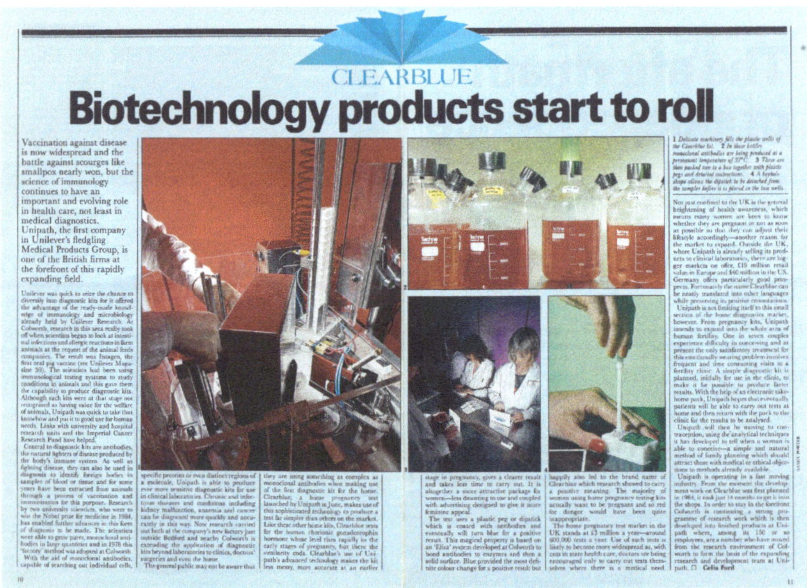

CLEARBLUE

Biotechnology products start to roll

Vaccination against disease is now widespread and the battle against scourges like smallpox nearly won, but the science of immunology continues to have an important and evolving role in health care, not least in medical diagnostics. Unipath, the first company in Unilever's fledgling Medical Products Group, is one of the British firms at the forefront of this rapidly expanding field.

Unilever was quick to seize the chance to diversify into diagnostic kits for it offered the advantage of the ready-made knowledge of immunology and microbiology already held by Unilever Research. At Colworth, research in this area really took off when scientists began to look at intestinal infections and allergic reactions in farm animals at the request of the animal feeds companies. The result was Intagen, the first oral pig vaccine (see Unilever Magazine 50). The scientists had been using immunological testing systems to study conditions in animals and this gave them the capability to produce diagnostic kits. Although such kits were at that stage not recognised as having value for the welfare of animals, Unipath was quick to take that knowhow and put it to good use for human needs. Links with university and hospital research units and the Imperial Cancer Research Fund have helped.

Central to diagnostic kits are antibodies, the natural fighters of disease produced by the body's immune system. As well as fighting disease, they can also be used in diagnosis to identify foreign bodies in samples of blood or tissue and for some years have been extracted from animals through a process of vaccination and immunisation for this purpose. Research by two university scientists, who were to win the Nobel prize for medicine in 1984, has enabled further advances in this form of diagnosis to be made. The scientists were able to grow pure, monoclonal antibodies in large quantities and in 1978 this 'factory' method was adopted at Colworth.

With the aid of monoclonal antibodies, capable of searching out individual cells, specific proteins or even distinct regions of a molecule, Unipath is able to produce ever more sensitive diagnostic kits for use in clinical laboratories. Chronic and infectious diseases and conditions including kidney malfunction, anaemia and cancer can be diagnosed more quickly and accurately in this way. Now research carried out both at the company's new factory just outside Bedford and nearby Colworth is extending the application of diagnostic kits beyond laboratories to clinics, doctors' surgeries and even the home.

The general public may not be aware that they are using something as complex as monoclonal antibodies when making use of the first diagnostic kit for the home. Clearblue, a home pregnancy test launched by Unipath in June, makes use of this sophisticated technology to produce a test far simpler than others on the market. Like these other home kits, Clearblue tests for the human chorionic gonadotrophin hormone whose level rises rapidly in the early stages of pregnancy, but there the similarity ends. Clearblue's use of Unipath's advanced technology makes the kit less messy, more accurate at an earlier stage in pregnancy, gives a clearer result and takes less time to carry out. It is altogether a more attractive package for women—less daunting to use and coupled with advertising designed to give it more feminine appeal.

The test uses a plastic peg or dipstick which is coated with antibodies and eventually will turn blue for a positive result. This magical property is based on an 'Elisa' system developed at Colworth to bond antibodies to enzymes and then a solid surface. Blue provided the most definite colour change for a positive result but happily also led to the brand name of Clearblue which research showed to carry a positive meaning. The majority of women using home pregnancy testing kits actually want to be pregnant and so red for danger would have been quite inappropriate.

The home pregnancy test market in the UK stands at £3 million a year—around 600,000 tests a year. Use of such tests is likely to become more widespread as, with cuts in state health care, doctors are being encouraged only to carry out tests themselves where there is a medical need.

1 Delicate machinery fills the plastic wells of the Clearblue kit. 2 In these bottles monoclonal antibodies are being produced at a permanent temperature of 37°C. 3 These are then packed two to a box together with plastic pegs and detailed instructions. 4 A keyhole shape allows the dipstick to be detached from the sampler before it is placed in the two wells.

Not just confined to the UK is the general heightening of health awareness, which means many women are keen to know whether they are pregnant or not as soon as possible so that they can adjust their lifestyle accordingly—another reason for the market to expand. Outside the UK, where Unipath is already selling its products to clinical laboratories, there are bigger markets on offer, £19 million retail value in Europe and $40 million in the US. Germany offers particularly good prospects. Fortunately the name Clearblue can be neatly translated into other languages while preserving its positive connotations.

Unipath is not limiting itself to this small section of the home diagnostics market, however. From pregnancy kits, Unipath intends to expand into the whole area of human fertility. One in seven couples experience difficulty in conceiving and at present the only satisfactory treatment for this emotionally wearing problem involves frequent and time consuming visits to a fertility clinic. A simple diagnostic kit is planned, initially for use in the clinic, to make it be possible to produce faster results. With the help of an electronic take-home pack, Unipath hopes that eventually patients will be able to carry out tests at home and then return with the pack to the clinic for the results to be analysed.

Unipath will then be moving to contraception, using the analytical techniques it has developed to tell when a woman is able to conceive—a simple and natural method of family planning which should attract those with medical or ethical objections to methods already available.

Unipath is operating in a fast moving industry. From the moment the development work on Clearblue was first planned in 1984, it took just 14 months to get it into the shops. In order to stay in the forefront Colworth is continuing a strong programme of research work which is then developed into finished products at Unipath where, among its 150 or so employees, are a number who have moved from the research environment of Colworth to form the basis of the expanding research and development team at Unipath. □ **Celia Ford**

10

11

11.2 Two-page spread in Unilever's in-house magazine offering a rare glimpse of the production line for Clearblue, set against boldly contrasting red. The caption reads: (1) "Delicate machinery fills the plastic wells of the Clearblue kit"; (2) "In these bottles monoclonal antibodies are being produced at a permanent temperature of 37°C"; (3) "These are then packed two to a box together with plastic pegs and detailed instructions"; (4) "A keyhole shape allows the dipstick to be detached from the sampler before it is placed in the two wells." Ford 1985, 10–11, with permission by Unilever PLC.

Organon (Teknika), the world's leading supplier of pregnancy tests to the professional market, launched Neo-Planotest Duoclon in 1984. An MAB-based slide test sensitive enough to detect low levels hCG as early as four days after a missed period, it combined the sensitivity of a tube test with the speed of a slide test.[24] In the same year, Monoclonal Antibodies Inc., a California startup founded in 1979, introduced Pregnastick, a plastic "dipstick" test, sensitive enough to detect still lower levels of hCG at about the time of the first missed period.[25] Also based on MABs, Pregnastick avoided cross-reactivity with luteinizing and follicle-stimulating hormones as well as interference from proteins such as hemoglobin. A color change from white to blue of the dipstick within twenty minutes indicated a positive result. Alpha Laboratories, a small family-run business in Hampshire, began distributing Pregnastick to British pharmacists in June 1984.[26] Clearblue, the first home pregnancy test to incorporate MABs, followed a year later.

Launched in June 1985 by Unipath, a subsidiary of the Anglo-Dutch conglomerate Unilever, Clearblue innovatively brought monoclonal, dipstick, and color-change technology to the consumer market.[27] Unilever had been developing the use of mouse MABs in laboratory diagnostics at its Colworth site since 1978 and had in 1983 set up Unipath in a converted warehouse on the outskirts of Bedford to begin work on a line of products for use in clinics and in homes. There, researchers developed an enzyme-linked immunosorbent assay (ELISA) system that would form the basis for Clearblue. The third home pregnancy test to enter the British market after Predictor and Discover 2, Clearblue uniquely eliminated the need for urinating into a cup. Instead, the user held a plastic sampling rod or "bucket" collection system in the urine flow for a few seconds and then incubated the rod in a plastic tray for three periods of ten minutes each, exposing the urine to reagents.

The rapid commercial success of Clearblue owed as much to innovative design and marketing, not to mention good timing, as to the incorporation of cutting-edge technologies. Market research had identified urine collection as a "problem for many women," and the high sensitivity of MABs allowed Unipath to promote Clearblue as reliable on the first day of a missed period, a significant improvement over the standard two weeks that polyclonal tests required.[28] Marketing literature boasted that Unipath's "advanced technology" made self-testing "less messy" and "more accurate at an earlier stage" and also that Clearblue gave a "clearer result" in "less time" than competitors.[29] A mixed review in the *Pharmaceutical Journal* judged it to be the "most complicated" of the three tests, albeit with "easy to follow" instructions and the "most easily read endpoint."[30]

Unipath spent the standard £200,000 on a consumer ad campaign in half a dozen women's magazines, including *Woman* and *Cosmopolitan*; organized seminars for family planning professionals; provided pharmacists with demonstration kits and display units; and set up a help line for consumers and pharmacists. Beyond technological sophistication, Clearblue came in a "more attractive package for women—less daunting to use and coupled with advertising designed to give it more feminine appeal." It avoided the "chemistry set" appearance of Predictor and Discover 2.[31] The brand name, fan-shaped logo, and blue color were carefully crafted to reinforce "positive" associations. As with all other home tests, Clearblue targeted women hoping to conceive, "so red for danger would have been quite inappropriate."[32]

Unilever correctly anticipated that the British market, already valued in 1985 at £3 million, or 600,000 tests a year, would expand as public-sector cutbacks pressured GPs to rein in hospital testing. As an article in the *Guardian* warned, the next time a woman asks her doctor for a pregnancy test, she could be told, "I'm sorry, but the National Health Service can't afford to find out if you're pregnant. You'll have to go to the chemist and buy a kit."[33] Beyond cuts, the company cited the "general heightening of health awareness," which increasingly compelled pregnant women to adjust "their lifestyle accordingly," as "another reason for the market to expand."[34] Seizing on the generally ascendent notion of "lifestyle," which pharmaceutical companies used to promote treatments for ever less serious conditions, Unipath's marketing director told the *Financial Times*, "It is very important to know as early as possible that you are pregnant so that you can change your lifestyle—for example stop smoking and drinking."[35] In the age of the "public fetus," the rapidly expanding market for home pregnancy tests fed off and reinforced an increasingly mainstream culture of pregnancy dos and don'ts for the British yuppie.[36]

Clearblue captured 30 percent of the British market in just three months. Chefaro and Carter Wallace quickly responded with their own

MAB-based tests: Predictor Colour and Discover Colour, respectively.[37] Both products were supported by major campaigns in women's magazines and with extensive "below-the-line" support, including consumer leaflets, telephone advice lines, shelf organizers, fact boards, and training sessions for pharmacy assistants. Chefaro also introduced Discretest, Britain's first home ovulation test kit, but Clearblue extended its lead over the competition, ending the year with over 40 percent of the market share, double that of Predictor, its nearest competitor.[38]

Competition was not a zero-sum game, and the rise of Clearblue had the effect of growing the market, valued at the end of 1986 at £4 million, by some 70 percent, including for other brands. Women, and nearly all purchasers were women, bought some 650,000 kits every year, but the number was even larger in some European countries, and analysts interpreted the two million professional pregnancy tests carried out annually in Britain as evidence of potential for further expansion.[39]

The start of 1987 saw the launch of three new home pregnancy tests: Evatest Blue 5, Evatest Rapid, and New Predictor.[40] Tambrands, the US company known for the Tampax brand of tampon, spent £1 million promoting its own home ovulation test, First Response, including with groundbreaking television commercials; the Independent Broadcasting Authority received no complaints.[41] Unilever launched Clearplan, its own home ovulation test, but the market for fertility monitoring was tiny in comparison to pregnancy detection, where the pace of change was reaching a peak.

July 1988 saw Tambrands launch a home pregnancy test, also called First Response (£6.95), a four-step product backed by a £600,000 campaign in women's magazines and cross-promoted with Tampax tampons (eight million packs of tampons carried a £1.50 off coupon for the pregnancy test).[42] More significantly, Unipath launched Clearblue One Step (£8.35 for a pack of two), the first one-step pregnancy test of the kind that is still dominant today.[43] By then, Unipath had captured half the British market, which had increased to £7 million and was growing at a rate of 15 percent a year; global sales approached £100 million annually.[44] Billed as the "most advanced pregnancy testing kit in the world," the new product, known as Clearblue Easy in the United States, incorporated the use of recently "humanized" MABs in a "novel format" based on Unipath's patented "rapid assay technology" (figure 11.3).[45]

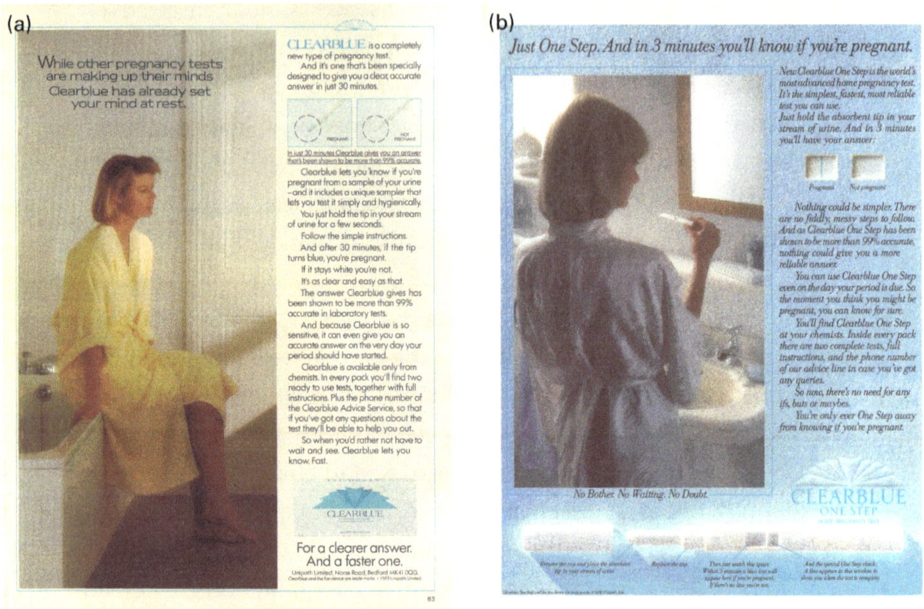

11.3 (a) Clearblue advertisement in *Woman*, Aug. 3, 1985, 63. (b) Clearblue One Step advertisement in *Woman*, Aug. 6, 1988, 68. Both advertisements are typical in depicting a white, middle-class woman in a bathrobe or nightgown waiting serenely in the comfort of her own home. For a contrasting experience: Agarwal 2021, 226–227. Cambridge University Library L448.a.11.97 and L448.a.11.103.

Winner of the 1989 British Design Award in Medical Equipment, Clearblue One Step dispensed with the messy handling of liquid reagents and introduced the now-familiar plastic stick (or wand) with an absorbent wick at one end and two display windows at the other.[46] Described as a "pen-like" device with a "cap," Clearblue One Step contained a porous membrane with three separate areas of antibody.[47] Urine rose up the membrane from the wick to the first area, where it picked up blue dye before seeping into the second area. Urine containing hCG caused a blue line to appear. The first window remained blank in the absence of hCG, but the urine continued to move up to the second, control window. Radically self-contained compared to previous designs, Clearblue One Step made competitors seem "cumbersome," much as tablets and test tubes had with toads.[48]

Unipath launched Clearview, a pregnancy test for professional use, the following year. By then, Clearblue One Step was already retailing in eighteen countries, including most of Western Europe as well as Australia, New Zealand, Canada, and the United States, and eyeing markets in South America, the Middle East, and Asia. The old polyclonal tests had "all but died out," with the few remaining products restricted to pharmacy use.[49] But progress came at a price.

CHEMICAL PREGNANCIES

Demand significantly shifted between 1971, when only a third of women who patronized commercial labs hoped for a positive, and 1989, when two-thirds of self-testers did.[50] Embedded in abortion protocols under the 1967 act, pregnancy testing likewise became a routine of assisted conception—one of the ten "stages of IVF" identified by anthropologist Sarah Franklin.[51] Although marginal in Britain following the birth of Louise Brown, IVF clinics proliferated in the late 1980s.[52] New treatment and monitoring regimes intensified the ambivalence of a positive result and the potential for loss in a process that typically ended in failure.

The combination of earlier tests and IVF research produced a new kind of experience: the "chemical pregnancy" (also "biochemical" or "occult"), as distinct from the "clinical pregnancy" (or "false" positive). Robert Edwards, the Cambridge physiologist and pioneer of IVF, reported "biochemical pregnancies" in between 10 and 15 percent of patients at Bourn Hall Clinic—significantly higher than the estimated 6.8 percent associated with "normal" fertilization.[53] Mary Chadwick, a patient undergoing IVF in Birmingham, where Franklin did her fieldwork, told her, "You know, I *was* pregnant, even it being just a chemical pregnancy and that, you were told it was positive."[54]

Some experts pushed back against the commercial drive for increased sensitivity and earlier detection made possible by MABs. A report in *New Scientist* of a professional test that could detect pregnancy five days *before* a first missed period provoked Bruce Hobson to remind readers that many conceptions did "not survive long enough to result in a missed period," and "even after implantation," there was "considerable embryonic loss." Accustomed to less sensitive tests, clinicians typically interpreted a "positive result to mean the patient is pregnant and can expect a child."[55] The new test was powered to transform the experience of menstruation into that of miscarriage, but a consensus had yet to form on the incidence of undetected (and previously undetectable) loss.

Writing in the *BMJ* in March 1988, a pair of Newcastle reproductive scientists questing for the "true incidence of unsuspected early pregnancy failure" asked, "How positive is a positive pregnancy test?" Clinical research that used immunoassays to measure the incidence produced results as low as 8 percent and as high as 57 percent.[56] A team from Dundee skeptically wondered whether "occult biochemical pregnancy" was "fact or fiction."[57] But an American study put the rate at 22 percent, lending credence to the "medical folklore" that "many women become pregnant only to abort before recognising it."[58] In the UK, women rapidly gained access to home tests that were *more* sensitive than the tests doctors had access to on the NHS. This had a destabilizing effect.

As Belfast pharmacist Terry Maguire warned in the *Pharmaceutical Journal*, the new generation of MAB-based tests could provoke new forms of anxiety. By Maguire's estimation, 60 percent of all implantations spontaneously aborted within twenty days. In the past, when pregnancy tests were not considered reliable until after a first missed cycle, women typically regarded such events as a "late period." The problem started when a woman "anxious to become pregnant" or with a history of miscarriage used a home test and obtained a positive result only to get her period in a few days. In this increasingly likely scenario, the event was experienced as a "'miscarriage' rather than a 'late period' with all the subsequent mental guilt and trauma." Pharmacists, Maguire concluded, would need all their "counselling skills to deal with the situation."[59] He wrote from experience.

Recently, a "very distressed" customer had tested positive using an MAB-based home test (probably Clearblue) marketed for use on the first day of a missed period. She then presented for confirmation of a urine specimen to her doctor, who informed her that it had tested negative. In "desperation," because she was "very anxious to conceive," she returned to Maguire's pharmacy for advice. He performed a third test and confirmed the initial positive result. He also contacted the woman's doctor, who told him he used Pregnosticon Planotest but had "failed to take into account the number of days from the woman's missed period."[60] The coexistence of polyclonal and monoclonal tests generated contradictory results that, in turn, produced confusion and, in some cases, distress.

Consumers soon gained access to tests marketed as reliable before the first missed period. Kent Pharmaceuticals promoted the Early Bird Hexagon hCG test for professional use by pharmacists up to five days before an anticipated period.[61] Such claims were "NOT necessarily a good thing" and could "cause much heartache," cautioned a report in *Community Pharmacy* that echoed Hobson's and Maguire's misgivings. Because "nearly two thirds" of all pregnancies spontaneously miscarried within twenty days of implantation, the Early Bird test forced women to choose between the anxiety of waiting two weeks and the risk of a "late period-miscarriage."[62]

Meanwhile, an expanding and diversifying market for pregnancy advice manuals, including books devoted to infertility, miscarriage, and assisted conception, cautioned patients and consumers against placing too much faith in an early positive result. For instance, *Getting Pregnant: The Complete Guide to Fertility and Infertility* (1989) warned that hCG injections used by some doctors to treat women "prone to miscarriage" or "having test-tube baby treatment" could produce false positives.[63] *The New Our Bodies, Ourselves* (1989) advised readers to "wait a few days to be more certain that a positive test means a viable pregnancy rather than be alarmed, or excited, prematurely."[64]

A Woman's Guide to Birth-Tech: Tests and Technology in Pregnancy and Birth (1989) explained that the high incidence of "failure" in early pregnancy might account for false positives and was "consistent with long-held theories surrounding natural selection and survival of the fittest," the idea being that a significant proportion of early miscarriages were associated with severe chromosomal anomalies.[65] *Tomorrow's Child: Reproductive Technologies in the 90s* (1990) defined "biochemical pregnancies" as "short-lived pregnancies, that show up as brief hormonal changes, but fail shortly after or during implantation."[66] And in an astonishing twist on a woman's right to know, *Miscarriage: Women's Experiences and Needs* (1990) gently suggested that "you may not want to know and sometimes very early miscarriage is easier to deal with by treating it as if it is a late period."[67] Choice had never been so fraught.

Researchers similarly interrogated the pros and cons of early testing. A Glasgow team reporting on the sensitivity of Unilever's Clearview acknowledged as a "possible drawback . . . the unnecessary detection of early complete abortion (often referred to as biochemical pregnancy), which might

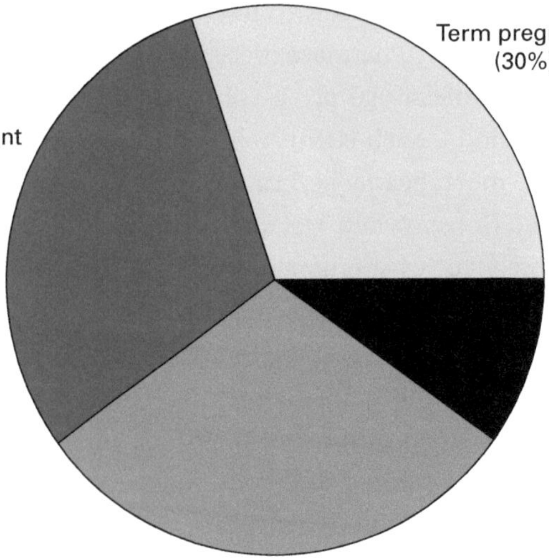

11.4 The fate of an early pregnancy. Chard's pie chart would later be adapted as the "pregnancy loss iceberg," an inverted triangle with live births (30 percent) and miscarriage (10 percent) as the clinically visible tip; preclinical implantation failure and early loss (60 percent) are hidden below: Macklon et al., 2002. Chard 1991, 186, with permission by Elsevier.

otherwise be regarded as a heavy period by the patient."[68] And Tim Chard, a reproductive physiologist at Barts, reminded fertility experts that it was more important than ever to "emphasize that a pregnancy and a positive test do not guarantee a baby 9 months later." A "substantial" number of pregnancies were lost "between conception and term," and the new, more sensitive tests made such losses more apparent (figure 11.4). The earlier a pregnancy was "diagnosed," the more likely it was to "abort." With the most sensitive tests increasingly detecting unviable implantations, technological progress seemed to be reaching a biological limit. Chard predicted the current generation of MAB-based tests was "unlikely to be surpassed either by better tests or alternative technology."[69]

YEARNERS AND DREADERS

In the early 1990s, sales of home pregnancy tests dwarfed the number of in-house tests performed by pharmacists.[70] Five companies marketed ten brands that ranged in cost between £5.95 and £8.75.[71] And the market was by no means static. First Response made history as the first home pregnancy test to be advertised on British television in a "second coup" that was set to "revolutionise" the Independent Broadcasting Authority's position on "sensitive" advertising.[72] Johnson & Johnson launched Fact Plus, which gave the result as a "plus" or "minus" sign in the test window.[73] Chefaro introduced its own pen-style test in June 1991, with a £400,000 ad campaign, the "biggest in the brand's history."[74] Unipath attempted to generate more discussion and interest, including among men, by advertising in the national press.[75] And Tesco supermarkets began selling Discover Today, a move that some pharmacists regarded as betrayal.[76]

Self-testing, reported *Community Pharmacy* in February 1992, had become an "accepted factor in women's conception and birth schedules, particularly now that home pregnancy tests are so simple and free GP tests seem to be becoming more restricted." But still many pharmacists refused to openly display the products, so the "final embarrassment barrier" of asking for a pregnancy test remained. In addition to display, stocking multiple brands appeared to boost sales because women often retested using another brand, reasoning that if two different brands agreed on the result, then it must be correct.[77]

Heteronormative marketing campaigns, meanwhile, did not prevent single women and lesbian couples from using the full range of products not intended for them. The revelation in the *London Evening Standard* that British lesbians were conceiving babies through artificial insemination by donor (AID) had shocked the nation in 1978.[78] But resourceful women continued to share information, for instance, about how to purchase an "inseminator" for seven pounds from John Bell and Croydon, a pharmacy on Wigmore Street, and cooperated with male friends and men's support groups to arrange semen transfer at a convenient tube station.[79] Although developed to support traditional heterosexual family formation, AID, IVF, and other fertility treatments inadvertently provided single women and lesbians with new ways of achieving motherhood.[80] *Challenging Conceptions* (1994), Lisa Saffron's guide to self-insemination, recommended any of the major brands of home pregnancy test, preferably a two-pack, for after the obligatory two-week wait, and also testing by doctor, which "cost nothing and will confirm what you may already know by this time." As recounted in the same book, the "idea of just weeing on a stick and looking for a colour change" (to self-monitor fertility) appealed to Eunice and Stella, who scoured Boots branches for discount vouchers to save money on the recently relaunched Clearplan ovulation test kit as a first step toward starting a family.[81]

Complacency was not an option in a competitive market that regularly saw all the major brands redesigned, repackaged, and relaunched. Even Clearblue One-Step, still a market leader, relaunched with "even clearer" blue lines in August 1992.[82] Continual renewal could be "confusing," observed trade journalist Judy Bargh, but it also "simplified things." The major brands had converged, and there was no longer any "great difference" between them. There was "simply no comparison" between the "miniature chemistry sets" of old, already a fading memory, and the new dipsticks that "simply need to be inserted into a urine sample to give a result in minutes." Women no longer needed to "set aside time to carry out the test or worry that they might have done it incorrectly." Bargh, like Chard, was hard-pressed to "imagine what future improvements there can possibly be."[83]

Yet despite all the innovation, "recessionary times" had slowed growth: potential customers had less disposable income to spend on what was still for many a "relatively expensive" commodity; financially struggling couples deferred starting or adding to their families; and women reverted to free NHS testing. Manufacturers nevertheless remained optimistic that the still unsaturated market would rebound and continue to expand.[84] Boots outperformed the market, with a growth of 10 percent compared with the 3 percent overall, possibly because of the "significant amount of money and shelf space" allocated to home pregnancy tests. As user confidence increased and belts tightened, customers tended to purchase (cheaper) single tests instead of packs of two. Communicating innovation to consumers remained a challenge and was another possible reason for "sluggish market performance." Potential users were "only in the market at the time they think they are pregnant," making it difficult for companies to sustain "brand awareness."[85]

Corporate help lines, meanwhile, had to some extent taken on the counseling role provided by women's centers in the 1970s. Unipath, for instance, recruited counselors from nursing and other "caring" professions to answer specific questions about Clearblue One Step and more generally about pregnancy. Most callers wanted "emotional reassurance" and some

11.5 Telephone helplines provided a crucial part of a marketing apparatus that anonymously provided information and reassurance to consumers, especially prior to rise of internet forum Mumsnet and other online resources in the early 2000s; see Pedersen 2020. *Unilever Magazine* 72, 1989, 36, with permission by Unilever PLC.

to help interpret a test result. The counselors fielded calls from "quite distressed individuals," including many from "women who may have found out they're pregnant, but are quite unprepared for the prospect." Other callers were "very anxious to become pregnant as soon as possible." The help lines referred callers as appropriate to BPAS (an abortion provider) or the National Fertility Association and offered customers access—during regular business hours—to an "impartial and sympathetic 'ear', particularly at a time when GP surgeries are becoming increasingly busy" (figure 11.5).[86]

A new Predictor, with a pink sponge tip, became the first pregnancy test to be advertised on radio—local and national, for five weeks—in May 1994.[87] Additionally supported by leaflets, including *So You Want to Have a Baby* and *Sex: What Next?*, it was—according to ads in various women's magazines—"easy peesy" and "so simple it shouldn't be called a test."[88] And yet, market research commissioned by Chefaro showed that consumers trusted the product but not themselves. The market had doubled in size between 1986 and 1994, but consumer awareness was "only inching up the learning curve." Just over half of potential customers surveyed for Unipath were "seasoned" users; 22 percent relied on a pharmacist, and 17 percent went to the GP even though one in five doctors no longer offered pregnancy testing.[89]

Televised commercials, a "shotgun approach," delivered a surge in sales, but manufacturers could only afford "short ad campaigns dotted round the country." Even Chefaro's pioneering television spot eschewed the "ultra-expensive London region," which accounted for about a third of national sales. Companies continued to rely on the "drip-drip method" of ads in women's magazines and newspapers. As a product category, pregnancy tests were "more generic" than brand-led goods like shampoo; women went into a pharmacy and asked for a home test kit, not a Predictor or Clearblue. In the marketing language of the day, consumers divided into two groups: "yearners" and "dreaders." Crucially, all advertising targeted "yearners," women who wanted to be pregnant. What about the "dreaders"? As the assistant product manager for First Response and Discover 2 explained in *Community Pharmacy*, "We don't need to target them[.]" "They will buy tests anyways."[90]

As with other established over-the-counter health care products, the market for home pregnancy tests diversified from a few "premium brands" to include "budget brands" and, finally, "own label." UniChem was the first, in 1995, to launch a store-brand pregnancy test, and Boots, which accounted for around 40 percent of all sales, followed in 1996. According to IMS Health data, own-label sales accounted in 1997 for 12.6 percent of the market, excluding Boots, and were increasing faster than the market. Reliability trumped cost, and first-time users were more likely to choose an established brand. But Boots was "virtually a brand name in its own right," and its pregnancy test appealed to the same customers who bought "branded products" and was "equally popular." In contrast to premium brands, own labels did not invest in "tech" or support phone lines, and a senior manager at Chefaro dismissed some of the budget brands as "little more than a litmus test."[91]

Trade journalist Sarah Purcell observed in November 1997 that home pregnancy testing had become a "part of modern life"; women who suspected they might be pregnant were now "more likely to buy a kit and wait nervously for the blue line to appear in the comfort of their own bathroom than make an appointment to see their GP to find out the news." Consumers spent around £27 million on some three million tests, with the market split evenly between single and double test packs. Half of women aged sixteen to forty had used a home test. Fewer women sought professional confirmation, and most GPs accepted the result of a home test.[92] More than a quarter century after the debut of Predictor, self-testing was only now becoming, for a younger generation of women, the new normal.

COMPUTER MAGIC

Media interest in "teenage pregnancy" peaked around 2000 under New Labour's high-profile effort to shift the terms of debate from sexual morality to health and socioeconomic disparities.[93] The narratives collected in *Tough Choices: Young Women Talk about Pregnancy* (1999), edited by Alison Hadley of Brook Advisory Centres, provide a window into the youthful experiences of mostly "dreaders" at the turn of the millennium. For instance, a positive home test confirms the "gut feeling" and "worse nightmare" of seventeen-year-old Julia, who stopped taking contraception because of the October 1995 "pill scare" and then split with her boyfriend. Samantha's "whole body went numb" as she "watched the bright blue line appear on the tester." Debbie "knew" she was pregnant when her period "stopped" and so "never did a pregnancy test." The appearance of a blue line filled Poppy, an atypically young "yearner," with the "most amazing feeling." Liz, on the other hand, was "absolutely devastated" when her "whole life fell apart."[94]

Meanwhile, pharmacy sales declined for the first time, as grocery sales increased.[95] To escape a slump, Predictor, the second-place brand with a 12 percent share, launched a £1 million television and press campaign supported by a telephone help line and, for the first time, a website: www.predictor.co.uk.[96] Unilever, meanwhile, sold Unipath for £103 million to Inverness Medical Innovations, a smaller, more specialist company that had been eyeing Clearblue since the late 1980s.[97] Clearblue Digital, the first pregnancy test to display the result as the words "Pregnant" or "Not Pregnant" on a liquid crystal display, debuted in 2003.

Initially sold in packs of three, Clearblue Digital retailed for £14.99. To use the test, a woman inserted a cartridge into a holder and held the absorbent sampler in the urine stream for a few seconds. A "test ready" symbol flashed to indicate the test was working, and the result remained on the display for one hour before automatically turning off.[98] Inverness relaunched Clearblue Digital in 2006 as a one-step test (£9.99 for one; £13.99 for two).[99] By then, a third of women self-tested before their period was due, and the Miscarriage Association supported their decision to do so even if it resulted, as a *Telegraph* headline proclaimed, in "unnecessary grief."[100] A newly added "egg-timer symbol" showed when the test was working, and taboo-breaking television advertising, on terrestrial, satellite, and digital channels, showed a "stream of liquid to represent urine."[101] And upgrade in 2008 introduced a "conception indicator" that additionally told the user whether she had conceived "1–2," "2–3," or "3+" weeks ago, a feature that harkened back to the graded results developed in Edinburgh in the 1930s.[102]

Today, home pregnancy tests seem to exist in a hierarchy of reputability and perceived trustworthiness that is directly proportional to cost. The Edinburgh women interviewed by sociologist Emily Ross, for example, disparaged the "cheapo Tesco" tests and praised Clearblue digital as the "Rolls Royce of pregnancy tests" (figure 11.6a).[103] When asked why she placed more faith in a digital result, one woman explained that "you probably just trust whatever computer magic is inside the test." Others reported "producing and deciphering their [analogue] test result in collaboration with others," for instance, by asking a partner to interpret a faint line or by Googling relevant pictures for comparison. Andrea, who had a history of early miscarriage, used "about thirty" cardboard test strips (cheaply available online in multipacks of fifty): "You kind of know if [the pregnancy'] s working because the line gets darker each day . . . that's why I kept on doing the test, cos it's like a reassurance thing."[104] Supported by online communities of users sharing photos and advice, even the cheapest analogue tests can be used creatively to mimic Clearblue Digital's advanced "conception indicator" feature, tracking the early progress of pregnancy and providing a kind of reassurance not advertised on the box.

Blue lines (parallel or crossed), meanwhile, seems to have a magic all their own. Possessing a ghostly permanence and almost religious symbolism that digital displays lack, they are a blessing to some, a curse to others. To the mixed-heritage granddaughter of a devout Jehovah's Witness in Zadie Smith's *White Teeth* (2000), they evoke the "face of the madonna in the zucchini of an Italian housewife."[105] To a distraught mother in Julie Bertagna's *The Opposite of Chocolate* (2003), the "bars of the prison . . . that her fourteen-year-old daughter had made of her own life."[106] As recounted in *Minus Nine to One: The Diary of an Honest Mum* (2006), Jools Oliver, wife of celebrity chef Jamie Oliver, kept the positive tests associated with their children as mementos "in a special box and, yes, the lines are still there just like magic!"[107] In *Keisha the Sket*, the viral creation of a Black inner-city schoolgirl first self-published on the social networking website Piczo in 2005, the eponymous hero sits "on the loo feeling cursed as the second line undeniably appear[s]"; her "deals with God" (sexual abstinence for a

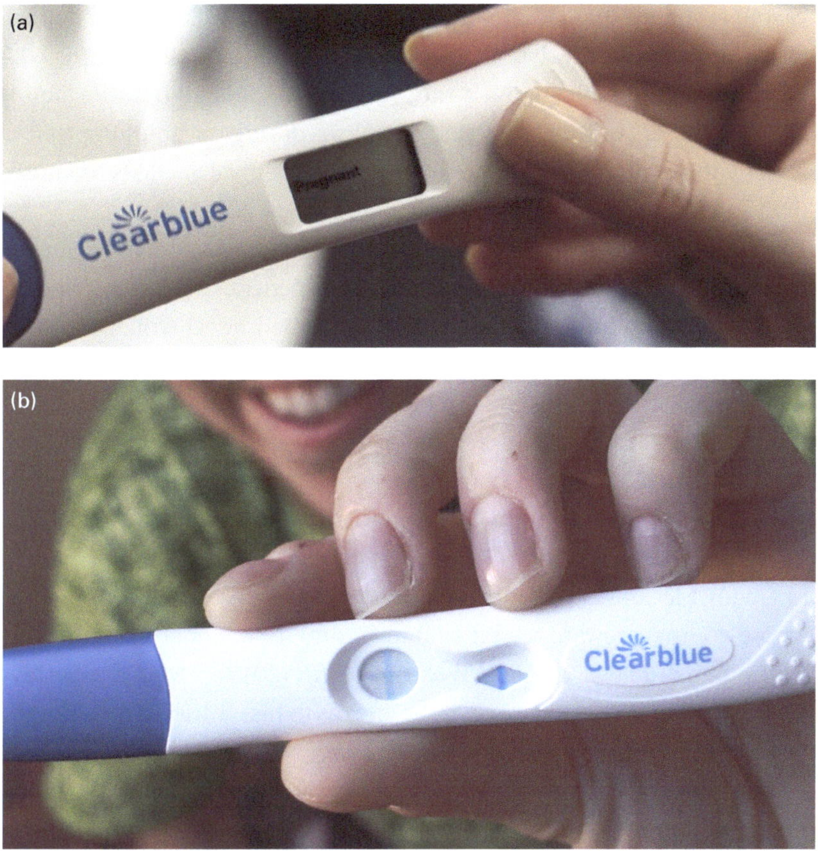

11.6 In the early twenty-first century, product placement (or "embedded marketing") consolidated the Clearblue close-up as a cinematic and televisual trope: Olszynko-Gryn 2017. (a) Clearblue (digital) close-up and example of product placement in *Bridget Jones's Baby* (Sharon Maguire, 2016): produced by Miramax, Perfect World, STUDIOCANAL, Universal and Working Title; distributed by Universal. (b) Clearblue (analogue) close-up (not product placement) in *Seahorse: The Dad Who Gave Birth* (Jeanie Finlay, 2019): produced by Andrea Cornwell, Jeanie Finlay, and Grain Media in association with Glimmer Films and The Guardian; distributed by Submarine Entertainment.

year in exchange for a negative) come to naught.[108] For Edinburgh journalist Chitra Ramaswamy, who had been "trying" for eighteen months with partner Claire, the blue cross is the "revelation [she] had been imagining for so long."[109] And for poet Holly McNish, during an hourlong wait at King's Cross station on her way to Glastonbury Festival, it elicits "confused then laughing sobs" and, back on the train, a new poem.[110]

Not only women, but men also invest the iconic lines with life-changing meaning and increasingly narrate their own experiences. Take, for example, the swell of emotion conveyed by Sheffield-based travel writer Peter Naldrett in *A Bun in the Oven: The Pregnancy Diary of a Dad-to-Be* (2007): "And sure enough, there it was. A little blue line has never meant so much. . . . It meant that I had helped to fertilise an embryo."[111] More rarely, although with increasing frequency, home tests are used by transmen to (privately or publicly) confirm their own impending fatherhood.[112] Two weeks after intrauterine insemination, a blue cross confirms successful implantation to Freddy McConnell, a journalist from the Kentish seaside town of Deal, whose journey toward gestational paternity is movingly documented in the film *Seahorse: The Dad Who Gave Birth* (2019): "That is a positive result . . . in more ways than one," he affirms to his mother as he holds the test up to the camera for the obligatory closeup (figure 11.6b).

Although typically and correctly seen as a significant turning point, this chapter has shown that the original Clearblue was less a sudden break with the past than a well-timed commercialization of technical and design elements that could be found already in professional tests. The marketing apparatus that created, maintained, and expanded demand, especially from women hoping for a baby, had as much to do with seemingly mundane technologies like counter display as with more sophisticated tools like monoclonal antibodies. Clearblue's rapid and sustained commercial success resulted from and contributed to a changing public culture of reproduction in Britain that made miscarriage, infertility, and assisted conception more visible than ever; the debate over "chemical pregnancies" signaled a biological limit to technological progress and called into question the ontological status of early pregnancy and its detection. The next and final chapter provides an epilogue to the story of supply and demand, production and consumption. It looks to the future of pregnancy testing and to the sometimes tenacious afterlives of obsolete practices.

12

FUTURES AND AFTERLIVES

Three minutes later, Tracey Emin, the artist who went to the bathroom at the start of this book, looked at the pregnancy test: "It's negative. Of course it's negative. Of course I'm not pregnant. I am relieved, relieved to know that at thirty-seven years of age, I am just a woman with a fucking good imagination."[1] Images of pregnancy tests now adorn the walls of galleries and other spaces with some regularity, part of a broader visual exploration of changing reproductive (and nonreproductive) experiences (figure 12.1).[2] From esoteric and taboo to a commonplace of everyday life and object of artistic contemplation, the humble pregnancy test has come a long way. Classified ads and posters in the London Underground once scandalized the BMA. Pharmacists kept the first test kits out of sight, and Boots, the largest chain, initially refused to stock them. Today, home test kits for sexually transmitted infections (STIs) and, most recently, for Covid-19, as well as preconception vitamins and morning-after pills have joined pregnancy and ovulation tests on pharmacy shelves. The tests have utterly and irreversibly lost the power to shock.

The social history I have constructed highlights significant changes and continuities that timelines of innovation, and histories of just the home test, miss. For instance, debates over the medical control of pregnancy testing came to a head not in the 1970s, with home tests, but in the 1960s, with the commercial laboratory and pharmacy services that served women

12.1 During a one-year residency in 2008 at the Guy's Hospital assisted conception unit, artist Gina Glover collaborated with staff and patients on *The Art in ART*, an award-winning collection of visual responses to the work of the unit. The ninety-five negative pregnancy tests in *Yes!* (from the collection) evoke the tally marks on a prison wall, counting down the days to freedom. See Franklin 2013: 275–296; courtesy of Glover.

directly as clients with consumer rights. Overturning the medical control of pregnancy testing, these services cleared the way for Predictor. The British welfare state, meanwhile, consistently refused to reign in commercialization, an approach that, at times, seemed to align with liberal feminism against medical paternalism. Despite petitions from the medical establishment, central government ceded control to the private sector. A generation of women more affluent than their mothers or grandmothers voted with their wallets.[3] The doctor–patient relationship, commercially mediated by the laboratory and pharmacy, would never be the same.

My account also complicates notions of a simple opposition between technological mediation, on the one hand, and self-knowledge, on the other. Successive generations of women did not discard old resources as they gained access to new ones. An online NHS video produced in 2012 explains that a positive result is probably correct, but a negative result should not be trusted if "you feel that you are pregnant or you continue to miss your period."[4] Feeling pregnant still counts. The venerable experience of protracted uncertainty and gradual realization, often mixed with fear, hope, and everything in between, was (and still is) coextensive with newer experiences of a technologically mediated diagnostic moment, which is not always as definitive as we might expect. Anxiety, such an important actor's category, took on new meanings as it shaped expectations for the social and technical innovations that brought pregnancy testing from the margins to the mainstream.

The liminal status of pregnancy, between the "normal" and the "pathological," has continuously challenged the legitimate boundaries of health care, including the provision of medical services for early detection.[5] Although male bodies have been medicalized too, debates over the medical surveillance of human procreation have overwhelmingly turned on the extent to which menstruation, menopause, pregnancy, childbirth, and miscarriage should be regarded as "natural" or disease-like.[6] In Britain, policymakers typically preferred to keep services related to sexual and reproductive health at arm's length. In the mid-1960s, the oral contraceptive pill became the only drug that GPs were allowed to prescribe privately to NHS patients for a fee. Similarly, doctors had the option of ordering nonmedical pregnancy tests for patients who were willing to pay out of pocket, a compromise that predated the NHS. Like contraception and menstruation, pregnancy testing was (and still is) big business.

Hormone tablets found a market in part because, as prescription drugs, they were covered by the NHS. For busy GPs in deprived areas, prescribing pills was cheaper and more convenient than ordering a laboratory test. But invasive pregnancy tests ran into trouble in the wake of the thalidomide disaster as suspicion fell on the use of synthetic hormone preparations in early pregnancy and the risk of birth defects. In 1978, around when Schering finally withdrew Primodos from the British market, a group of parents formed the Association for Children Damaged by Hormone Pregnancy Tests to take legal action against the company. The action was discontinued

Victims start 'new thalidomide' fight

Lois Rogers

VICTIMS of a "forgotten thalidomide-style drug scandal" have begun a fresh attempt for compensation for serious birth defects linked to a pregnancy-test pill taken by 1.5m women.

Some women given Primodos, a drug based on a super-strength version of hormones later used in the "morning-after" contraceptive pill, suffered instant miscarriages in the 1960s and 1970s. Thousands more gave birth to babies with missing limbs, abnormalities of internal organs, or other handicaps.

The recent discovery of documents at the National Archives in London has prompted victims to demand a government investigation and a review of a legal action that failed more than 30 years ago. It has now emerged that 26 studies from 1960 onwards suggested the drug caused miscarriages and birth defects, yet it remained on the market.

The documents include studies suggesting Primodos caused abortions in pregnant rats, as well as evidence pointing to its manufacturer, Schering, offering incentives to GPs to give the unprescribed drug to women.

The AGM in Birmingham yesterday of the Association for Children Damaged by Hormone Pregnancy Testing heard more than 40 MPs support calls for a public inquiry. Dan Poulter, the health minister and a gynaecologist, met the group last month and agreed to investigate.

Primodos, which predated modern urine pregnancy test kits, contained a combination of 10mg of norethisterone and 0.2mg of ethinylestradiol, 13 times the strength of the "morning after" pill. Women were told to take two tablets 12 hours apart, and if they did not bleed afterwards they were told to assume they were pregnant.

The Sunday Times raised concerns about Primodos in 1975. The drug was withdrawn in 1978 without explanation. In 1986 this newspaper reported an admission by Michael Briggs, Schering's medical director, that he had fabricated evidence relating to its contraceptive pill.

Many of the affected children are believed to have died in early adulthood from heart or organ defects or from cancer. Of those still alive, many have missing limbs, fingers or toes, or defects such as cleft lips. Others have been left deaf or blind. A survivors' group is also campaigning for compensation for thousands of affected children in Germany.

Marie Lyon, 67, a retired building society manager from Wigan, Lancashire, whose daughter Sarah, 43, was born without her left forearm, leads the British campaigners. "Many of the parents are concerned about what will happen to their badly disabled adult children after they die, and they want funds made available to care for them," she said.

Nicola Williams, 42, who has had more than 20 operations to correct digestive and spinal abnormalities, said she had found documents at the National Archives in which doctors likened defects associated with Primodos to those caused by thalidomide. "I couldn't believe how much the government and senior doctors knew about this at the time," she said.

A spokesman for Bayer, the drug company that took over Schering in 2006, said: "Bayer denies that Primodos was responsible for causing any deformities in children. [We] are not aware of any contention that evidence of alleged harmful effects from Primodos was suppressed. We have no knowledge of any payments made by Schering to doctors in respect of their use of Primodos."

Yasmin Qureshi, the Labour MP leading the parliamentary campaign, said: "The fact is Schering knew the drug dose could cause miscarriage, they knew it affected development of the early embryo, and they knew there was a correlation with increased risk of birth defects, yet it stayed on the market."

12.2 A bold headline and defiant group photograph in the same newspaper that campaigned for thalidomide survivors here announces the new fight over a "forgotten thalidomide-style drug scandal." The smaller photograph of hands reinforces the sense of solidarity and similarity to thalidomide. The "recent discovery of documents at the National Archives in London" is cited as having prompted calls for a government investigation and a review of the failed legal action. Lois Rogers, "Victims start 'new thalidomide' fight," *Sunday Times*, Feb. 23, 2014, 9, Cambridge University Library NPR.B.2026, with permission by Andrew Fox/The Sunday Times/News Licensing.

in 1982 but on terms that left the plaintiffs free to proceed again pending further evidence that Primodos caused birth defects. Today, after a long period of dormancy, the association is once again campaigning for financial redress and regulatory reform, this time armed with previously inaccessible archival records (figure 12.2).[7] The final chapter in the history of Primodos and related products remains to be written.

The same goes for Predictor, although in a happier mode. Described in 2016 as a "real-life Peggy Olson," the secretary promoted to copywriter in

12.3 The Osler Library of the History of Medicine at McGill University, Montreal, acquired an original Predictor from Crane in 2019: Olszynko-Gryn 2020b. Photograph of Crane proudly displaying her invention at the unveiling, which featured a roundtable discussion with historians and gender studies scholars. In back, left to right: Jenna Healey, Alanna Thain, Mary Yearl, and Christabelle Sethna. Photo: Cynthia Tang, May 7, 2019.

the hit series *Mad Men*, Meg Crane may yet become a household name.[8] Complete with romance, the invention story of Predictor seems made for Hollywood. A film company is working on the story and a stage play recently premiered.[9] To be sure, Predictor was a significant milestone and Crane deserves a place in the history books.[10] Alongside the typically female "invisible technician" or "human computer," not to mention researchers like Rosalind Franklin and research subjects like Henrietta Lacks, historians of science, technology, and medicine should also consider the invisible designer, invisible no more (figure 12.3).[11] Yet, as with the pill or any "revolutionary" invention, it is important to resist ascribing too much disruptive potential to the advent of home pregnancy testing.

Predictor did not suddenly sweep away alternatives any more than the pill made other forms of birth control obsolete. Twentisec had damaged the reputation of the sector, and most women preferred to rely on familiar arrangements. It would take not only technological improvement but also cultural change to normalize self-testing. Embarrassing and fiddly home tests coexisted with a palimpsest of new and old diagnostic services and resources, including women's (mediated) knowledge about their own bodies.[12] Only when Unilever launched Clearblue Easy in 1988 did the home pregnancy test achieve its present-day form. And even then, Unilever provided a full phone-in service to advise users. Mass change started around 1990 and was slower outside London. Only around 2000 did self-testing become a ubiquitous rite of passage for a younger generation of consumers. Feminism helped to make Predictor thinkable in the 1960s, but the market Predictor made is more indebted to neoliberalism.

Gone but not forgotten, the *Xenopus* test continues to appear regularly on science quiz shows. Since the early 2000s, it has perennially made headlines as the culprit behind the global spread of a lethal fungus held responsible for the mass die-off of amphibians. When anthropologist Eben Kirksey attempted publicly to revive the test in 2012, he was compelled by People for the Ethical Treatment of Animals (PETA) to change the location of the "performative experiment" from a Brooklyn art gallery to a friend's house (figure 12.3).[13] Today, visitors to the medical galleries at the Science Museum in London can see on display a vintage 1802 *Xenopus*, on loan from the adjacent Natural History Museum. Placed near apparently sophisticated devices for blood and urine analysis, the preserved zoological specimen turns heads. Ordinarily, however, pregnancy testing doesn't raise an eyebrow.

Home tests have become so thoroughly taken for granted that it is hard to believe they ever caused trouble. Their ubiquity makes them invisible. What, then, can we expect regarding future access to today's contested technologies of sexual and reproductive health? Does the long-term normalization I have recounted in this book have implications for home STI tests, morning-after pills, or abortion pills? Will they, like pregnancy tests before them, become normalized and socially accepted despite decades of medical and moral resistance?

HIV testing provides a suggestive comparison since the virus, like an unplanned pregnancy, can also be the consequence of unprotected sex and has a history of social stigmatization. In 2012, the FDA approved OraQuick, the first over-the-counter home test for HIV and a version of the swab test

12.4 The event promised free pregnancy testing, pet frogs, and potential co-authorship of an academic publication. (a) Kirksey catches a *Xenopus* frog to test it for fungal spores at the Proteus Gowanus Gallery in Brooklyn; (b) Albino *Xenopus* frogs, obtained for $2.88 apiece from local pet shops, on display in iconic Ball mason jars at the gallery. Photos courtesy of the Multispecies Salon; see Kirksey 2014.

that health care professionals had been using a version since 2002. The FDA advised that OraQuick was not 100 percent reliable and that additional testing by medical professionals was needed to confirm results. OraSure, the manufacturer, claimed a 99 percent accuracy for negative results but only 92 percent for positives.[14] As with pregnancy testing in the 1960s, users were encouraged to see a doctor if they test positive and to retest if they tested negative. OraQuick followed a similar trajectory of demedicalization and commercialization, as well as raised a familiar set of questions about anxiety, privacy, reliability, and consumer responsibility.[15]

The distinction between nonmedical "entertainment" or "keepsake" ultrasound, on the one hand, and diagnostic imaging, on the other, is reminiscent of past categories of "curiosity" and "pathological" cases for pregnancy testing. Since the mid-1990s, entrepreneurs have bought or leased equipment that has previously been confined to a professional clinical setting.[16] Some doctors and medical organizations have condemned nonmedical ultrasound for risking false diagnoses, whereas its defenders have argued that the application of consumer safety regulations, not a ban, would adequately ensure the responsible provision of a low-risk, in-demand service to the general public.[17] The parallels with pregnancy testing are striking. In both cases, we can see the commercial appropriation and marketing of a medical technology for ends that are not strictly medical, and in both cases, opposition is framed in terms of state regulation, the risk of false results, and doctors' responsibility to control a reproductive technology that had found a nonmedical market.

For some, the "ideal" pregnancy test—still a thing of the future—would be "100% accurate," "very cheap," and "able to diagnose pregnancy immediately following conception."[18] Such perfect accuracy is elusive. And the criteria of cost, speed, and convenience are not the only ones that have mattered. Anthropologist Linda Layne has proposed that home tests could better serve users by indicating not only the presence or absence of hCG but also the level, which could then be tracked over time. Women would be able to use these tests not only to confirm or exclude the possibility of pregnancy but also to ascertain "whether a pregnancy is likely to end in miscarriage, is likely a multiple gestation (twins, triplets), or is likely to be an ectopic." Placed "in women's hands," this information could "save lives."[19] But "semi-quantitative" or "multilevel" pregnancy tests, as they are called, have yet to materialize as a consumer product, and the frontier of pregnancy testing may have moved on.

Today, the latest in pregnancy testing is Lia, the first biodegradable, flushable, and plastic-free home pregnancy test (figure 12.5). Developed by Bethany Edwards and Anna Couturier Simpson when they were graduate students at the University of Pennsylvania's Integrated Product Design program, Lia is well timed to ride the waves of "zero waste" and of "femtech" that are boosting a range of products, including redesigned menstrual cups, compostable tampons, "smart" breast pumps, and fertility-tracking apps. Pitched to glowing reviews as a more ecological and more discrete alternative to the conventional plastic wand, Lia received FDA approval in 2017

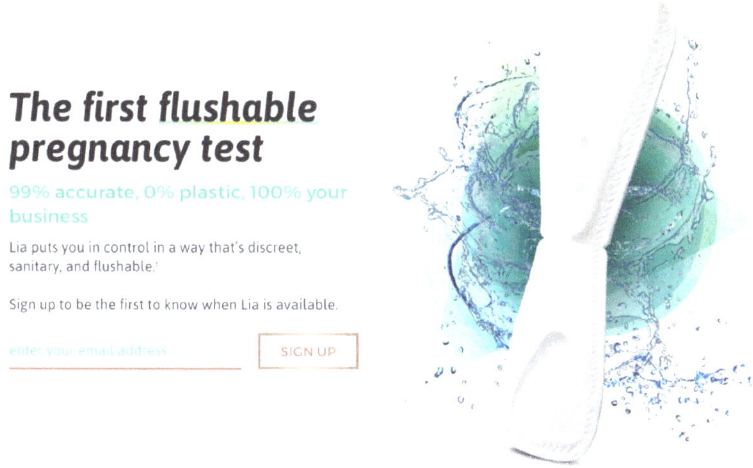

12.5 The Lia pregnancy test, which resembles a menstrual pad, may represent the first major redesign of a home pregnancy test since digitization. Detail from the product website, March 31, 2019, https://meetlia.com.

and launched in March 2021 with a direct-to-consumer website offering two-packs for $13.99 each.

In the late 1970s, Ann Oakley identified a "tension" in the process of pregnancy determination "between a desire to defer to medical authority and a feeling that the body's own signals should be trusted."[20] It "may seem obvious," she elaborated, that the "person whose body the baby is in should know about it first. But, on the other hand, people are used to going to doctors to have their symptoms interpreted. Why should pregnancy be a special case?"[21] Yet in many ways, pregnancy is a special case. Pregnancy is not a disease in any straightforward sense, and historically, many doctors rejected the demand by women to take responsibility for its medical verification. In Britain, pregnancy testing is technically free on the NHS, but most women will shell out for a home test, and most doctors will accept the result. Momentum has gathered behind the effort to tackle "period poverty" by making sanitary products freely available to all.[22] Should a similar case to be made for pregnancy testing?

In the ascent of private, for-profit pregnancy testing, for better or for worse, the market won out over alternatives. This victory was not inevitable but resisted at every turn with alternatives proposed along the way. Many GPs wished they could offer a free, while-you-wait service, but their hands were tied. And for nearly two decades, feminists did just that. Central government rejected paternalistic requests to ban or regulate pregnancy testing, but they also rejected progressive calls to make rapid, convenient tests directly available to GPs. Health authorities encouraged doctors to direct NHS patients to pharmacies, where they could pay for a home test. From abortion to IVF, the British welfare state has a record of allowing the private sector to meet demand for sexual and reproductive health services. The humble pregnancy test is no exception. But it didn't have to be that way.

A final takeaway from my analysis is that women have not always needed home pregnancy tests. Demand from married women, in particular, had to be constructed. Today, doctors are no longer in control. But the purchase of a home test is often embedded in an elaborate and stressful regime of medical surveillance and self-discipline, especially for women trying to become pregnant. Without arguing that we should turn back the clock, I do want to underscore that technological progress has resulted in losses as well as gains. For many women, including IVF patients, pregnancy tests have transformed the experience of a late period into that of an early miscarriage. The effect, as others have observed, can be more distress and grief.[23] Meanwhile, social pressure continues to mount on women to strive for the "perfect" pregnancy.[24] The stakes, often constructed in terms of the fetal origins of adult health, could not be higher.[25] That said, I want to conclude by suggesting that the past may yet contain resources for the present and future. Home pregnancy tests may be here to stay, but the broader culture they help to sustain is resistible.

ACKNOWLEDGMENTS

This book began in 2009, with an MPhil essay and dissertation and then a PhD thesis in the Department of History and Philosophy of Science at the University of Cambridge, supervised by Nick Hopwood. Evidence of Nick's guidance can be found on every page, not least in the range of sources he encouraged me to consult. The research and writing benefited immeasurably from being embedded for nearly a decade within the Wellcome-funded Generation to Reproduction project, led by Nick. For memorable conversations and constructive criticisms while at Cambridge and beyond, I thank Salim Al-Gailani, Daniel Anderson, Leah Astbury, Jenny Bangham, Christina Benninghaus, Dominic Berry, Fran Bigman, Mary Brazelton, Polly Brownlee, Sarah Bull, Miquel Carandell Baruzzi, Margaret Carlyle, Helen Curry, Lucy Delap, Will Dunlop, Patrick Ellis, Lukas Engelmann, Mary Fissell, Rebecca Flemming, Clara Florensa, the late John Forrester, Sarah Franklin, Caitjan Gainty, Richard Harding, Petter Hellström, Maggie Jack, Alex Jacobs, Jelena Jovanovic, Lauren Kassell, Tim Lewens, Lavinia Maddaluno, Lara Marks, Richard McKay, Dmitriy Myelnikov, Ayesha Nathoo, Francis Neary, Eóin Phillips, Sadiah Qureshi, Joanna Radin, Caroline Rusterholz, Jim Secord, Susanne Schmidt, Kathryn Schoefert, Kathryn Ticehurst, Linnea Tillema, Natalie Thomlinson, Caitlin Wylie, and, for impeccable administrative support, Tamara Hug, Aga Lanucha, Louisa Russell, and David Thompson. My Cambridge students, including Alona Bach, Julia Davies, Aleka Gürel, and Amirah Sequeira, taught me a great deal.

In 2013 and 2017, when my children were born, work on the manuscript was greatly facilitated by the hospitality of then director Philippe Lautesse and members of the Science, Society, Historicity, Education, and Practices (S2HEP) laboratory at Claude Bernard University Lyon 1, including Baptiste Baylac-Paouly, Lucie Dalibert, Claire Grino, Ronald Guilloux, Nicolas Lechopier, Thibaut Serviant-Fine, and Jonathan Simon. I completed revisions in 2022 as a visiting fellow of the Institute for Advanced Studies at the University of Lyon, in the stimulating company of Jumana Bayeh, Michael Daniel, Candise Delmas, Olga Dror, Glenn Dutcher, Aparecida Maria Fontes, Madhi Kazempour, Max Kramer, Amine Lahyani, Melissa Marschke, Plinio Smith, Joe Wheaton, and Verónica Zubillaga, an opportunity for which I thank current S2HEP director Catherine Bruguière and successive Institute directors Hervé Joly and Jean-Michel Roy.

In 2016, Solveig Jülich graciously hosted me in the Department of History of Science and Ideas at Uppsala University, where I had the pleasure of also getting to know Eva Åhrén, Annelie Drakman, Helena Franzén, Francis Lee, Lena Lennerhed, Hampus Östh Gustafsson, Morag Ramsey, Mike Sappol, Petter Tistedt, and Sven Widmalm. Since launching the Risky Hormones project with Birgit Nemec in 2020, the Institute for the History of Medicine and Ethics in Medicine at the Charité in Berlin has served as a base of operations, and I am grateful to members and friends of the project, including Samuël Coghe, Kübra Göksel, Axel Hüntelmann, Andreas Jüttemann, Miriam Koppehl, Jennise Krusche, Lisa Malich, Tamar Novick, Martina Schlünder, Anja Suter, Dóra Vargha, Sophia Wagemann, Ludger Wimmelbücker, and Hanna Worliczek. Joanna Dietzel, Peter Dietzel, Maria Dietzel-Papakyriakou, and Glenn Silver deserve special mention for their enduring hospitality and generosity of spirit.

In the School of Humanities at the University of Strathclyde, I am particularly indebted to successive heads of school Kristie Blair and David Murphy and support team members Ann Bartlett, David Cunningham, Jacqueline Daly, Pablo de Castro, Colin Fyfe, Margaret Jolly, Caroline Marley, Maureen McDonald, and Maureen Noor. Since moving to Glasgow in 2018, I have learned much from colleagues, students, and friends, including Daisy Abbott, Patricia Barton, Kirstie Blair, Karen Boyle, Maud Bracke, Tanja Bueltmann, Monica Callaghan, Grant Carpenter, Peder Clark, Hannah-Louise Clarke, Phil Cooke, Fabiola Creed, Petya Eckler, Matt Eisler, Catriona Ellis, Mark Ellis, Erica Fudge, Brian Garvey, Chi Chi Huang, Yun Huang, Sophie

Jones, Laura Kelly, Chloe Khan, Wendy Kline, Poppy Lamberton, Wankun Li, Emma Macrae, Churnjeet Mahn, Holly Marley, Toni Marzal, Yvonne McFadden, Anna McFarlane, Reuben Message, Neil McGarvey, Arthur McIvor, Hussein Medhi, Jim Mills, Esperanza Miyake, Colin Moore, Cédric Moreau, Perside Ndandu, Emma Newlands, Laura Newman, Victoria Peters, Lucy Pickering, Meha Priyadarshini, Hannah Proctor, Nicky Reeves, Stefanie Reher, Elsa Richardson, Luc Richert, Daniela Sime, Matt Smith, Hanaa Taloub, Natasha Telepneva, Angie Turner, Caroline Verdier, Niall Whelehan, Manuela Williams, Enlu Yang, Zang (Jeff) Yifan, and Xioaye Zhu.

It was a magical moment when Katie Helke offered to look at my manuscript, and I remain grateful that she and the MIT Press took a chance on my book. I am likewise obliged to her assistants Suraiya Jetha and Laura Keeler, the series editors Wiebe Bijker and Rebecca Slayton, project manager Madhulika Jain, production manager Emma Martin, as well as Virginia Crossman, Gillian Dickens, and Nitesh Singh for meticulous copyediting, and Edwin Durbin for indexing. I am deeply indebted to the three anonymous reviewers for carefully reading the entire manuscript and offering much thoughtful, practical advice. I take full responsibility for any remaining errors.

A version of chapter 3 first appeared in *Studies in History and Philosophy of Biological and Biomedical Sciences* 47 (2014), 233–247. Parts of chapter 7 repurpose *Reproductive BioMedicine and Society Online* 6 (2018), 34–44. Chapter 10 is summarized in *Journal of British Studies* 59 (2020), 638–642; its final section draws on *Women's History Review* 28 (2019), 869–894. A few paragraphs in chapters 8 and 11 use examples discussed at length in the *British Journal for the History of Science* 50 (2017), 495–520. The most concise articulation of the argument elaborated in this book is "Exhibit 32" in *Reproduction: Antiquity to the Present Day*, edited by Nick Hopwood, Rebecca Flemming, and Lauren Kassell (Cambridge: Cambridge University Press, 2018). A very brief survey of the Canadian history can be found in the *Osler Library Newsletter* 132 (2020), 7–8.

A series of generous grants, especially from the Wellcome Trust (0887 08/Z/09/A, 106553/Z/14/Z), which also paid for open access, sustained the research and writing. An award jointly from the UK Arts and Humanities Research Council (AH/T013281/1) and the German Research Foundation (NE 2419/1-1) funded additional work on chapter 7 and helped pay for the images. Postgraduate scholarships from Wellcome, the Fonds québécois de la recherche sur la société et la culture, Cambridge Commonwealth Trust, and Robinson College, Cambridge, provided additional support in the initial phases. At Wellcome, I thank Tom Bray, Chris Hassan, Margaret Hurley, Sophie Hutchison, Ross MacFarlane, Dan O'Connor, and Angela Saward and, at AHRC, Emily Bultitude and Jamie Davies. I am grateful to the committee members who interviewed for the Trust, especially Gayle Davis, Sarah Hodges, and Stuart Murray.

I owe much to the expertise and generosity of archivists, librarians, and curators, including Michele Beckerman (Rockefeller Archive Center); Tim Boon, Robert Bud, and Katie Dabin (Science Museum); Sue Breakell (University of Brighton); Clare Button, Rachel Hosker, and Stephen Willis (University of Edinburgh); Sophie Clapp (Boots); Philipp Gafner (Novartis); Thore Grimm (Schering); Helen Knowles (Birth Rites Collection); Hannah Lowery (Feminist Archive South); Gemma Read (LSE); Rose Reeve (Feminist Archive North); Arlene Shaner (New York Academy of Medicine); Claire Tunstall (Unilever); Hannah Westhall (Girton College); Anne Wicks (SWWEC); and especially Lesley Hall (Wellcome Collection). Ann Hale, Max Long, and Dmitriy Myelnikov came to my rescue when I was unable to get to a library myself. For imaging, I thank Daniel Hirons (British Library), Domniki Papadimitriou and Johanna Ward (Cambridge University Library), and Holly Peel (Wellcome Collection). Wendy Phillips (Wits University) designed the map and charts.

For opportunities to communicate my research to diverse audiences across media, I thank Eddie Boldger, Helen Castor, John and Margaret Cooper, Isabel Davis, Nadja Durbach, Jason Farrell, Amy Gastelum, Liz Lane, Ina Linge, Clare Murphy, Anne Noyes Saini, Charlotte Sleigh, Rose Stokes, Hermione Wiltshire, and Mary Yearl. I presented parts and versions of the book at numerous seminars. Memorable experiences include post-talk discussion at King's College London and University College Dublin, and I wish to single out a fruitful visit to the University of Sussex, where I benefited from the insightful feedback of Carol Dyhouse, Jim Enderby, and Claire Langhammer. Anne Hardy suggested the title at a seminar convened by Gareth Millward at the London School of Hygiene and Tropical Medicine, and Judith Leavitt made me feel welcome at a crucial juncture.

For generously sharing knowledge and experience, and responding to my many emails, I thank John Abraham, Stuart Anderson, Alison Bashford, Roberta Bivins, Lucy Bland, Fabrice Cahen, Angela Davis, Kate Fisher,

Lara Freidenfelds, Jean-Paul Gaudillière, Carlos Gracia Zamacona, Fiona Hackney, Anne Hanley, Jenna Healey, Volker Hess, Claire Jones, Nils Kessel, Laura King, Rob Kirk, Alison Kraft, Linda Layne, Sarah Leavitt, Jan Moje, Alex Mold, Malcolm Nicolson, Bibia Pavard, Anne Rasmussen, Emily Ross, Christabelle Sethna, Sally Sheldon, Steve Sturdy, Sezin Topçu, Noémi Tousignant, Neil Vargesson, Paul Weindling, Michael Worboys, and especially Ilana Löwy, who with Simon Szreter also examined my thesis. For reading and commenting on a near-final draft of the framing chapters, I thank Salim Al-Gailani, Maud Bracke, Nick Hopwood, Camilla Røstvik, Caroline Rusterholz, and Susanne Schmidt. Feedback on various parts of the book and proposal was provided by fellow members of the History of Reproduction Virtual Circle: Salim Al-Gailani, Jenny Bangham, Branka Bogdan, Heather Dron, Charlotte Greenhalgh, Janet Greenlees, and Birgit Nemec. Karen Weingarten's work on pregnancy testing challenged my thinking and spurred me to submit the manuscript.

A Woman's Right to Know would not have been thinkable without the pioneering work of Ann Oakley and a generation of feminist scholars. It is greatly enriched by the memories, documents, photographs, and other artifacts generously shared by a range of extraordinary people, including Sheila Abdullah, Kraig Adler, Liesel Adler-Kastner, Mary Bernard, Linda Brice, Jenny Corrigall, Meg Crane, Delia Davin, Pat Fincham, Kathy Gal, Gina Glover, Jana Goldsmith, John Hey, Naomi Hull, Lucy King, Eben Kirksey, Bernard Leddy, Marie Lyon, Audrey Peattie, David Petts, Naomi Pfeffer, Rose Reeve, Martin Richards, Roy Summerfield, Maureen Symons, Cynthia Tang, Leif Wide, and Jonathan Wiesner.

I will always be from Montreal, where I was encouraged by Vanier teachers Richard Lock, Chandra Prakash, and Paul Yeargans and McGill historians Nick Dew, Elsbeth Heaman, Thomas Schlich, Frank Stahnisch, Andrea Tone, and George Weisz. Friends who led by example and helped in countless other ways include Nina Chepurniy, Nick Cowan, Isabeau Doucet, Daryl Haggard, Jocelyne Lamarche, Kim Penney, Anna Sheftel, Frank Suerich-Gulick, and Jenn Woo. For childcare in Glasgow and Lyon, I thank Ruth Ni Chumhaill, Marianna Golinucci, Victoria Moundounga, Zoe Richard, and Becci Webster. Above all, I thank my parents, to whom this book is dedicated, as well as Ndollo and the Eboumbou family, for the love and warmth they've all shown to Vera-May and Maxime, the joys of my life.

ABBREVIATIONS

ACDHPT	Association for Children Damaged by Hormone Pregnancy Tests
ALRA	Abortion Law Reform Association
ASA	Advertising Standards Authority
BMA	British Medical Association
BMJ	*British Medical Journal*
BPAS	British Pregnancy Advisory Service
CA	Consumers' Association
CPA	Cape Provincial Administration
CRA	Clinical Research Association
CSD	Committee on Safety of Drugs
DHSS	Department of Health and Social Security
FPA	Family Planning Association
GMC	General Medical Council
HMSO	Her Majesty's Stationery Office
JAMA	*Journal of the American Medical Association*
LSE	London School of Economics
MDU	Medical Defence Union
MJA	Medical Journalists' Association
NHS	National Health Service
RCOG	Royal College of Obstetricians and Gynaecologists

RCPE Royal College of Physicians of Edinburgh
UFAW Universities Federation for Animal Welfare
WNC Women's National Commission

NOTES

CHAPTER 1

1. Leavitt 2006, 339–343; Olszynko-Gryn 2017.

2. Emin 2005, 164.

3. Rothman 1986; Petchesky 1987; Morgan and Michaels 1999; Rapp 1999; Cowan 2008; Morgan 2009; Dubow 2011; Jülich 2011; Franklin and Roberts 2006; Taylor 2008; Roberts 2012; Nicolson and Fleming 2013; Löwy 2014; Löwy 2017; Löwy 2018a; Löwy 2018b.

4. For an overview: Olszynko-Gryn 2018; Drucker 2020.

5. Classic studies of reproductive technologies: Corea 1985; Rowland 1993; Squier 1994; Raymond 1995; Holmes 1994; Stanworth 1987; Lublin 1998; Saetnan et al. 2000; Henig 2004.

6. Duden 1991.

7. McClive 2002; McClive 2015; Gowing 2003; Astbury 2015.

8. Sánchez 2015.

9. Rautelin 2015.

10. Fissell 2003, 64–65.

11. Stolberg 2015, 82–90; Stolberg 2020.

12. Withycombe 2019, 45–46.

13. The practice was abolished in 1931: Butler 2019; Crosby 2019.

14. Jalland 1988, 139; Withycombe 2019, 34–35.

15. Fisher 1998, 35.

16. Jüttemann 2020.

17. Gurdon and Hopwood 2000; Shah 2004; Leavitt 2006; Layne 2010; Tone 2012; Amster 2013, 168–170; Childerhose and MacDonald 2013; Olszynko-Gryn 2014b; Olszynko-Gryn 2017; Olszynko-Gryn 2019; Olszynko-Gryn 2020a; Olszynko-Gryn 2020b; Robinson 2016; Robinson 2022; Kevin 2017; Ross 2018; Cahen 2019; Bärnreuther 2021, 33–34; Lynn 2022; Weingarten 2023.

18. Mathieu 1929, 363.

19. Bruehl 1952, 591.

20. Marcus 2011, 43.

21. For user- and use-based approaches: Cowan 1983; Oudshoorn and Pinch 2003; Edgerton 2011; Olszynko-Gryn 2014a; gender and technology: Wajcman 1991; Cockburn and Ormrod 1993; Horowitz and Mohun 1998; MacKenzie and Wajcman 1999; Lerman et al. 2003; Hicks 2017.

22. Social histories (of birth control): Watkins 1998; Tone 2001; Fisher 2006; Jones 2020; science and technology studies (of sex hormones): Oudshoorn 1994; Clarke 1998; Gaudillière 2018.

23. Newspapers: Bingham 2004; Bingham 2009; Conboy 2006; Bingham and Conboy 2015; magazines: Clay et al. 2017; Forster and Hollows 2020; media histories of medicine and reproduction: Nathoo 2009; Hopwood et al. 2015.

24. Visual histories: Cartwright 1995; Hansen 2009; Serlin 2010; Ostherr 2013; Hopwood 2015; Olszynko-Gryn and Ellis 2017; Bonah et al. 2018.

25. On the importance of infrastructure in histories of science, technology, and medicine: Creager 2004; Creager and Landecker 2009; Bangham 2020; Olszynko-Gryn 2021.

26. Hopwood 2015, 146.

27. Tone 1996; Tone 2000; Tone 2001; recent histories have similarly examined the business of contraception and menstruation: Borge 2020; Jones 2020; Røstvik 2022.

28. The NHS: Webster 2002; Crane and Hand 2022; Seaton 2023; neoliberalism: Davies et al. 2021; Stoller 2023; consumer culture: Hilton 2003; Mold 2015; Tomes 2016; Gurney 2017.

29. Campaigns: Leathard 1980; Brooke 2011; Debenham 2014; Rusterholz 2020; financial independence: Hilton 2002; McCarthy 2020; women's liberation: Jolly 2019; Delap 2020.

30. King 2017; see also Leavitt 2009; King 2015.

31. Oakley 1984; see also Arney 1982; Leavitt 1988; Hiddinga 1995; Eisenberg 2013; Al-Gailani 2018; Wolf 2018; Kline 2019.

32. Leathard 1980; Cook 2004; Marks 2010; Brooke 2011; Hopwood 2018; Dow 2019; Rusterholz 2020; Beers 2022; Sheldon et al. 2022.

33. For comparison, in the same period, his home deliveries declined from 71.3 percent to 1.3 percent: Oakley 1984, 230.

34. Keenan 1971.

35. Voss 1992.

36. Purcell 1997.

37. Hess 2005; Vostral 2008; Freidenfelds 2009; Bivins and Marland 2016.

38. Olszynko-Gryn et al. 2018; Nemec and Olszynko-Gryn 2022.

39. Freidenfelds 2020, 186.

CHAPTER 2

1. Stopes 1929, 29–31; excerpted in Hall 1978, 37–38.

2. On uroscopy in pregnancy diagnosis: Bayon 1939; Forbes 1957; Fissell 2003, 64–65; Stolberg 2015, 82–90; Stolberg 2020; clairvoyant pregnancy detection in the 1920s: Bland 2012; earlier astrological determination: Kassell 2018.

3. Ross 1993, 108.

4. Jacyna 2001.

5. On British debates over the appropriate use of the speculum: Moscucci 1990; Nicolson 2011; Yeniyurt 2014; earlier debates over forceps: Wilson 1995; King 2007.

6. Nicolson 2011.

7. Calder 1906, 56–61.

8. Jardine 1905, 15–16.

9. Jardine 1905, 15. See Rautelin 2015.

10. Oakley 1984, 18–20.

11. Eden 1897, 687–688.

12. William Fetherstone Montgomery's *An Exposition of the Signs and Symptoms of Pregnancy*, first published in 1838, was entirely devoted to the subject: Fleming 1966.

13. Playfair 1893, 154.

14. Galabin 1900, 118.

15. Jardine 1905, 14.

16. Oakley 1984, 18.

17. Galabin 1900, 118–119.

18. Jardine 1905, 7.

19. Galabin 1900, 118; for morning sickness in nineteenth-century Britain: Russell 2012, 126–162.

20. Jardine 1905, 7.

21. Ballantyne 1914, 151.

22. Oakley 1984, 25.

23. Eden 1897, 692.

24. Galabin 1900, 123.

25. Eden 1906, 57.

26. To confirm pregnancy, Simpson used a stethoscope to listen for the fetal heartbeat or placental souffle: Nicolson 2011, 59.

27. Dakin 1897, 61.

28. Ballantyne 1914, 160.

29. Dakin 1897, 64.

30. Herschkorn-Barnu 2002.

31. Galabin 1900, 130.

32. Galabin and Blacker 1910, 186–187.

33. Fothergill 1900, 48.

34. Porter and Hall 1995, 45; Fissell 2003.

35. Rose 2010, 207.

36. *The Works of the Famous Philosopher* (London: Smith, c. 1850s), 81–82.

37. Rosenberg 2003.

38. Bull 1877, 50. See Al-Gailani 2010, 31.

39. Allbutt 1888, 5; original emphasis; on the adverts: Stevenson 1984, 154.

40. Allbutt 1888, 7.

41. Glasson 1901; "Library table," *Lancet*, December 7, 1901, 1592–1594.

42. Dodd 1914, 129.

43. Dodd 1914, 134.

44. The second posthumously revised edition, published in 1914, removed this reference to the German expression; on Ballin's magazine for mothers and baby foods: D'Antonio 2008.

45. Ballin 1904, 1–4; see also Stephens 1910, 88.

46. Ballin 1904, 6–7; see also Dodd 1914, 216; for the pay scale of nurses before World War I: Daniel 1925, 12.

47. On *Expectant Motherhood*: Al-Gailani 2010, 238–241; Seigel 2014, 48–50.

48. Ballantyne 1914, 147–148; *Dr. Chavasse's Advice to a Wife* also condemned the "heinous sin" of abortion before and after quickening: Dodd 1914, 134.

49. Ballantyne 1914, 148–150.

50. Fleck 1979; Löwy 1993; Löwy 2004; Brunton 2004; van den Belt 2011; Hanley 2017.

51. Bröer 2004, 135; on serology in the decades around 1900: Bangham 2022, 27–31.

52. Ballantyne 1913, 362–363.

53. King 1913, 296.

54. King 1913, 298.

55. Ballantyne 1913, 363–366.

56. Ballantyne 1913, 367.

57. Ballantyne 1913, 363; on this debate: Hanson 2004, 9, 95; Al-Gailani 2010, 240.

58. Williamson 1913, 211.

59. Mackenzie Wallis 1913, 256; on Mackenzie Wallis: "Obituary," *BMJ*, April 13, 1929, 710–711; his commercialized pancreatic extract: Edwards 2007, 42–46.

60. Brockman 1913, 1387.

61. King 1913, 296.

62. Mackenzie Wallis 1914, 55.

63. Mackenzie Wallis 1914, 71.

64. Ballantyne 1914, 352.

65. "Review and notices of books," *Lancet*, 181, January 4, 1913, 39–40.

66. "Library table," *Lancet*, 182, November 8, 1913, 1328.

67. "Review and notices of books," *Lancet*, 183, May 16, 1914, 1399.

68. Abderhalden 1914b, viii.

69. Gavronsky 1915, 121.

70. On Michaelis: Deichmann 2007.

71. "Berlin," *BMJ*, February 28, 1914, 508–509.

72. "The serum reaction for pregnancy," *BMJ*, March 7, 1914, 555.

73. "Berlin," *BMJ*, March 21, 1914, 683–684.

74. Leitch 1914, 332; Leitch later became director of the institute: J. A. Murray, "Prof. A. Leitch," *Nature*, 127, February 7, 1931, 206–207.

75. Bullock 1915, 223, 228.

76. Gavronsky 1915, 119.

77. Gavronsky 1915, 120–123.

78. Mackenzie Wallis 1916, 148–149.

79. Mackenzie Wallis 1916, 160–161.

80. Gavronsky 1918, 830.

81. "The Abderhalden reaction," *Lancet*, February 12, 1921, 343; on Abderhalden's reaction as a blood test for madness: Kaasch 2000; Noll 2006.

82. Topley 1917, 213–214.

83. Eden 1919, 75–77.

84. Berkeley and Bonney 1919, 192, 425.

85. Robertson 1921, 216.

86. Munro Kerr et al. 1923, 147–148.

87. Johnstone 1923, 93.

88. Cameron et al. 1924, 50.

89. Smith 1925, 243.

90. Stockham 1918, 35. Derived from the Greek word *τόκος* (*tokos*) meaning "birth," Americans referred generically to pregnancy manuals as "tokologies": Brumberg 1993, 111; on Stockham: Versluis 2008; Silberman 2009.

91. Stopes 1920, 123. See also Geppert 1998.

92. Stopes 1920, 239.

93. Lovat 1921, 63.

94. *For Women Only* (London: Cecil Palmer, 1924), 50, 54.

95. Dorland and Hubeny 1926, 259; Oakley 1984, 100; Howell 1995, 149–150.

96. Claye 1936; Roberts 1938; Oakley 1984, 98.

97. Long and Hirst 1923.

98. On this laboratory: Young et al. 1997.

99. Hirst and Long 1926, 846. See also Kamnitzer and Joseph 1921a; Kamnitzer and Joseph 1921b; Milnor and Fennel 1924, 538; on the importance of scientific articles in the marketing of medical products: Hüntelmann 2013.

100. Hirst and Long 1926, 849–852.

101. Hirst and Long 1926, 853.

102. Quoted in Marsh and Ronner 2008, 56–57.

103. Allan and Doisy 1923.

104. Frank 1929, 247.

105. Siddall 1928a; on Siddall: Myers 2013.

106. Siddall 1928a, 381.

107. "A test for pregnancy," *BMJ*, June 2, 1928, 952–953.

108. Siddall 1928b, 779.

109. Siddall 1928b, 781–782.

110. "Diagnosis of early pregnancy," *BMJ*, February 9, 1929, 259.

111. Finkelstein and Zondek 1966.

112. Schneck 1997.

113. Schneck 1997.

114. Frank 1929, 48.

115. Henriksen 1941, 572.

CHAPTER 3

1. Oakley 1984, 97; Clarke 1998, 320; Gurdon and Hopwood 2000, 45–46; Hanson 2004, 136; Wilmot 2007, 433; McLaren 2012, 100–101.

2. Oakley 1984, 98.

3. Usborne 2007, 180.

4. Brookes 1988, 62–63.

5. Oakley 1984, 97–98.

6. Jones 2007, 135.

7. Fisher 1999, 221–222; Jones 2007, 134.

8. Sengoopta 2006, 281; men with certain testicular tumors also tested positive: Finkelstein and Zondek 1966, 9; Leavitt 2006, 321; Han 2013, 12.

9. McLaren 2012, 101.

10. For a discussion of the ultracentrifuge as a versatile tool: Rheinberger 2010, 130.

11. Clarke 1998, 149.

12. On the merits of use-based history of technology: Edgerton 2006; Edgerton 2010.

13. Rosenberg 1990; Chen 1992; Crenner 2006; Creager 2008; Close-Koening 2011; Worboys 2018.

14. Sturdy and Cooter 1998; Kohler 2008.

15. Singleton and Michael 1993; Clarke and Casper 1996; Casper and Clarke 1998; Löwy 2010.

16. Worboys 2018.

17. Gooday 2008, 786.

18. O'Dowd and Philipp 2000, 21.

19. Digby and Bosanquet 1988; Hardy 2001, 80.

20. Brookes 1988, 62.

21. Foster 1961; Foster 1983; Cunningham 1992; Prüll 1998; Prüll 2003.

22. Stevens 1966.

23. Amsterdamska and Hiddinga 2003.

24. Hutt 1931; Marie 2004; Clarke 2007; Button 2018. See also Margaret Deacon, "The Institute of Animal Genetics at Edinburgh: The first twenty years," typescript, c. 1971, IN1/ACU/A1, University of Edinburgh.

25. Wilmot 2007, 433.

26. Hogben 1974; Porter 1997; Richmond 2007; Ha 2011.

27. Borell 1987; Soloway 1995; Löwy 2009; Löwy 2011.

28. On Macaulay: Li 2003; Smith 2009; the sex congress: Marcuse 1927; Riddle 1927; Dose 2003.

29. On glandular rejuvenation: Hamilton 1986; Sengoopta 2006; Stark 2020; the Vivarium: Mèuller 2017.

30. Interview with Hugh Paterson Donald, July 1971, CA16/1/5, University of Edinburgh.

31. Fletcher to Crew, May 8 and 15, 1928, in "Sexual hormones: research by Dr. B. P. Wiesner, Edinburgh University, 1928–32," FD 1/2816, TNA; on Johnstone: Peel 1976, 221–223.

32. Fletcher to Wiesner, March 13, 1929, FD 1/2816, TNA; on the MRC's limited support for sex physiology: Thomson 1975, 42–43; Austoker and Bryder 1989, 27, 49.

33. As a weakly institutionalized field, sex physiology relied heavily on interdisciplinary collaborations and material exchanges: Oudshoorn 1994; Clarke 1987; Clarke 1998; on lab–clinic collaboration more generally: Hull 2007; Sturdy 2011.

34. Parkes 1966; Borell 1985; Oudshoorn 1994; Gaudillière 2005; Ratmoko 2010; Stoff 2012.

35. See Gaudillière 2010.

36. See Valier 2002; Chen 1992; Oudshoorn 1994, 97; Rader 2004, 171.

37. See Johnstone et al. 1932; "Sex hormone therapy," *Lancet*, September 3, 1932, 525.

38. Sharpey-Schafer 1930, 31; Lawrence 2005, 36.

39. Rogers 1937; on the standardization of therapeutic agents: Gradmann and Simon 2010; von Schwerin et al. 2013; of laboratory animals: Kohler 1994; Rader 2004; Kirk 2008; Kirk 2010; Myelnikov 2015; Bolman 2022.

40. Tansey 1994.

41. Stone 1932, 383; Lawrence 2005, 66.

42. Johnstone 1947, 11.

43. Kirk 2008, 285.

44. Quoted in Bayly 1936, 25.

45. Crew 1930.

46. "Biochemical diagnosis of pregnancy," *Lancet*, October 20, 1928, 834–835.

47. Johnstone 1929a; Johnstone 1929b.

48. "Diagnosis of early pregnancy," *BMJ*, February 9, 1929, 259.

49. Crew 1929a; Crew 1929b. Prior to decimalization on February 15, 1971, the British pound (£) was divided into twenty shillings (s) or 240 pence (d); a guinea (gn) was equivalent to one pound and one shilling. See Cook 2022.

50. Griffiths 1997, 678.

51. Worboys 2018; Newman 2020.

52. Cunningham 1992, 311.

53. *The Practitioner's Guide to Clinical Research: A Review of Recent Advances, Followed by Alphabetically Arranged General Sections* (London: CRA, 1929), 168; on the regulation of packaging in the case of radium: Rentetzi 2011; and more generally: Bauer et al. 2020.

54. Dupree 1997, 100; Worboys 2018; Newman 2020.

55. Foster 1983, 32–37.

56. In Edinburgh, a small hand press was eventually built to extract the urine from the contents (correspondence and all) of broken packets: Crew 1937b, 996.

57. On the technical and secretarial staff of medical research laboratories in mid-twentieth-century Britain: Tansey 2008.

58. Hannan 1930, 637–638.

59. Allan and Dickens 1930.

60. "Biological diagnosis of pregnancy," *Lancet*, January 11, 1930, 94.

61. "Reports of societies," *BMJ*, January 25, 1930, 150–153; on the importance of trust in MRC clinical trials in this period: Cox-Maksimov 1997, 70–80.

62. On molar pregnancies in early modern Europe: McClive 2002; McClive and King 2007; Blanarsch 2009.

63. Johnstone 1934, 267.

64. Aschheim 1929.

65. The incidence at a large teaching hospital was on average only three cases of hydatidiform mole every year and one case of chorioepithelioma every two years: Brews 1939, 814.

66. Austoker 1988; Medina-Domenech and Castañeda 2007; Cantor 2008; Löwy 2009; Moscucci 2009.

67. Donaldson et al. 1936.

68. Wright and Wolf 1930.

69. Blair-Bell 1930, 221; on Blair-Bell: Peel 1986.

70. Robertson 1930, 131.

71. Wiesner to Thomson, February 8, 1930, FD 1/2816, TNA.

72. Wiesner to MRC, February 18, 1931, FD 1/2816, TNA.

73. Johnstone 1930, 175.

74. Crew 1930, 662.

75. On calibration: Collins 1985; Pinch 1993.

76. On medical understandings of pregnancy loss in the first half of the twentieth century: Elliot 2014.

77. Quoted in Lawrence 1998, 99.

78. Dukes 1936; Cameron 1945, 59; Newman 2020.

79. "Accuracy of laboratory diagnosis," *Journal of Clinical Research*, 20, October 1934, 141–142.

80. "New and valuable tests," *Journal of Clinical Research*, 19, April 1933, 63–65.

81. Green-Armytage 1934; "Laboratory tests for pregnancy," *Journal of Clinical Research*, 19, July 1933, 88–90.

82. "Reports of societies," *BMJ*, February 28, 1931, 351–356.

83. "A reliable test for pregnancy," *Lancet*, January 4, 1930, 36–37.

84. Green-Armytage 1934, 53.

85. Stallworthy 1951, 119.

86. Chisolm 1930; for a discussion of similar in Victorian practice: Nicolson 2011.

87. Haultain and Fahmy 1931, 31.

88. Johnstone 1932, 83.

89. Blair-Bell 1934, 149.

90. Bourne 1935, 68–69.

91. Macleod 1936; Davis 2008.

92. Sharpey-Schafer 1930, 31; "Professor's defence of vivisection," *Scotsman*, July 1, 1930, 6; Americans blamed antivivisectionists for the centralization of pregnancy testing in Britain: Harding 1931; Rock 1932.

93. M. J. A. Giles, "The position of pregnancy tests under the Cruelty to Animal Act, 1876," "Vivisection, pregnancy tests: Legal option as to the necessity for licence coverage, 1943–63," HO 45/25145, "Vivisection: Crewe [*sic*], Dr. F. A. E.: Pioneer in animal genetics and pregnancy testing techniques," HO 45/24715, TNA.

94. "Laboratory methods for the diagnosis of pregnancy," *BMJ*, May 24, 1930, 962.

95. Voge 1929; Hannan 1930; Dodds 1930; on Voge's subsequent research on spermicides in Crew's department: Borell 1987; Soloway 1995; Clarke 1998; Clarke 2000; Löwy 2010; Szuhan 2018; Szuhan 2022.

96. Friedman and Lapham 1931.

97. Leavitt 2006.

98. Wiesner 1931; Wiesner 1932.

99. On the London hospitals: Rivett 1986.

100. Ford 1935, 6; "Friedman's pregnancy test," *BMJ*, February 2, 1935, 211.

101. Bamforth 1936, 132.

102. Ralph 1934, 57.

103. Bishop 1932; Bishop 1933; Bishop 1934.

104. Crew 1936a, 993.

105. Crew 1937b, 990.

106. Crew 1936b, 1093.

107. Johnstone 1933, 557.

108. Deacon, "The Institute of Animal Genetics."

109. Li 2003.

110. Interviews with F. A. E. Crew, 1969–1971, CA16, University of Edinburgh; on the research: Wiesner and Sheard 1933.

111. Lane-Roberts et al. 1939; Lane-Roberts et al. 1948; Pfeffer 1987; Pfeffer 1993.

112. Richards 2008; today Wiesner is best known as a profligate sperm donor: Davis 2017.

113. Crew 1934, 531.

114. "The female sex hormones," *BMJ*, July 18, 1936, 126; also Robson 1934a; Robson 1934b.

115. Cameron 1936, 331.

116. Donaldson et al. 1936, 28–29.

117. Hannan 1933, 1047.

118. Crew 1936b, 1092.

119. Dodds 1936, 225.

120. Patterson 1937, 524.

121. "Another test for pregnancy," *BMJ*, October 16, 1937, 757.

122. Gill and Howkins 1937, 1069; on Parke, Davis's commercial growth hormone: Medeiros 2016, 63–68.

123. Keevil 1937; Sharman 1937; Sharman 1938.

124. Kanter et al. 1934, 2027.

125. Way 1937, 1143.

126. Crew 1939, 767.

127. Brews 1935, 1225.

128. "Hydatidiform mole," *Lancet*, April 6, 1935, 823.

129. "Hydatidiform mole and chorionepithelioma," *Lancet*, February 4, 1939, 283–285; also Brews 1939.

130. Crew 1936a, 993.

131. Jeffries 1935.

132. Crew 1937b, 994; on the widespread use of abortifacients: Brown 1977; Knight 1977; Weeks 2018, 75–76.

133. "Evidence of Professor F. A. E. Crew, Inter-Departmental Committee on Abortion, Tuesday, 21 June 1938," A.C. Paper 53, MH 71/27, "Memorandum of Professor Crew," A.C. Paper 138, MH 71/26, TNA.

134. Birth control clinics in the United States used the Aschheim–Zondek test to reject pregnant patients: Hajo 2010, 56, 203.

135. B. M. Hobson, "Facilities for pregnancy diagnosis in Britain," 1966, HH 102/858, NRS.

136. "The eleventh British Congress of Obstetrics and Gynaecology," *Journal of Obstetrics and Gynaecology of the British Empire*, 46, June 1939, 582–589, 585.

137. Adler-Kastner 1998; on medical refugees in Britain: Collins 2009; Weindling 2009a; Weindling 2009b.

138. Crew 1939, 768.

CHAPTER 4

1. Interview with Hugh Paterson Donald, July 1971, CA16/1/5, University of Edinburgh.

2. Quoted in Hall 2000, 290.

3. Brooke 2011, 102.

4. "Miss Browne's meetings," *New Generation*, 10, March 1931, 29; Hall 2011, 178.

5. For example, Brookes 1988, 75; McIntosh 2000, 88; Usborne 2007, 134, 149.

6. Geppert 1998.

7. Fisher 2006, 27; see also Fisher 2018.

8. Szreter and Fisher 2010, 90; see also Fisher 2018.

9. Leap and Hunter 1993, 72–82; Alexander 1996; Vincent 1998; Szreter and Fisher 2010, 64–65; for a survey of sex manuals: Cook 2004, 341–354; see also Porter and Hall 1995.

10. Sutton 1992, 58.

11. Glucksmann 1990, 108; on pregnancy in the workplace: McCarthy 2020; Millward 2022.

12. Sutton 1992, 57–58.

13. Humphries 1988, 75.

14. Kate Fisher to author, October 18, 2010.

15. Brittain 1957, 54; Longford 1986, 140–141.

16. Lehmann 1936, 228–229.

17. "New novels," *Times*, July 7, 1936, 11.

18. Beauman 1983, 3.

19. McAleer 1992.

20. Beauman 1983, 10.

21. Ingman 1998; Humble 2001; Hinds 2009; Brown and Grover 2012; Jouannou 2013.

22. Macaulay 1926, 132–138.

23. Braybrooke 1927, 45–47; see Bensen 1969, 98; Passty 1988, 14–15; Beauman 1983, 155; Macdonald 2018, 220.

24. Jesse 1934, 273.

25. Blumenfeld 1935, 200–203.

26. Bourke 2003; Overy 2009; Jackson 2013.

27. Mitchison 1935, 487–492; see Mitchison 1979, 14; Chan 2019; Dougall 2021.

28. Fisher 1998, 35.

29. Hackney 2008, 115–116.

30. See contributions especially to part III of Clay et al. 2017.

31. Barrell and Braithwaite 1988, 13; Hackney 2008, 117–118.

32. *Woman's Own* innovatively coordinated its commercial and editorial messages about domesticity, motherhood, and shopping: Greenfield and Reid 1998, 161.

33. White 1970; Tinkler 1995; Barrell and Braithwaite 1988; Greenfield and Reid 1998; Hackney 2007; Hackney 2008; Hackney 2010; Clay et al. 2017; Forster and Hollows 2020.

34. Greenfield and Reid 1998, 170.

35. For example, *Woman's Illustrated*, August 21, 1937, 16; September 26, 1937, 26.

36. Newspapers invested considerable resources in responding to hundreds of real letters every week: Bingham 2012, 52.

37. *Mother*, May 1938, 40.

38. *Woman's Own*, November 11, 1933, 138; on hormone therapy in menopause: Watkins 2007.

39. *Mother*, December 1936, 53.

40. *Mother*, September 1939, 36–37.

41. *Mother*, October 1937, 55.

42. *Mother*, March 1938, 54.

43. *Mother*, June 1938, 49.

44. *Mother*, September 1939, 36–37.

45. *Woman's Illustrated*, February 6, 1937, 45.

46. *Woman's Illustrated*, July 17, 1937, 20.

47. *Woman's Illustrated*, April 17, 1937, 46.

48. *Woman*, July 9, 1938, 41.

49. Hackney 2008, 115–116; Bingham 2004, 102–105.

50. Bingham 2009, 129.

51. Bingham 2004, 178.

52. Sharpey-Schafer 1930, 31.

53. "Professor's defence of vivisection," *Scotsman*, July 1, 1930, 6.

54. "Defence of vivisection," *Times*, July 1, 1930, 18; on the public debate over vivisection in the late 1920s: Tansey 1994.

55. "Jews in Germany," *Times*, April 26, 1933, 12.

56. "A Jewish token of gratitude," *Manchester Guardian*, January 22, 1934, 11; see also "Manchester Jewish Hospital new appointments," *Manchester Guardian*, September 30, 1933, 13, "The Zondek brothers," *Manchester Guardian*, July 10, 1934, 10; Aschheim and Zondek were shortlisted for the Nobel Prize four times in the 1930s: Björk 2001, 399.

57. *The Daily Mirror*, December 1, 1937, 27.

58. Keown 1988, 49–59; Brookes and Roth, 1994.

59. "Surgeon who 'deliberately defied the law,'" *Daily Telegraph*, July 2, 1938; "Surgeon for trial," *Hull Daily Mail*, July 1, 1938, 14.

60. "Surgeon found 'not guilty,'" *Times*, July 20, 1938, 9.

61. Sir Bernard Spilsbury, "Rex v. Dr. Bourne. Report. Respecting the diagnosis of early pregnancy," July 16, 1938, DPP 2/564, TNA; on Spilsbury: Burney and Pemberton 2011.

62. Mechling 1975; K. Brown 2003; Sauerteig 2009; Seigel 2014.

63. Porter and Hall 1995, 259–260.

64. On the importance of gay-oriented mail-order catalogues in the 1960s: Johnson 2019; on mail-order retailing in Britain: Coopey et al. 2005.

65. Davin 1978, 47–48; Hardyment 1983, 167–179; Thomas 1997; Raphael 2010; McIntosh 2012, 60; Davis 2012, 115–119.

66. Porter and Hall 1995; Rose 2010, 214.

67. Comrie 1931, 292; on Comrie: "Obituary," *BMJ*, October 14, 1939, 789; Pitman 1993.

68. *The Concise Home Doctor* (London: Amalgamated Press, 1932), 1055–1056.

69. Cox 1933, 292.

70. *The Universal Home Doctor Illustrated* (London: Odhams Press, 1934), 567.

71. Erskine 1929, 131.

72. Pink 1930, 13–14.

73. Arbuthnot Lane 1934, 490; on the society: Zweiniger-Bargielowska 2005; Zweiniger-Bargielowska 2007.

74. Meredith 1935, 350.

75. Scott 1936, 2135–2137.

76. Wells et al. 1935; Crew 1932; Crew 1933; see Bowler 2006, 173; Bowler 2009, 106.

77. Crew 1930; Crew 1939; on the BBC: Nicholas 1995; radio science: LaFollette, 2008; Hall 2021; Long 2022.

78. Wiesner 1936, 62; on the library: Bowler 2009, 114–142.

79. Van de Velde 1931, 133, 283–285; see Bullough 1994, 601; Melody and Peterson 1999; on Heinemann: St. John 1990.

80. Thesing 1935, 116.

81. "The Listener's Book Chronicle," *Listener*, June 19, 1935, 1068; on *The Listener*: Kavanagh 1999, 83–84.

82. Griffith 1939, 98–99.

83. Stone and Stone 1936, 116–117; on the book club: Samuels 1966; Neavill 1971.

84. Batten 1939, 21.

85. Pink 1941, 3; on Abrams: Lawrence 1999, 439–441.

86. Moscucci 2003, 170.

87. Fisher 2006.

88. Lewis 1980, 209; Brooke 2011, 95.

89. Crew 1937b, 989.

90. On democratization: LeMahieu 1988; Hilliard 2006.

91. "Memorandum of Professor Crew," A.C. Paper 138, MH 71/26.

92. "Evidence of Professor F. A. E. Crew," MH 71/27.

93. See "Death of Lady Ruth Balfour," *Glasgow Herald*, August 31, 1967, 7.

94. "Memorandum of Professor Crew."

95. "Memorandum of Professor Crew."

96. Ministry of Health and Home Office 1939, 110.

97. Brooke 2001a.

98. Brooke 2001a, 150; on abortifacients: Brown 1977; Knight 1977; Weeks 2018, 75–76.

99. Malleson 1939, 366–367.

100. Browne 1939, 478.

101. Blacker 1939, 93.

102. Thurtle 1940a, 15–16.

103. Thurtle 1940b, 75–76.

104. Thurtle 1940a, 15–16.

105. de Chadaravian 2002, 20–49.

106. See Crew to Chief Medical Officer, Scotland, January 21, 1946, HH 102/858, NRS; Sher 1946.

107. On the EMS: Dunn 1952–1953; blood transfusion during World War II: Whitfield 2013.

108. On the ELS: "Public health in the war years: Ministry of Health retrospect," *BMJ*, September 21, 1946, 432–434; Valier and Timmermann, 2008, 497; on the Scottish system: "Scotland: Emergency Bacteriological Service," *BMJ*, February 10, 1940, 228; Dunn 1952–1953, Vol. 2, 64–68; Balmer 2001, 22–24.

109. Schuster 1983, 21; on rabbit as human food: Zweiniger-Bargielowska 2000; Zweiniger-Bargielowska 2011; Martin 2010, 271; supply issues: "The University of Edinburgh Pregnancy Diagnosis Laboratory," HH 102/858, NRS.

110. Schuster 1983, 21.

111. Giles, "The position of pregnancy tests under the Cruelty to Animals Act, 1876," February 2, 1944, 2, "Vivisection: Pregnancy tests: legal opinion as to the necessity for licence coverage," HO 45/25145.

112. Summerfield and Crockett 1992; Summerfield 1998; McCarthy 2020.

113. Hall 2013.

114. Riley 1981.

115. Soloway 1990, 312.

116. Oakley 1984, 125.

117. "Extra foods for the expectant mother," *BMJ*, July 24, 1943, 15.

118. Zweiniger-Bargielowska 2000, 131, 144.

119. *Mother*, September 1941, 32.

120. Soloway 1990, 313.

121. *Mother*, October 1939, 23, 83.

122. *Mother*, July 1940, 34.

123. *Mother*, July 1940.

124. On women and antivivisectionism: Lansbury 1985; Elston 1987; Ferguson 1998.

125. *Mother*, July 1940, 34.

126. *Mother*, January 1941, 35.

127. *Mother*, February 1941, 12–13.

128. "Motherhood ahead," *Mother*, June 1942, 71–72.

129. *Mother*, July 1942, 71–72; see McCarthy 2020, 122.

130. "What are the first signs?," *Mother*, April 1943, 79–80.

131. Dick-Read 1942, 131; see also Thomas 1997; Michaels 2014.

132. Diarist 5247, "Mass Observation Diaries 1939–1945," SxMOA1/4; on Mass Observation and women: Sheridan 1994; Langhamer 2007; see also *The Mass-Observation Diaries: An Introduction* (Brighton: University of Sussex, 1991).

133. *Mother*, April 1945, 63.

134. Quoted in Oakley 1984, 97–98.

CHAPTER 5

1. "'The frog' becomes 'the squeaker,'" *Daily Mirror*, February 8, 1938, 2.

2. "Frogs that 'say mother,'" *Daily Mirror*, August 30, 1938, 25.

3. Via pregnancy testing, *Xenopus* eggs and embryos would also rise to international prominence as an experimental system in developmental biology: Gurdon and Hopwood 2000; Crowe 2021, 105–112.

4. Crew 1939.

5. Hogben 1936; Erlingsson 2016.

6. Wells 1978, 205–206; Zwarenstein 1985; Hogben and Hogben 1998, 101; Bröer 2004, 140–141; Haarburger and Pillay 2011; Sleigh 2012, 136.

7. On the university and department: Phillips 1993; A. C. Brown 2003.

8. Hogben and Slome 1931.

9. Hogben 1930; it was standard practice in comparative physiology to remove a gland from a test animal and then attempt to restore the action of the removed gland with injections of glandular extract: Schlich 2010.

10. Gurdon and Hopwood 2000, 45.

11. For social biology at the LSE: Renwick 2013.

12. Bellerby 1933, 616.

13. Bellerby 1934; on *Nature*: Baldwin 2015.

14. Bellerby 1934, 494.

15. Robson 1934a, 1064.

16. Schonland 1934, lxxv; on Zwarenstein and Shapiro: "In memoriam," *South African Medical Journal*, 66, November 24, 1984, 824; "In memoriam," *South African Medical Journal*, 86, August 1996, 989; Zwarenstein 1986; Gevers 1998; the physiology department: Sloan 1979; the royal society: Carruthers, 2010.

17. "Societies and academies," *Nature* 133, March 3, 1934, 338–340.

18. Shapiro and Zwarenstein 1934; on Golberg: "In memoriam," *South African Medical Journal*, 69, March 29, 1986, 461.

19. Shapiro and Zwarenstein 1934.

20. Shapiro and Zwarenstein 1935.

21. Zwarenstein and Shapiro 1933, 372.

22. Shapiro and Zwarenstein 1933, 189.

23. Alexander and Bellerby 1935, 312; on pathologically crowded rats: Ramsden 2011.

24. Crew 1936b.

25. Crew 1939, 766; Landgrebe 1939, 94.

26. See "In memoriam," *South African Medical Journal*, December 13, 1947, 922–923.

27. Bosman 1937.

28. Crew 1937a.

29. Crew 1939, 768.

30. Crew acknowledged the assistance of Miss Marie Mackay and Miss Janet Meikle; on "invisible" technicians: Shapin 1989; Tansey 2008; Bangham et al. 2022.

31. Crew 1939, 768.

32. Crew 1939, 766–768.

33. Crew 1939, 767; mice were still needed to produce graded results in differential diagnosis (see chapter 3).

34. Gunn 1939a.

35. Hogben 1939.

36. Gunn 1939b.

37. "Letters, notes, and answers," *BMJ*, February 16, 1946, 262–264.

38. Milton 1946.

39. Hogben 1946a.

40. Shapiro and Zwarenstein 1946.

41. Hogben 1946b; on scientific patriotism in South Africa: Dubow 2006; Dubow 2020; Beinart and Dubow 2021.

42. Landgrebe 1946, 963.

43. See "Obituary," *BMJ*, October 8, 1983, 1071; Edward Elkan, "Sketches from my life," typescript, 1983, MS.9151, WCL; Proctor 1988, 271–272; Cooper 1989; Williams 1994; Adler 2007, 193–194; von Villiez 2009, 258–259.

44. Elkan, "Sketches," 51–52; see also Grossmann 1995, 24, 119, 134–135, 153, 180, 186; on Hodan: Lennerhed 2009, 60–61; Sauerteig 2009, 133–136.

45. Elkan, "Sketches," 51; on medical refugees in Britain: Weindling 2009a; Weindling 2009b.

46. On Ottesen-Jensen: Lennerhed 2009, 59–62.

47. On Wright: Evans 1984; Szuhan 2022, 85–143; on the Zurich conference: Grossmann 1995, 41–43.

48. Elkan, "Sketches," 52–53. On refugee physicians in Scotland: Collins 2009.

49. Elkan, "Sketches," 54–55.

50. Jackson 2013, 115–116; on the Papworth experiment: Bryder 1984.

51. Obermer 1935a; Obermer 1935b.

52. Elkan, "Sketches," 54.

53. Elkan 1938b, 1253.

54. Elkan 1938a, 313.

55. Stevenson and Cook 2010, 18; on electrification in Britain: Luckin 1990; Gooday 2015; Weightman 2011.

56. Hodge 1927, 11.

57. Hamera 2012, 193; Klee 2003, 152–160; Grier 2006, 255.

58. Quoted in Naomi Hull to author, March 9, 2014.

59. See Elkan 1938a; Elkan 1938b; Elkan 1939a; Elkan 1939b Elkan 1939c; Elkan 1940; Elkan 1946a; Elkan 1946b; Elkan 1947a; Elkan 1947b.

60. Elkan 1938a, 315.

61. Elkan 1939b; on Vichy mineral water: Gordon 2012.

62. Elkan 1947a, 104.

63. Elkan 1938b, 1255.

64. Elkan 1946b.

65. Elkan 1938b.

66. Elkan 1938b.

67. Elkan 1947a, 255.

68. Elkan, "Sketches," 56.

69. Survival rate of tadpoles: Elkan 1938b, 1253; frog farming prediction: Elkan 1939b; pathologist's pet: Elkan 1939c, 95.

70. Elkan 1940; promotion of rabbits as food: Zweiniger-Bargielowska 2000; Zweiniger-Bargielowska 2011; Martin 2010.

71. On the internment of enemy aliens: Cesarani and Kushner 1993.

72. Elkan, "Sketches," 58.

73. Ministry of Health vivisection minutes, February 1, 1944, HO 45/25145, TNA.

74. Kennedy to Giles, December 31, 1943, HO 45/25145; see "Obituary," *BMJ*, 284, May 1, 1982, 1339–1340.

75. Obermer 1947, 1948.

76. "The position of pregnancy tests under the Cruelty to Animals Act, 1876," February 2, 1944, 2, HO 45/25145, TNA; on Giles: "Obituary," *BMJ*, July 22, 1961, 248; see also Lawrence 1987.

77. Ministry of Health vivisection minutes, February 7, 1944, HO 45/25145, TNA.

78. "The position of pregnancy tests."

79. "The position of pregnancy tests."

80. Ministry of Health vivisection minutes, February 1, 1944, HO 45/25145, TNA.

81. Hume 1947, 7; on the handbook: Kirk 2009, 528–529; as late as the early 1960s, the Home Office was still clarifying to pathologists that pregnancy tests fell outside the 1876 act: Hume 1962, 2; Beedle to Calman, "Cruelty to Animals Act, 1876," July 11, 1963, HO 45/25145, TNA.

82. Gurdon and Hopwood 2000, 47.

83. Elkan, "Sketches," 58.

84. Parkes 1947; on the students: Bangham 1999, 57.

85. Bury 1953, 145.

86. Wilson 1948.

87. Nottingham 2000; McCrae 2003; Stewart 2003; McTavish 2005.

88. DHS Circular No. 97/1947, Graham, "Bacteriological service," October 6, 1947, HH 102/858, NRS.

89. Crew 1971, 293.

90. Crew's draft letter to Davidson, No. 11, undated, HH 102/858, NRS.

91. Import difficulties: "Letters, notes, and answers," *BMJ*, February 16, 1946, 262–264; mice: B. M. Hobson to R. W. Johnstone, November 13, 1953, GD11; see also Johnstone 1954, 92.

92. Maclean 2004, 59; on Waddington: Peterson 2016; Squier 2017; Park 2019.

93. Wells 1978, 206.

94. Hobson 1952, 352.

95. Crew to Davidson, June 20, 1947, Crew's draft letter to Davidson, 1947, Crew to Davidson, June 20, 1947, HH 102/858, NRS.

96. Crew to Davidson, March 25, 1947, HH 102/858, NRS.

97. See "Obituary," *BMJ*, March 25, 1962, 880–881; Davidson 1948.

98. Crew to Davidson, January 21, 1946, Crew to Sir Andrew, November 25, 1946, HH 102/858, NRS.

99. Stewart 2003, 399–400.

100. Hobson to Callan Wilson, August 17, 1949, notes of meeting with Dr. Bowman, December 25, 1949, "Pregnancy diagnosis," December 30, 1949, HH 102/858, NRS.

101. "Pregnancy diagnosis service," June 7, 1951, Pringle to Callan Wilson, June 1, 1950, HH 102/858, NRS.

102. DHS Minutes, June 16, 1947, Hughes to Taylor, May 28, 1951, Pringle to Callan Wilson, June 1, 1950, HH 102/858, NRS; on Douglas: "Obituary," *BMJ*, March 29, 1980, 948.

103. Beek to Watson, June 10, 1947, Crew to Davidson, December 9, 1947, HH 102/858, NRS; on Panton: "Obituary," *Journal of Clinical Pathology*, 4, February 1951, 109–111; Panton 1951.

104. Schwabacher 1950; Elkan, "Sketches," 58; on Schwabacher: "Obituary," *BMJ*, November 10, 1984, 1319.

105. Schwabacher 1951, 84.

106. In the 1950s, parents continued to send unmarried pregnant daughters to workhouses until childbirth, at which point newborns would be put up for adoption: Spensky 1992; Kiernan et al. 1998; Keating 2009, 31; Gallwey 2011; Thane and Evans 2012.

107. Schwabacher 1951, 84; this was a generalized concern in the early years of the NHS, most famously regarding dentures and spectacles but also X-rays and other diagnostic tests: Loudon and Drury 1988, 108–109.

108. Schwabacher 1951, 84.

109. N.E. Metropolitan Regional Hospital Board Circular No. (50) 150, HA/NT/117/000702, LMA.

110. Shea and Warrack 1963, 581.

111. "Toad supply gives out: Hospital centre closed," *Manchester Guardian*, November 6, 1954, 1; "Toads still at work: Edinburgh aids Sheffield," *Manchester Guardian*, November 10, 1954, 8; see also "Pregnancy diagnosis," *Lancet*, November 13, 1954, 1025; "Toad supply," *BMJ*, November 20, 1954, 1229.

112. Shea and Warrack 1963, 581.

113. Hitchens and Lowe 1966, 144.

114. Johnstone 1954, 92.

115. Quoted in Wells 1978, 206.

116. Schwabacher 1956, 392.

CHAPTER 6

1. Campion 1950, 7.

2. Campion 1950, 5; "Consumer's end," *Manchester Guardian*, November 17, 1950, 4; see also Hanson 2004, 132; compare with Knott 2019.

3. Campion 1950, 8.

4. Campion 1950, 10–11.

5. Campion 1950, 20.

6. Campion 1950, 23.

7. Sage 2000, 236.

8. Interview by Amanda Kennet with Sheila Walker, April 4, 1999, C900/16008 C1, Millennium Memory Bank, British Library; on the project: Gallwey 2013.

9. Interview by Angela Davis with Hope, CO11; for details: Davis 2012, 219.

10. Rayner started out writing scripts for BBC Radio 4's *Women's Hour*, which were later published as educational leaflets: Loughlin, 2005, 306.

11. Rayner 2003, 311–312.

12. Maureen Symons to author, April 16, 2013.

13. Interview with Audrey Peattie, August 1, 2011; see Olszynko-Gryn 2013.

14. "Twenty years of pregnancy tests," *Lancet*, December 18, 1948, 977–979; see further Cowie 1948.

15. Galli Mainini 1947a; Galli Mainini 1947b; Galli Mainini 1948a; Galli Mainini 1948b; Galli Mainini 1948c; Bröer 2004, 141; Haarburger and Pillay 2011, 546; on endocrinology in Latin America: Eraso 2007; Rustoyburu and Eraso 2018; Haines: Young 2009, 151; Chelsea Hospital for Women: Blaikley and Chamberlain 1971.

16. Haines 1948, 923.

17. Haines 1948, 924.

18. "Twenty years of pregnancy tests."

19. Lane-Petter 1961, 24–25.

20. Haines 1948, 926.

21. Frazer 1950; Frazer and Wohlzogen 1949, 1950.

22. Klopper and Frank 1949, 9–10.

23. Bishop 1951, 101.

24. Barton 1953.

25. Jeffree 1953, 151.

26. Ferreira 1954.

27. Ferreira 1954, 358.

28. Russell and Burch 1959, 53.

29. Allison 1957, 786.

30. Allison 1955, 282.

31. Elkan 1960.

32. Pat Fincham to author, April 19, 2013.

33. David Petts to author, March 7, 2013.

34. Linda Brice to author, May 1, 2013.

35. Nathoo 2009, 38–48.

36. "Family doctor," *BMJ*, March 17, 1951, 571–572; Nathoo 2009, 207n26; see also Bartrip 1996, 285.

37. Kemp 1959; see "Obituary," *BMJ*, March 10, 1990, 673–674.

38. Kemp 1959.

39. Some American gynecologists used blood serum for the *Xenopus* test when they wanted to conceal the true nature of the test: Rosenfeld and Rosenfeld 1944, 529.

40. Kemp 1959.

41. Kemp 1959.

42. Greene 1999; Dry 2006.

43. Kemp 1959.

44. Kemp 1959.

45. Pumphrey to FPA, June 28, 1949, FPA to Pumphrey, June 30, 1949, SA/FPA/A3/11, WCL.

46. Leathard 1980, 73–74.

47. Pfeffer 1993, 109.

48. Brookes 1988, 136.

49. Leathard 1980, 79–80; see further Rusterholz 2020; Szuhan 2022.

50. Leathard 1980, 73; see Pfeffer 1993, 59–60; Szuhan 2018; Szuhan 2022; Beers 2022.

51. Pfeffer 1993, 105; for an earlier history of sperm analysis: Benninghaus 2012.

52. Davidson 1949; Davidson 1953.

53. F. Robertson, "Should a family be planned?," *Picture Post*, June 12, 1948, 11–13.

54. Leathard 1980, 81.

55. Oakley 1984, 97.

56. Nicholson 1995, 98; Martin Richards to author, February 14, 2013.

57. Wright to Schwabacher, February 15, 1949, SA/FPA/A3/11, WCL.

58. Wright to Medical and Sub-fertility Committee, January 20, 1949, SA/FPA/A3/11, WCL.

59. Wright to Guy, January 21, 1949, Wright to Crew, July 1, 1949, SA/FPA/A3/11, WCL.

60. Executive minutes, March 12, 1949, SA/FPA/A3/11; on Helena and Beric Wright: Evans 1984; Rusterholz 2020.

61. Wright, April 13, 1949, SA/FPA/A3/11, WCL.

62. E. K. Bowman Ltd., to Wright, February 23, 1949, Wright to Guy, January 21, 1949, SA/FPA/A3/11, WCL.

63. Wright to Schwabacher, February 15, 1949, Schwabacher to Wright, February 18, 1949, Forbes to Wright, March 1, 1948, SA/FPA/A3/11, WCL; on the union: Forbes 1948.

64. Unsigned letters to R. Christie Brown, J. Carrie, G. Graff and H. I. Winner, March 29, 1949, Robinson to Godber, April 1, 1949, Godber to Robinson, April 2, 1949, SA/FPA/A3/11, WCL; on Godber: Craft 1975.

65. Robinson to Godber, April 6, 1949, Godber to Robinson, April 6, 1949, SA/FPA/A3/11, WCL.

66. Wright to Crew, July 1, 1949, letter to Wright, July 5, 1949, SA/FPA/A3/11, WCL.

67. On Thomas Cook: Brendon 1991.

68. *Inland Fisheries Department Report* 1, 1944, 31; the now contested term "coloured" referred under apartheid to a heterogeneous group of people with mixed European, African, and Asian ancestry: Adhikari 2005; compare with rural yam pickers in Mexico, who supplied the raw materials for the first oral contraceptives: Soto Laveaga 2009.

69. *Inland Fisheries Department Report* 1, 1944, 32.

70. Hey 1949, 45.

71. Hey 1949.

72. Hey 1977, 109.

73. Hey 1977, 108.

74. Hey 1949, 52.

75. Weldon et al. 2004; van Sittert and Measey 2016.

76. Hey 1949, 54–55.

77. Hey 1949.

78. "Seminological centre," *Lancet*, April 16, 1949, 676; "Family Planning Association," *BMJ*, April 30, 1949, 790.

79. "Diagnosis of pregnancy," *Chemist & Druggist*, July 2, 1949, 11.

80. Wright 1949a.

81. Commercial laboratories have left little trace, and it is difficult to recover the scale. Welbeck Laboratories and Camden Chemical Co. are occasionally mentioned in passing but not discussed at any length.

82. van Abbé 1949, 78.

83. Anderson and Berridge 2000, 64–68.

84. Wright 1949b, 115; on postwar austerity: Zweiniger-Bargielowska 2000; Kynaston 2010.

85. Draft letter, SA/FPA/A3/11, WCL.

86. Robinson to Lloyd, August 15, 1949, General Secretary to Hadfield, December 1, 1949, SA/FPA/A3/11, WCL.

87. Secretary to *Pharmaceutical Journal*, August 2, 1949; "'Radar' for the stork," *Reveille*, August 27, 1949, SA/FPA/A3/11, WCL; on *Reveille*: Hoggart 1957; Owen 2005.

88. "'Radar' for the stork."

89. "New science aids family planning," *South London Advertiser*, September 16, 1949, SA/FPA/A3/11, WCL.

90. "Trial by toads," *News Review*, December 22, 1949, SA/FPA/A3/11, WCL; on postwar businesswomen: Perriton 2007; women's leisure: Langhamer 2000; more generally: Brooke 2001b; Thomas 2008; Tinkler et al. 2017.

91. On the appeal: Colpus 2011.

92. Wright to Pyke, August 20, 1949, Pyke to Wright, August 24, 1949, Wright to Pyke, August 28, 1949, General Secretary to Wright, September 19, 1949, SA/FPA/A3/11, WCL.

93. General Secretary to Mr. Avory, February 27, 1950, letter to chemists, January 31, 1950, SA/FPA/A3/11, WCL.

94. Wright to Finance Committee, A. Restall, Ltd. to Wright, May 11, 1950, Wright to Finance Committee, May 19, 1950, SA/FPA/A3/11, WCL.

95. Wright to Leach, May 26, 1950, Leach to Thos. Cook, May 31, 1950, Wright to Leach, July 19, 1950, Wright to Pyke, July 23, 1950, General Secretary to Wright, October 20, 1950, Wright to Finance Committee, October 23, 1950, SA/FPA/A3/11, WCL.

96. Wright to Finance Committee, January 26, 1951, Wright, "Report 1950/51 and budget 1951/2," February 17, 1951, SA/FPA/A3/11, WCL.

97. "Pregnancy test," *Woman's World*, January 6, 1951.

98. *Mother*, February 1952, 69.

99. Shepherd 1994, 94.

100. Interviews, Beric Wright, PP/PRE/J.1/18, WCL.

101. Florence 1956, 33.

CHAPTER 7

1. Greer 1984, 169–170; on the RCGP evidence: Heneghan and Aronson 2021; on the regulatory history, including in Sweden and Finland: Olszynko-Gryn et al. 2018.

2. Henriksen 1941.

3. For example, "The skin test for pregnancy," *JAMA*, 93, August 17, 1929, 559.

4. Soskin et al. 1940.

5. Weisman 1940; Crew's assistant, Janet Meikle, had a small batch shipped to New York, where Weisman enlisted Christopher Coates, an aquarist at the Bronx Zoo, to domesticate *Xenopus* for use in pregnancy diagnosis: Weisman and Coates, 1944, xiii; "Frog test: Egg-laying by African frog yields quick diagnosis of human pregnancy," *Life*, April 24, 1944, 87–88; "Christopher W. Coates, 74, dies; headed New York Aquarium," *New York Times*, January 1, 1974, 22; "A. I. Weisman, 81, fertility specialist, professor and writer," *New York Times*, December 24, 1990, 48.

6. Winkelstein 1942, 239.

7. Zondek 1942. See Finkelstein and Zondek 1966; Novick 2023.

8. Zondek and Rozin 1948.

9. Matthew and Hobson 1953; on Matthew: "Obituary," *BMJ*, 292, 1678.

10. Matthew 1956, 979.

11. "Diagnosis of early pregnancy," *BMJ*, October 27, 1956, 987–989.

12. "Diagnosis of early pregnancy."

13. Britton 1956a, 419.

14. Britton 1956b, 1118.

15. Nixon 1956, 314, 321.

16. Lambert 1956, 118.

17. On Organon: Tausk 1984; Oudshoorn 1994.

18. *An Introduction to Endocrinology* (London: Organon Laboratories, 1957), 35; see also *Everyday Treatment of Endocrine Disorders* (London: Organon Laboratories, 1959), 83.

19. Hogg 1959.

20. Bew 1960.

21. Forster 1959.

22. Hobson 1959.

23. Clow 2003; Stern 2012; Löwy 2017; Löwy 2018a; de Chadarevian 2020.

24. "Woman herself is best guinea pig for test," *Science News Letter* 77, March 5, 1960, 151.

25. I thank Isabel Davis for this point.

26. *Squibb Product Reference for the Medical Profession* (New York: Squibb, 1962), 76–77.

27. Abraham 2009, 936; Logan 2022.

28. Kenton 1959, 409–410.

29. Davis 1963; ethnicity usually goes unmentioned in studies of this kind, so Davis is highly unusual in noting that his patients include "women of Northern Mediterranean and African genotypes"; on Davis: "Obituary," *BMJ*, February 21, 2004, 467.

30. Jaffé 1960, 654; for a critique of Jaffé's "therapeutic enthusiasm": Read 1960.

31. Higgins and Sadler 1960, 677–678.

32. Lister 1962, 86.

33. Seaward, "Pregnancy test . . . a modern scientific achievement," *Woman*, June 24, 1961, 27, SA/FPA/A3/12, WCL.

34. Dubowitz 1962; see further Dubowitz 2005; on chromosomes in clinical practice: de Chadarevian 2020; genetic understandings of sex: Richardson 2013; clinical approaches to sexual ambiguity: Eder 2022.

35. Smithells 1965; see further Al-Gailani 2014.

36. On pediatricians and children's hospitals in Budapest: Vargha 2018.

37. Mike Montuschi, "Isabel Gal obituary," *Guardian*, March 13, 2018.

38. Gal 1979, vii; on fetal surgery: Casper 1998, 2007.

39. Al-Gailani 2018.

40. Gal and Gal 1975, 161–162.

41. Gal and Gal 1975, viii–ix.

42. Gal and Gal 1975, 162–163.

43. Herrald to Richter, June 27, 1967, FD 23/127, TNA.

44. Kirman to Farrat, May 30, 1967, MH 71/39, TNA.

45. Abraham 1995, 66–70; Tansey and Reynolds 1997, 103–132.

46. Inman 1999; Inman 2006; Marks 1999; Marks 2002.

47. Tansey and Reynolds 1997, 118.

48. Inman 1999, 28.

49. Inman 1999, 29.

50. Tansey and Reynolds 1997, 118.

51. Tansey and Reynolds 1997, 125.

52. Rawlins 1995.

53. Inman 1999, 118.

54. Inman to Farrat, June 2, 1967, MH 171/39, TNA.

55. Witts to Inman, June 6, 1967, MH 171/39, TNA.

56. Jeffcoate to Inman, June 13, 1967, MH 171/39, TNA; for a contemporary view of the components of Primodos: Brown et al. 2018.

57. Jeffcoate to Inman, June 13, 1967, MH 171/39, TNA.

58. Inman to Gal, June 23, 1967, MH 171/39, TNA.

59. Inman to Jeffcoate, June 26, 1967, MH 171/39, TNA.

60. Jeffcoate to Inman, July 7, 1967, MH 171/39, TNA.

61. Inman to Jeffcoate, July 11, 1967, MH 171/39, TNA.

62. Smithells to Inman, July 14, 1967, MH 171/39, TNA.

63. Gal et al. 1967.

64. "Test pills may injure foetus," *Medical News*, October 13, 1967.

65. Gal to Inman, December 5, 1967, MH 171/39, TNA.

66. Inman to Gal, December 11, 1967, MH 171/39, TNA.

CHAPTER 8

1. Delia Davin, "Using commercial pregnancy test in the 1960s," typescript, January 19, 2014, quoted with permission of Davin; on the demonstration: Thomas 2002.

2. Edgerton 2006, 8.

3. Donnelly 2005; Weeks 2018, 272–301; Rusterholz 2022.

4. On consumerism choice and public interest: O'Hara 2013; Mold 2015; reproductive freedoms: Davis and Davidson 2006; Bashford 2014, 344–351; Bracke 2022a; Bracke 2022b; somewhat later developments in journalism: Schmidt 2020, 53–55.

5. On the hormonalization of pregnancy: Malich 2017; postwar biomedicine: Löwy 1996; Gaudillière 2002; Quirke and Gaudillière 2008; immunology: Mazumdar 1989; Mazumdar 2002; Cambrosio and Keating 1995; Kroker et al. 2008; Silverstein 2009; Anderson and Mackay 2014; Marks 2015.

6. Boyden 1951; on the Rockefeller Institute and its flagship journal, in which Boyden reported his method: Lederer 1992.

7. For histories of related (ABO) blood group agglutination tests in paternity determination and human genetics: Milanich 2019; Bangham 2021.

8. Wide 2005, 194–195.

9. Read and Stone 1958; on Gemzell: Pfeffer 1993; Pfeffer 2000; Nordlund 2011.

10. Wide 2005, 195.

11. Greaves 1968; Meryman 1976; Schneider 2003; Radin 2017; Radin and Kowal 2017; Bangham 2022.

12. Wide 2005, 197.

13. Tausk 1984, 236.

14. Wide and Gemzell 1960.

15. Wide 2005, 198.

16. Tausk 1984, 236; on Polderman: Gerritsma 1980.

17. On Uppsala University: Lindroth 1976.

18. Wide 2005, 199; see also Wide 1962; Wide and Gemzell 1962.

19. Interview by Naomi Pfeffer with Carl Axel Gemzell, c. 1990s, OH4, WCL.

20. Tausk 1984, 237.

21. On Burroughs Wellcome: Macdonald 1980; Tansey 1989; Tansey 1990; Church and Tansey 2007; the pharmaceutical industry in twentieth-century Britain: Slinn 1995; Slinn 2005; Quirke 2008; Quirke 2009.

22. Fulthorpe et al. 1963; see also Anderson 2005; Church and Tansey 2007.

23. Barr 1963a, 556.

24. Shea and Warrack 1963, 581–582.

25. Shuttleworth 1963, 699; "Obituary notices," *BMJ*, October 26, 1974, 234–235.

26. Oliver 1963, 840.

27. Simpson to Organon Laboratories, January 8, 1963, SA/FPA/A7/90, WCL.

28. *Pregnosticon* (Morden: Organon Laboratories, c. 1963); "Contraceptive testing manufacturers: Organon 1962–1965," SA/FPA/A7/90, WCL.

29. *"Prepuerin": New Pregnancy Test Combines Accuracy with Test-Tube Simplicity* (London: Burroughs Wellcome, 1963), Prepuerin "Dear Doctor" letter, WF/M/PL/247, WCL.

30. Hobson 1966; J&J created Ortho in the early 1930s to market spermicidal creams and jellies: Marks 2010, 36; "Ortho: A her-story of achievement in women's health,"

Women's Health Care 5, 2006, 16–20; see Keating and Cambrosio 2003, 133, 171; Löwy 2010, 2012.

31. Rasmussen 1997, 204, 235–236.

32. Singer and Plotz 1956.

33. Henry and Little 1962.

34. "Immunological tests for pregnancy," *BMJ*, December 22, 1962, 1668–1669.

35. "A better test for pregnancy," *New Scientist*, 323, January 24, 1963, 190; on *New Scientist*: Nathoo 2009, 210.

36. Hobson 1963a.

37. Duncan 1952; Larkin 1955.

38. Garden et al. 1963, 480.

39. On the fellowship: Maisin 1968.

40. Hobson 1963b.

41. Garden et al. 1963, 482.

42. Hobson 1963c, 749.

43. Warrack 1963, 869.

44. Hobson 1963d.

45. "Obituary," *BMJ*, March 7, 1970, 634–636; Macdonald and Sharman 1959; Davis 2013, 116–117; Davis 2017, 368–372.

46. Sharman and Pearston 1964; Sharman 1965.

47. "'Exciting results' from new pregnancy test," *Medical News*, 97, October 25, 1963; on *Medical News*: Dopson 1971, 145; Thistlethwaite 1997, 11; Nathoo 2009, 50–51.

48. Sharman to Department of Health for Scotland, November 11, 1963, HH 102/858, NRS; original emphasis. On the origins and uses of the newspaper clipping: te Heesen 2014.

49. Barr 1963b.

50. Sharman 1963.

51. Butt 1963.

52. Ennis 1963, 1379.

53. H. McLeave, "British doctors find quick way to test for pregnancy," *Scottish Daily Mail*, December 9, 1963, 3.

54. Sharman 1964, 291.

55. For example, "2,000 toads in a hole," *Daily Mail*, July 10, 1964; "Redundant toads," *Birmingham Post*, July 10, 1964, SA/FPA/A3/13, WCL.

56. "Science makes 2,000 toads redundant," *Daily Telegraph*, July 10, 1964, SA/FPA/A3/13, WCL.

57. "Pregnancy Diagnosis Laboratory, 1949–1965," SA/FPA/A3/13, WCL.

58. Brodie and Mellis 1966.

59. Hobson 1968, 722.

60. Warrack 1965, 731.

61. Hilary Haywood, "Pregnancy testing," *New Statesman*, March 25, 1966, 416.

62. "Pregnancy testing services," *BMJ Supplement*, June 17, 1967, 174.

63. On Kilburn in the 1960s: Wilson 1996.

64. Haywood, "Pregnancy testing"; see also Janice Solomons, "Pregnancy testing," *New Statesman*, April 15, 1966, 538.

65. On the ASA: *Celebrating 35 Years of Effective Self-Regulation, 1962–1997* (London: ASA, 1997).

66. *The Third Report of the Advertising Standards Authority, 1 April 1965–31 March 1966*, May 3, 1966, 23.

67. On the earlier history of the magazine, the circulation of which peaked at over 90,000 in the mid-1960s: Hyams 1963; Smith 1996.

68. "The facts about pregnancy testing," *Times*, August 30, 1967, 9.

69. Irene Black, "Pregnancy tests," *Sunday Citizen*, January 1, 1967.

70. "Doctors differ on new test," *Lincoln Echo*, May 2, 1967.

71. "The facts about pregnancy testing," *Times*, August 30, 1967, 9.

72. *The Fourth Report of the Advertising Standards Authority, 1 April 1966–31 March 1967*, 26; on the council: Hilton 2003, 228–241.

73. "Advertising and pregnancy tests," *Guardian*, August 17, 1966, 6.

74. "Immunological tests for pregnancy," MoH circular no. 130, August 7, 1967; "Pregnancy tests," *BMJ*, August 19, 1967, 504.

75. "Tests for pregnancy," *BMJ*, August 19, 1967, 448.

76. "The facts about pregnancy testing," *Times*, August 30, 1967, 9.

77. See T. Clifton, "The tube drops an advert," *Sunday Times*, August 20, 1967; P. Deschampsneufs, "Tube and rail differ on pregnancy posters," *Sunday Times*, October 1967; "Liberties of the subject," Hansard Commons Debates, vol. 755, December 1, 1967, cc808–909.

78. Balfour 1967, 16.

79. Webster 2002, 132–133.

80. Cook 2004, 268.

81. Brooke 2011, 149.

82. Rusterholz 2021; Rusterholz 2022; Sheldon et al. 2022.

83. Leathard 1980, 109; Latham 2002, 34.

84. Cook 2004, 279–281.

85. "Immunological pregnancy test," Hansard Commons Debates, vol. 693, April 16, 1964, c107W.

86. "Pregnancy tests: Oral answers to questions—Ministry of Health," Hansard Commons Debates, vol. 743, March 20, 1967, c1049.

87. "Pregnancy tests (advertisements)," Hansard Commons Debates, vol. 744, April 7, 1967, cc106–7W; "Parliament," *Lancet*, April 1, 1967, 732.

88. "The facts about pregnancy testing," *Times*, August 30, 1967, 9; on Jeger: Brooke 2011, 165.

89. The code vetted certain classes of advertisement for preclearance, including cigarettes, pregnancy testing and counseling, abortion, vasectomy, and sterilization: Nevett 1982, 202.

90. "Liberties of the subject."

91. Stacey 1992; Smith 1994; Morrice 1994.

92. Brown 1991, 51.

93. Smith 1993, 59; Jenkinson 2012, 28.

94. Nathoo 2009, 35–36; Morrice 1994.

95. "Pregnancy testing services," *BMJ Supplement*, June 17, 1967, 174.

96. "Pregnancy testing services"; see also "Denial of profit aim by doctor," *Guardian*, May 27, 1967, 2; "Doctor accused over pregnancy test firm," *Daily Mail*, May 27, 1967, 7; "Pregnancy-test doctor warned," *Daily Mirror*, May 27, 1967, 7.

97. "Pregnancy-testing service," *BMJ Supplement*, June 15, 1968, 206–207.

98. Nathoo 2009, 33–42, 55.

99. Haywood, "Pregnancy testing."

100. Haywood, "Pregnancy testing"; on *Born Free*: Beinart 2007; Beinart and McKeown 2009.

101. On personal responsibility in matters of individual reproduction choice: Bracke 2022a; Bracke 2022b; Lambe 2022.

102. Haywood, "Pregnancy testing."

103. Claire Rayner, "Pregnancy testing," *New Statesman*, April 1, 1966, 466; on the global rubella pandemic of the early 1960s: Reagan 2010.

104. Claire Rayner, "Pregnancy testing."

105. Derek Stevenson, "Pregnancy testing," *New Statesman*, April 8, 1966, 501–502.

106. Hilary Haywood, "Pregnancy testing," *New Statesman*, April 8, 1966, 502.

107. Alan Massam, "Pregnancy testing," *New Statesman*, April 8, 1966, 502; on the MJA: Thistlethwaite 1997.

108. Solomons, "Pregnancy testing"; see also *Medical News*, April 28, 1966.

109. "Pregnancy testing services," *Medical News*, July 1, 1966.

110. Peter Gladstone Smith, "Pregnancy tests," *Sunday Telegraph*, March 27, 1996.

111. "ASA stresses danger of pregnancy testing advertisements," *Guardian*, May 3, 1966, 3.

112. "Pregnancy tests," *BMJ Supplement*, July 16, 1966, 22, 45; on McCallum: "Obituary," *BMJ*, September 4, 1971, 586.

113. "Advertising and pregnancy tests: Letters," *Guardian*, August 30, 1966, 6.

114. Edward Connolly, "A woman's right to know," *Daily Sketch*, May 5, 1966, 8; see Conboy 2006; Bingham and Conboy 2015.

115. Grigg 1995, 24–25.

116. "The facts about pregnancy testing," *Times*, August 30, 1967, 9.

117. Bigman 2014; Bigman 2017; Schaefer 2014; Comella 2017; Lieberman 2017; Olszynko-Gryn and Ellis 2017; Cofield et al. 2022.

118. Banks 1960, 33.

119. Lourie 1962, 13.

120. Lodge 1965, 77.

121. Newman 1966, 166, 20.

122. Hines 1966, 201.

123. See Olszynko-Gryn 2017, 500–502. For *A Taste of Honey*, see further Todd 2019; Williams 2023.

124. Johnstone and Kellar 1968, 103.

125. Hobson to Macdonald, November 20, 1969, HH 102/858, NRS.

126. P. Clarke, "Hogben test's last croak," *Guardian*, February 7, 1970, 18.

CHAPTER 9

1. Anne Robinson, "Chemists attack pregnancy test kits," *Sunday Times*, March 16, 1969, 4.

2. Gilmour to Shaw, November 23, 1967, MH 159/77, TNA.

3. "Doctors want to give 5s pregnancy test," *Guardian*, January 4, 1968; see also "Pregnancy tests urged by B.M.A.," *Greenock Telegraph*, January 5, 1968; "NHS pregnancy tests proposed," *Times*, January 5, 1968.

4. Shaw to Hedgcock, January 4, 1968, MH 159/77, TNA.

5. Franklin to Robinson, July 27, 1968, MH 159/78, TNA.

6. Smith et al. 1973; Preece 1974; on prescriptions: Greene and Watkins 2012.

7. Gullick to Godber, February 17, 1969, MH 159/78, TNA.

8. Carr to Ralph, February 19, 1969, MH 159/78, TNA.

9. Hobson to Scarff, March 18, 1969, MH 159/78, TNA.

10. Shaw to Gullick, August 11, 1969, MH 159/78, TNA.

11. Thomson to Hobson, February 3, 1970, MH 159/78, TNA.

12. Hobson to Thomson, March 2, 1970, MH 159/78, TNA.

13. Shaw to Gullick, January 1, 1970, MH 159/78, TNA.

14. McKee to Giddon, April 23, 1970, MA 159/78, TNA.

15. Murray to Gidden, June 11, 1970, MH 159/78, TNA.

16. Gullick to Godbar, September 20, 1968, MH 159/78, TNA.

17. Ralph to Croft, October 3, 1968, MH 159/78, TNA.

18. Godber to Gullick, November 5, 1968, MH 159/78, TNA.

19. Anderson 2001, 26.

20. Stuart Anderson to author, August 3, 2017.

21. *Times*, January 1, 1969, 12; John Crosby, "No seats at the soda fountain," *Observer*, July 14, 1968, 22.

22. Langer 1969, 90.

23. "Pregnancy testing," *Pharmaceutical Journal*, July 26, 1969, 87–88.

24. Benz et al. 1969.

25. Langer 1969, 138–139.

26. "Pregnancy tests alarm," *Times*, July 28, 1969, 2.

27. "Pregnancy testing," *Pharmaceutical Journal*, October 16, 1969, 161.

28. "Boom for the positive-negative tester," *Scarborough Evening News*, August 2, 1969; on women's travel for abortion, including to London: Sethna and Davis 2019.

29. "So you want to know quickly if you are or you aren't . . . ," *Chingford Express*, August 22, 1969; although no mention is made of ethnicity, the perception that Britain was becoming a "nation of Asian shopkeepers" accompanied conspicuous demographic change in the late 1960s: Jones 1981, 467; I thank reviewer 1 for the point about cards.

30. "Council statement: Pregnancy testing," *Pharmaceutical Journal*, August 16, 1969, 162.

31. "Pregnancy testing kits distributed through pharmacists," *Pharmaceutical Journal*, September 13, 1969, 309.

32. "Pharmacists might give pregnancy test results," *Medical News*, August 29, 1969.

33. "Boots reject test offer," *Guardian*, September 16, 1969.

34. "Pregnancy testing kits distributed through pharmacists."

35. "Pregnancy testing kits distributed through pharmacists."

36. "Pregnancy testing—opposition by B.M.A.," *Pharmaceutical Journal*, September 20, 1969, 319.

37. "Pregnancy testing—opposition by B.M.A."

38. "Risk to life in pregnancy tests—doctor," *Morning Telegraph*, September 19, 1969.

39. "Pregnancy tests at chemists," *Times*, September 15, 1969.

40. Honor Owen, "Paging the ladies," *Bellshill Speaker*, September 19, 1969.

41. "Pregnancy testing: Doctors must show 'greater willingness,'" *Pharmaceutical Journal*, September 27, 1969, 359.

42. "Pregnancy tests over the counter," *BMJ*, September 20, 1969, 667.

43. "Over the counter," *Economist*, September 20, 1969, 54.

44. "A paradox," *Pharmaceutical Journal*, September 27, 1969, 357.

45. Wilcox and Block 1969.

46. Millington and Smith 1969.

47. Hunter 1969; "Doctor defends new pregnancy testing scheme," *Pharmaceutical Journal*, October 4, 1969, 392.

48. "President defends pregnancy testing," *Pharmaceutical Journal*, October 11, 1969, 442–444; on the production of public opinion through polls and surveys: Igo 2007; Mold 2015.

49. "President defends pregnancy testing."

50. "Pharmacist's changing role," *Pharmaceutical Journal*, October 11, 1969, 443–444.

51. "Pregnancy testing through pharmacies—doctors opposed but public demand increases," *Pharmaceutical Journal*, November 1, 1969, 525.

52. "The notice on my door," *Pharmaceutical Journal*, November 22, 1969, 601–602; on the lack of privacy in pharmacies: Anderson 2002, 81.

53. "The notice on my door"; Organon's proposal: "Pregnancy test equipment," *Pharmaceutical Journal*, October 11, 1969, 441.

54. "Pregnancy testing: Society accused of trespassing," *Pharmaceutical Journal*, November 22, 1969, 606.

55. "Support for commercial pregnancy tests," *Pharmaceutical Journal*, January 10, 1970, 29.

56. Lunay 1969.

57. "Pharmacy 'no place' for pregnancy tests," *Pharmaceutical Journal*, December 13, 1969, 699.

58. Block 1970.

59. Pembleton 1969.

60. "'Which?' looks at pregnancy tests," *Pharmaceutical Journal*, January 17, 1970, 49.

61. Anderson 2001, 26; the Pharmaceutical Society also permitted retail chemists to openly display condoms in 1970: Mechen 2017, 221.

62. "Pharmacists' big role in pregnancy testing service," *Pharmaceutical Journal*, June 13, 1970, 680.

63. "Survey of women using pregnancy test," *Pharmaceutical Journal*, January 30, 1971, February 6, 13, 20, 27, 1971, March 6 and 13, 1971, 94.

CHAPTER 10

1. Margaret Crane to author, May 18, 2015.

2. Crane to author, May 16, 2015.

3. Pagan Kennedy, "Who made that? (Home pregnancy test)," *New York Times Magazine*, July 29, 2012, MM16.

4. Crane to author, July 7, 2017.

5. Crane to author, May 18, 2015; "Lot 37," Bonhams, June 16, 2015, http://www.bonhams.com/auctions/22407/lot/37/.

6. Roger Catlin, "The unknown designer of the first home pregnancy test is finally getting her due," *Smithsonian Magazine*, September 21, 2015, http://www.smithsonianmag.com/smithsonian-institution/unknown-designer-first-home-pregnancy-test-getting-her-due-180956684/.

7. On the women's liberation movement: Jolly 2018; "Sisterhood and after," British Library, March 8, 2013, https://www.bl.uk/sisterhood; on the permissive society: Collins 2007; Mort 2010; Mort 2011; Payling 2022; as a phrase in journalism: Bingham 2009, 121; the 1970s more generally: Black et al. 2013.

8. Crane to author, July 27, 2017.

9. Margaret Crane, "Development of the home pregnancy test," typescript, April 2015.

10. Crane, "Development of the home pregnancy test."

11. Crane, "Development of the home pregnancy test."

12. Crane, "Development of the home pregnancy test."

13. Crane, "Development of the home pregnancy test."

14. See Edward B. Fiske, "Clergymen offer abortion advice," *New York Times*, May 22, 1967, 1, 36; Carmen and Moody 1973.

15. Crane to author, May 18, 2015.

16. Crane to author, August 12, 2016.

17. On pregnancy testing in Canada: Olszynko-Gryn 2020b.

18. "Home pregnancy test simple to use, reasonably accurate, especially if result is positive," *Family Planning Perspectives*, 11, May–June 1979, 190–191; Leavitt 2006; Robinson 2016; Crane to author, July 21, 2017.

19. "Pregnancy test kit due for home use," *Vancouver Sun*, November 27, 1970, 70; Roger Kenneth Field, "Automated medicine," *New York Times*, January 31, 1971, 8.

20. Johnson et al. 1976, 129.

21. Green 1991, 71.

22. Olszynko-Gryn 2019, 39.

23. Market research by Survey Computing Ltd. in cooperation with Belmont Laboratories disclosed that just over half (55 percent) of clients were unmarried; 72 percent of these planned to abort if positive, and the proportion plummeted to 45 percent of married women: "Survey of women using pregnancy test," *Pharmaceutical Journal*, January 30, February 6, 13, 20, 27, March 6 and 13, 1971, 94.

24. Christine Doyle, "Testing pregnancy tests," *Observer*, September 12, 1971, 23; "Do-it-yourself pregnancy test," *Times*, September 15, 1971, 16; on decimalization: Cook 2022.

25. Joan Williams, "Pregnancy tests," *Mother*, February 1971, 101–102.

26. Saroja Ramaswamy, "Critical review of pregnancy tests," *Medical Newsletter* (London: FPA, 1972), FD 23/4486, TNA.

27. "Do-it-yourself pregnancy test," *Chemist & Druggist*, June 19, 1971, 770.

28. John Ezard, "Pause over pregnancy," *Guardian*, June 23, 1971, 11; on Schmid: Tone 2002.

29. "Growing point," *Times*, September 15, 1971, 16.

30. "Pregnancy test gets medical backing," *Chemist & Druggist*, June 26, 1971, 793.

31. Ronald Bedford, "Pregnant? First D.I.Y. testing kit," *Daily Mirror*, June 9, 1971, 11.

32. James Wilkinson, "Home pregnancy test kits . . . Doctors give a negative," *Daily Express* (Glasgow), June 21, 1971.

33. "Do-it-yourself pregnancy test to be launched," *Chemist & Druggist*, June 19, 1971, 765; on the fear of pregnancy while on the pill, perhaps having missed one or more, and the distress of amenorrhea after coming off the pill: Blackwell and Walker 1969; Jones 2011; Olszynko-Gryn 2019; packaging and inserts: Gossel 1999; Watkins 2012a.

34. Ezard, "Pause."

35. Ezard, "Pause."

36. "Pregnancy check," *Penthouse*, 6, November 15, 1971, 15.

37. Doyle, "Testing."

38. "Do-it-yourself pregnancy test," *Times*, September 15, 1971, 16.

39. "Growing point," *Times*, September 15, 1971, 16; on NHS waiting rooms: Moore 2022.

40. "Company profile," *Chemist & Druggist*, September 11, 1971, 344.

41. "Pregnancy testing," MH 156/633, TNA; on the CA: Hilton 2003; Black 2004.

42. "Doubt on home pregnancy test accuracy," *Chemist & Druggist*, December 4, 1971, 829.

43. Fairweather and Cremer 1972.

44. Ramaswamy, "Critical review of pregnancy tests."

45. Thursz 1971.

46. Winton 1971.

47. Benson et al. 1971.

48. Benson et al. 1971.

49. Benson et al. 1971; on the "social clause": Keown 1988, 114–119.

50. Benson et al. 1971.

51. Benson et al. 1971.

52. Brodie and Porter 1971.

53. Holland and Whitehead 1973; Gorsky 2013, 89.

54. "Pregnancy testing," *BMJ*, November 20, 1971, 444–445.

55. "Pregnancy testing."

56. "Pregnancy testing."

57. Dennis Newson, "Kits that tell if a baby's on the way," *Daily Record*, May 26, 1971, 5.

58. Letter to Graham, November 9, 1971, "Predictor home pregnancy test," PP/GRA/B.4, WCL; Giles, "Report to the Lane Committee on the Working of the Abortion Act 1967," c. 1974, MH 156/633, TNA.

59. "Home pregnancy test kit," *Pharmaceutical Journal*, November 13, 1971, 475.

60. L'Aimable 1996.

61. "Predictor na twee jaar," *Farmakopij*, 16, September 5, 1973, 3.

62. "Home pregnancy test kit."

63. *Chemist & Druggist*, November 13, 1971, 723.

64. *Daily Express*, December 7, 1971, 14.

65. Giles 1971.

66. L'Aimable 1996.

67. Blake 1971; on "subnormality": Burt 1977; Welshman 1996; Bakker 2015.

68. Wivel 1998.

69. "Pregnancy kits ban urged," *Times*, February 22, 1972, 4; "Women call for ban on birth-test kits," *Daily Mirror*, February 22, 1972, 3.

70. Giles, "Report"; original emphasis.

71. Grob et al. 1973.

72. *Sex with Health: The Which? Guide to Contraception, Abortion and Sex-Related Diseases* (London: CA, 1974), 53–54.

73. L'Aimable 1996.

74. "Large increase in Predictor sales," *Chemist & Druggist*, March 19, 1977, 371.

75. *Chemist & Druggist*, May 28, 1977, 777.

76. Holland to Royle, July 7, 1976, MH 156/633, TNA.

77. Clarke 1987.

78. Jeffreys 2001.

79. Abraham 2009, 938–939.

80. Fogden to Robertson, July 2, 1976, Grimston's minute on PO(MIN-H) 1374/9, undated response to letter of July 13, 1976 addressed to Sir Anthony Clarke, Robertson, August 20, 1976, MH 156/633, TNA.

81. Robertson, April 22, 1976, Robertson, August 11, 1976, MH 156/633, NA.

82. Robertson, August 11, 1976, MH 156/633, NA.

83. Minute addressed to Grimston, Private Office, August 1976, Brereton, "Pregnancy testing kits," August 16, 1976, MH 156/633, TNA.

84. Owen to Royle, September 8, 1976, MH 156/633, TNA.

85. Abraham 2009.

86. Holland to Butterfield, October 14, 1976, Butterfield to Holland, October 20, 1976, MH 156/633, TNA.

87. Butterfield to Holland, November 24, 1976, MH 156/633, TNA.

88. Handwritten note, January 18, 1977, MH 156/633, TNA.

89. Oakley 1979, 27–28.

90. Olszynko-Gryn 2019.

91. "The free pregnancy testing service," November 1974, University of Bristol Library Special Collections, Bristol, DM2123/8/115.

CHAPTER 11

1. Brooke 2014; Sutcliffe-Braithwaite 2014; Hilton et al. 2017.

2. Löwy 2018.

3. Mulkay 1997; Wilson 2014; Hopwood 2018; Dow 2019; Beers 2022.

4. Durham 1991; Franklin et al. 1991; Dee 2019; Amery 2020; Sheldon et al. 2022.

5. Al-Gailani 2014; Nemec and Dron 2022; Nemec and Olszynko-Gryn 2022.

6. Healey 2016; van de Wiel 2020.

7. "Two-test pregnancy kit," *Chemist & Druggist*, October 13, 1979, 564; "Pregnancy testing," *Which?*, February 1982, 106–107.

8. Millar and Jarvis 1980.

9. Millar and Jarvis 1980.

10. Smith 1980, 125.

11. Kitzinger 1980, 27–28.

12. Thom 1981, 37.

13. Oakley et al. 1990, 89; she may have been alerted to the possible risk of prescription hormones in pregnancy by intensive press coverage of Primodos in the late 1970s: Olszynko-Gryn et al. 2018.

14. "Pregnancy testing," *Which?*, February 1982, 106–107.

15. "Pregnancy testing in pharmacies," *Pharmaceutical Journal*, November 27, 1982, 631.

16. "Pregnancy test reports 'dumped,'" *Pharmaceutical Journal*, May 21, 1983, 573.

17. Evens 1983.

18. "Predictions for prediction," *Chemist & Druggist*, December 1, 1984, 991–994.

19. "Predictions for prediction."

20. Larbie 1985, 10; on race and health activism in Britain: Bivins 2015; Redhead 2021.

21. On the production of aspirational middle-class whiteness through menstrual marketing: Røstvik 2022, 24.

22. "Predictions for prediction."

23. On Britain's biotech boom: Green 1991; Cambrosio and Keating 1995; de Chadarevian 2011; Marks 2015; Myelnikov 2017; the lucrative diagnostics sector: Creager, 2008, 216.

24. "Predictions for prediction"; on the leaflet in the communication of reproductive health: Nemec 2023.

25. Marks 2015, 134.

26. "Predictions for prediction."

27. Jones and Kraft 2004; Jones 2005, 294–296.

28. Zajicek 1986.

29. Ford 1985; "Monoclonal pregnancy test," *New Scientist*, 106, May 30, 1985, 22.

30. Donaldson 1986.

31. "Colour change pregnancy test with sampling device," *Chemist & Druggist*, June 1, 1985, 1135.

32. Ford 1985.

33. Lindsay Cook, "How the cuts cause a pregnant pause," *Guardian*, November 4, 1985, 21.

34. Ford 1985; on "lifestyle drugs": Watkins 2012b.

35. Clive Cookson, "Two new babies on the way," *Financial Times*, July 7, 1988, 13; "New pregnancy test cuts out the chemistry lesson," *New Scientist*, 119, July 21, 1988, 39; on campaigns against smoking and drinking in pregnancy: Berridge 2001; Armstrong 2003; Golden 2005; Al-Gailani 2014.

36. On the public fetus: Petchesky 1987; Jülich and Björklund 2024; British yuppies: Edwards 2022.

37. "Discover in colour too!," *Chemist & Druggist*, September 27, 1986, 493.

38. *Community Pharmacy*, November/December 1986, 3.

39. Bargh 1986.

40. "A year for new arrivals," *Chemist & Druggist*, February 7, 1987, 210–215.

41. "Birth of anther market," *Chemist & Druggist*, February 7, 1987, 216; on Tampax: Røstvik 2022.

42. "Tambrands add pregnancy test to First Response," *Chemist & Druggist*, June 18, 1988, 1194.

43. "After Clearblue comes Clearplan," *Unilever Magazine* 65, 1987, 7–8.

44. Cookson, "Two new babies."

45. Conway 1989.

46. "Clearblue one step by Unipath Limited," DES/DCA/30/7/1989/5/1, Design Archives, University of Brighton.

47. "Unipath launches OTC pregnancy test 'pen,'" *Pharmaceutical Journal*, July 2, 1988, 18.

48. Conway 1989, 36.

49. Maguire 1989, 533.

50. "Testing times," *Community Pharmacy*, September 1989, 12–13.

51. Abortion protocols: Olszynko-Gryn 2019; assisted conception: Franklin 1997, 109.

52. Franklin 1997, 78.

53. Edwards 1985; Birke et al. 1990, 143–144.

54. Franklin 1997, 181.

55. Hobson 1987; the report: "Pregnancy test gets the positive touch," *New Scientist*, January 29, 1987, 38; on loss: Layne 2002; Freidenfelds 2020.

56. Lind and Whittaker 1988.

57. Walker et al. 1988.

58. Brattebø 1988.

59. Maguire 1989, 533.

60. Maguire 1989, 534.

61. *Chemist & Druggist*, April 29, 1989, 718.

62. "Testing times," *Community Pharmacy*, September 1989, 12–13.

63. Winston 1989, 53–54.

64. Philips and Rakusen 1989, 310.

65. Charlish 1989, 29.

66. Birke et al. 1990, 143–144.

67. Moulder 1990, 5.

68. Kingdom et al. 1991, 1311.

69. Chard 1992, 710.

70. Pye and Thomas 1990; Voss 1992; *Community Pharmacy*, February 1992, 11.

71. "Testing time," *Over the Counter*, 2, September 21, 1990, 4–5.

72. "First Response: A breakthrough," *Chemist & Druggist*, September 15, 1990, 432.

73. Bargh 1992.

74. "Into the pink," *Chemist & Druggist*, June 1991, 6.

75. "Home help," *Chemist & Druggist*, February 1992, 13.

76. "Carter-Wallace in Tesco," *Chemist & Druggist*, March 7, 1992, 350.

77. "How to grow the market," *Community Pharmacy*, February 1992, 12.

78. Klein 1984; Jennings 2017.

79. Wikler and Wikler 1991, 18–19.

80. Mamo 2007; Claes 2022.

81. Saffron 1994, 67–69.

82. "Clearer step," *Community Pharmacy*, August 1992, 8.

83. Bargh 1992.

84. Bargh 1992.

85. "Pregnant pause," *Community Pharmacy*, February 1993, 14.

86. "Home pregnancy testing: The inside story," *Community Pharmacy*, March 1993, 26–27.

87. "Test drive," *Community Pharmacy*, May 1994, 40.

88. "With new Predictor sales are 'easy peesy,'" *Community Pharmacy*, July 1994, 23.

89. Platt 1995.

90. Platt 1995.

91. Purcell 1997.

92. Purcell 1997.

93. Arai 2009.

94. Hadley 1999, 5, 40, 63, 92, 98; a regular though infrequent occurrence since 1969, pill scares regarding thrombosis or cancer constituted yet another source of demand for pregnancy tests; on the 1995 scare: Furedi 1999.

95. "Grocery tests pharmacy's patients," *Community Pharmacy*, August 2000, 12.

96. "Product round-up," *Community Pharmacy*, August 2001, 24.

97. Jones and Kraft 2004, 119; Holt and Cameron 2010, 205.

98. "Product round-up," *Community Pharmacy*, July 2004, 30–31; "Go digital for clearer result," *Chemist & Druggist*, August 9, 2003, 22.

99. "Spelling it out with Clearblue," *Chemist & Druggist*, April 8, 2006, 28.

100. Charlotte Edwardes, "Early pregnancy tests 'causing unnecessary miscarriage grief,'" *Telegraph*, October 4, 2005; Hodgekiss 2005.

101. "Clearblue predicts sales surge," *Chemist & Druggist*, September 9, 2006, 36; see Holt and Cameron 2012, 202–219.

102. "Clearblue's just in time," *Chemist & Druggist*, September 20, 2008, 22.

103. Ross 2015, 116–117.

104. Ross 2018, 94–96.

105. Smith 2000, 515.

106. Bertagna 2003, 25.

107. Oliver 2006, 35.

108. LB 2021.

109. Ramaswamy 2016.

110. McNish 2016.

111. Naldrett 2007, 14–15.

112. Love 2022.

CHAPTER 12

1. Emin 2005, 166.

2. See Matthews and Wexler 2000; Liss 2009; Buller 2012; Bright 2013; Birth Rites, a collection of contemporary art on childbirth, is currently housed at the University of Kent: http://birthritescollection.org.uk.

3. On "following the money" to revisionist histories of twentieth-century science, technology, and business: Gaudillière 2001; Edgerton 2012.

4. "Doing a pregnancy test," National Health Service, February 2, 2023, https://www.nhs.uk/pregnancy/trying-for-a-baby/doing-a-pregnancy-test/; thanks to Kathryn Schoefert for the link.

5. On this distinction more generally: Canguilhem 1991; Méthot with Sholl 2020; Arminjon et al. 2022.

6. On the medicalization of male reproductive health: Almeling 2020.

7. For the latest developments: ACDHPT, https://primodos.org; Risky Hormones, https://riskyhormones.org.

8. Pagan Kennedy, "The pregnancy test scandal," *New York Times*, July 31, 2016, 1, 4–5.

9. There is also a comic strip: Morgane Pellennec and Cécily de Villepoix, "Inconsciences," *La Revue Dessinée* 33, Autumn 2021, 208–221.

10. For example, Fisher and Winick 2021, 81–82.

11. Shapin 1989; Gaudillière and Löwy 1998; Maddox 2002; Grier 2005; Skloot 2010; Shetterly 2016; Berner and Björkman 2017; Hicks 2017; Bangham et al. 2022.

12. On self-diagnosis in the 1970s: Jordan 1977; more recent ethnographic accounts: Gregg 1995; Papen 2008; Ross 2018; Robinson 2020.

13. Kirksey et al. 2016.

14. "US approves first over-the-counter HIV home-use test," *BBC News*, July 4, 2012, http://www.bbc.co.uk/news/world-us-canada-18700121.

15. See Banda 2015.

16. Taylor 2008; Roberts 2012.

17. Burlbaw 2004; Green and Platt 2005; Doublilet 2005; Alexander 2007.

18. Haarburger and Pillay 2011, 548.

19. Layne 2010, 108.

20. Oakley 1979, 27–28.

21. Oakley 1979, 30.

22. For a historical perspective on the menstrual product industry up to the present-day equity movement: Røstvik 2022.

23. Layne 2002; Freidenfelds 2020.

24. Karpin and Kristin 2012; Waggoner 2017.

25. Buklijas 2014; Arni 2018; Richardson 2021.

REFERENCES

ARCHIVES

London Metropolitan Archives (LMA)
London School of Economics Special Collections (LSE)
National Records of Scotland (NRS)
Rockefeller Archive Center (RAC)
Schering Archives, Berlin
Second World War Experience Centre (SWWEC)
The National Archives (TNA)
University of Brighton Design Archives
University of Bristol Library Special Collections
University of Edinburgh Special Collections
University of Sussex Special Collections
Wellcome Collection, London (WCL)

BOOKS AND ACADEMIC ARTICLES

Note: Newspaper and magazine articles, reviews, editorials, obituaries, unsigned articles, typescripts, films, websites, and personal communications are cited fully in the notes and not included here.

Abderhalden, E. 1914a. *Abwehrfermente das Auftreten blutfremder Substrate und Fermente im tierischen Organismus unter experimentellen physiologischen und pathologischen Bedingungen*. 4th ed. Berlin: Springer.

Abderhalden, E. 1914b. *Defensive Ferments of the Animal Organism*. London: Bale & Danielsson.

Abraham, J. 1995. *Science, Politics and the Pharmaceutical Industry: Controversy and Bias in Drug Regulation*. London: UCL Press.

Abraham, J. 2009. "Partial progress: Governing the pharmaceutical industry and the NHS, 1948–2008." *Journal of Health Politics, Policy and Law* 34: 931–977.

Adhikari, M. 2005. *Not White Enough, Not Black Enough: Racial Identity in the South African Coloured Community*. Athens: Ohio University Press.

Adler, K. 2007. *Contributions to the History of Herpetology*. Vol. 2. St. Louis, MO: Society for the Study of Amphibians and Reptiles.

Adler-Kastner, L. 1998. "From *personae non gratae* in Vienna 1938 to respected citizens of Edinburgh: A vignette of my parents, Dr. Ernst Adler and Dr. Regina Kapeller-Adler." *Wiener klinische Wochenschrift* 110: 174–180.

Agarwal, P. 2021. *(M)otherhood: On the Choices of Being a Woman*. Edinburgh: Canongate.

Al-Gailani, S. 2010. *Teratology and the Clinic: Monsters, Obstetrics, and the Making of Antenatal Life in Edinburgh, c.1900*. PhD diss., University of Cambridge.

Al-Gailani, S. 2014. "Making birth defects 'preventable': Pre-conceptional vitamin supplements and the politics of risk reduction." *Studies in History and Philosophy of Biological and Biomedical Sciences* 47: 278–289.

Al-Gailani, S. 2018. "Hospital birth." In *Reproduction: Antiquity to the Present Day*, edited by N. Hopwood, R. Flemming, and L. Kassell, 553–566. Cambridge: Cambridge University Press.

Alexander, A. A. 2007. "Just scanning around with diagnostic medical ultrasound: Should state regulate the nondiagnostic uses of this technology?" *Annals of Health Law* 16: 1–41.

Alexander, S. 1996. "The mysteries and secrets of women's bodies: Sexual knowledge in the first half of the twentieth century." In *Modern Times: Reflections on a Century of English Modernity*, edited by M. Nava and A. O'Shea, 161–175. London: Routledge.

Alexander, S. S., and C. W. Bellerby. 1935. "The effect of captivity upon the reproductive cycle of the South African clawed toad (*Xenopus laevis*)." *Journal of Experimental Biology* 12: 306–314.

Allan, H., and F. Dickens. 1930. "The Zondek and Aschheim test for pregnancy." *Lancet* 215: 39–41.

Allbutt, H. A. 1888. *The Wife's Handbook*. 7th ed. London: R. Forder.

Allen, E., and E. A. Doisy. 1923. "An ovarian hormone: Preliminary report on its localization, extraction and partial purification, and action in test animals." *JAMA* 81: 819–821.

Allison, R. M. 1955. "The merits of some English and foreign male toads and frogs used for the diagnosis of pregnancy and measures to overcome their disadvantages." *Laboratory Practice* 4: 229–234, 277–282.

Allison, R. M. 1957. "The care and maintenance of male toads and frogs used in the bioassay of chorionic gonadotrophins." In *The UFAW Handbook on the Care and Management of Laboratory Animals*, 2nd ed., edited by A. N. Worden and W. Lane-Petter, 703–803. London: UFAW.

Almeling, R. 2020. *Guynecology: The Missing Science of Men's Reproductive Health*. Oakland: University of California Press.

Amery, F. 2020. *Beyond Pro-Life and Pro-Choice: The Changing Politics of Abortion in Britain*. Bristol: Bristol University Press.

Amster, E. J. 2013. *Medicine and the Saints: Science, Islam, and the Colonial Encounter in Morocco, 1877–1956*. Austin: University of Texas Press.

Amsterdamska, O., and A. Hiddinga. 2003. "The analyzed body." In *Companion to Medicine in the Twentieth Century*, edited by R. Cooter and J. Pickstone, 417–433. London: Routledge.

Anderson, S. 2001. "'The most important place in the history of British birth control': Community pharmacy and sexual health in 20th century Britain." *Pharmaceutical Journal* 266: 23–29.

Anderson, S. 2002. "Community pharmacy in Great Britain: Mediation at the boundary between professional and lay care, 1920 to 1995." In *Biographies of Remedies: Drugs, Medicines, and Contraceptives in Dutch and Anglo-American Healing Cultures*, edited by M. Gijswijt-Hofstra, G. M. van Heteren, and E. M. Tansey, 75–97. Amsterdam: Rodopi.

Anderson, S. 2005. *Making Medicines: A Brief History of Pharmacy and Pharmaceuticals*. London: Pharmaceutical Press.

Anderson, S. C., and V. S. Berridge. 2000. "The role of the community pharmacist in health and welfare, 1911–1986." In *Oral History, Health and Welfare*, edited by J. Bornat, R. B. Perks, P. Thompson, and J. Walmsley, 48–74. London: Routledge.

Anderson, W., and I. R. Mackay. 2014. *Intolerant Bodies: A Short History of Autoimmunity*. Baltimore: Johns Hopkins University Press.

Arai, L. 2009. *Teenage Pregnancy: The Making and Unmaking of a Problem*. Bristol: Policy.

Arbuthnot Lane, W. 1934. *Everywoman's Home Doctor*. London: Odhams.

Arminjon, M., C. Cherici, and P. O. Méthot, eds. 2022. *Le normal et le pathologique: des categories pérmiées?* Paris: Éditions Matériologiques.

Armstrong, E. M. 2003. *Conceiving Risk, Bearing Responsibility Fetal Alcohol Syndrome and the Diagnosis of Moral Disorder*. Baltimore: Johns Hopkins University Press.

Arney, W. R. 1982. *Power and the Profession of Obstetrics*. Chicago: University of Chicago Press.

Arni, C. 2018. *Pränatale Zeiten: das Ungeborene und die Humanwissenschaften (1800–1950)*. Berlin: Schwabe.

Aschheim, S. 1929. "The early diagnosis of pregnancy, chorion-epithelioma and hydatidiform mole by the Aschheim-Zondek test." *American Journal of Obstetrics and Gynecology* 28: 335–342.

Astbury, L. 2015. *Breeding Women and Lusty Infants in Seventeenth-Century England.* PhD diss., University of Cambridge.

Austoker, J. 1988. *A History of the Imperial Cancer Research Fund, 1902–1986.* Oxford: Oxford University Press.

Austoker, J., and L. Bryder, eds. 1989. *Historical Perspectives on the Role of the MRC.* Oxford: Oxford University Press.

Bakker, N. 2015. "Identifying the 'subnormal' child in an age of expansion of special education and child science in the Netherlands (*c.*1945–1965)." *History of Education* 44: 460–479.

Baldwin, M. 2015. *Making Nature: The History of a Scientific Journal.* Chicago: University of Chicago Press.

Balfour, M. 1967. *Help Yourself in London: A Guide to Services, Facilities and Things to Do.* London: Garnstone.

Ballantyne, J. W. 1913. "Blood tests, especially Abderhalden's, in pregnancy." *Edinburgh Medical Journal* 11: 362–367.

Ballantyne, J. W. 1914. *Expect Motherhood: Its Supervision and Hygiene.* London: Cassell.

Ballin, A. S. 1904. *The Expectant Mother: A Complete Guide to the Care of Her Health and the Preparations for the Infant.* London: Office of "Baby": The Mother's Magazine.

Balmer, B. 2001. *Britain and Biological Warfare: Expert Advice and Science Policy, 1930–65.* Basingstoke: Palgrave.

Bamforth, J. 1936. "Experiences of the Friedman test for pregnancy." *St. Thomas's Hospital Reports* 1: 132–139.

Banda, J. 2015. "Rapid home HIV testing: Risk and the moral imperatives of biological citizenship." *Body & Society* 21: 24–47.

Bangham, D. R. 1999. *A History of Biological Standardization: The Characterization and Measurement of Complex Molecules Important in Clinical and Research Medicine.* Bristol: Society for Endocrinology.

Bangham, J. 2020. *Blood Relations: Transfusion and the Making of Human Genetics.* Chicago: University of Chicago Press.

Bangham, J., X. Chacko, and J. Kaplan, eds. 2022. *Invisible Labour in Modern Science.* Lanham, MD: Rowman & Littlefield.

Banks, L. R. 1960. *The L-Shaped Room.* London: Chatto & Windus.

Bargh, J. 1986. "To be or not to be: The advent of technology." *Community Pharmacy,* November/December, 4–6.

Bargh, J. 1992. "Dipping into a testing market." *Community Pharmacy*, September, 12–14.

Bärnreuther, S. 2021. *Substantial Relations: Making Global Reproductive Medicine in Postcolonial India*. Ithaca, NY: Cornell University Press.

Barr, W. A. 1963a. "A comparison of the Hogben pregnancy test with an immunological method." *Journal of Obstetrics and Gynaecology of the British Commonwealth* 70: 551–556.

Barr, W. A. 1963b. "Immunological methods used in pregnancy diagnosis." *Medical News* 56, November 1, 8.

Barrell, J., and B. Braithwaite. 1988. *The Business of Women's Magazines*. 2nd ed. London: Kogan Page.

Barton, G. M. 1953. "Comparison of the male toad and Hogben pregnancy tests." *BMJ* 2: 868.

Bartrip, P. W. J. 1996. *Themselves Writ Large: The British Medical Association 1832–1966*. London: BMA.

Bashford, A. 2014. *Global Population: History, Geopolitics, and Life on Earth*. New York: Columbia University Press.

Batten, L. W. 1939. *The Single-Handed Mother*. London: George Allen & Unwin.

Bauer, S., M. Schlünder, and M. Rentetzi, eds. 2020. *Boxes: A Field Guide*. Manchester: Mattering.

Bayly, M. B. 1936. *Cancer: The Failure of Modern Research*. London: Health Education and Research Council.

Bayon, H. P. 1939. "Ancient pregnancy tests in light of contemporary knowledge." *Proceedings of the Royal Society of Medicine* 32: 1527–1538.

Beauman, N. 1983. *A Very Great Profession: The Woman's Novel, 1914–39*. London: Virago.

Beers, L. 2022. "Not a priority: Infertile women and the symbolic politics of IVF in 1980s Britain." *Gender & History*, online ahead of print: July 20.

Beinart, W. 2007. "The lion queen." *Environmental History* 12: 383–386.

Beinart, W., and S. Dubow. 2021. *The Scientific Imagination in South Africa: 1700 to Present*. Cambridge: Cambridge University Press.

Beinart, W., and K. McKeown. 2009. "Wildlife media and representations of Africa, 1950s to the 1970s." *Environmental History* 14: 429–452.

Bellerby, C. W. 1933. "The endocrine factors concerned in the control of the ovarian cycle. I. *Xenopus laevis* as a test animal." *Biochemical Journal* 27: 615–620.

Bellerby, C. W. 1934. "A rapid test for the diagnosis of pregnancy." *Nature* 133: 494–495.

Benninghaus, C. 2012. "Beyond constructivism? Gender, medicine and the early history of sperm analysis, Germany 1870–1900." *Gender & History* 24: 647–676.

Bensen, A. R. 1969. *Rose Macaulay*. New York: Twayne.

Benson, W. G., G. Grant, M. Knowles, J. Segall, and S. L. Henderson Smith. 1971. "Pregnancy testing." *BMJ* 4: 425.

Benz, R., M. J. Morris, A. Norman, and P. J. Wildblood. 1969. "Pregnancy testing." *Pharmaceutical Journal* 203: 113–114.

Berkeley, C., and V. Bonney. 1919. *A Guide to Gynaecology in General Practice*. 2nd ed. London: Frowde.

Berner, B., and M. Björkman. 2017. "Modernizing the flow of blood: Biomedical technicians, working knowledge and the transformation of Swedish blood centre practices." *Social Studies of Science* 47: 485–510.

Berridge, V. 2001. "Constructing women and smoking as a public health problem in Britain 1950–1990s." *Gender & History* 13: 328–348.

Bertagna, J. 2003. *The Opposite of Chocolate*. London: Picador.

Bew, M. 1960. "'Orasecron' and pregnancy." *Practitioner* 184: 372–373.

Bigman, F. 2014. *"Nature's a Wily Dame": Abortion in British Literature and Film, 1907–1967*. PhD diss., University of Cambridge.

Bigman, F. 2017. "Babies without husbands: Unmarried pregnancy in 1960s British fiction." In *Perceptions of Pregnancy from the Seventeenth to the Twentieth Century*, edited by J. Evans and C. Meehan, 161–185. Cham: Palgrave Macmillan.

Bingham, A. 2004. *Gender, Modernity, and the Popular Press in Inter-War Britain*. Oxford: Clarendon.

Bingham, A. 2009. *Family Newspapers? Sex, Private Life, and the British Popular Press 1918–1978*. Oxford: Oxford University Press.

Bingham, A. 2012. "Newspaper problem pages and British sexual culture since 1918." *Media History* 18: 51–63.

Bingham, A., and M. Conboy. 2015. *Tabloid Century: The Popular Press in Britain, 1896 to the Present*. Oxford: Peter Lang.

Birke, L., S. Himmelweit, and G. Vines. 1990. *Tomorrow's Child: Reproductive Technologies in the 90s*. London: Virago.

Bishop, P. M. F. 1932. "The Zondek-Asch[h]eim reaction and the early diagnosis of pregnancy." *Guy's Hospital Gazette* 46: 89–90.

Bishop, P. M. F. 1933. "The Friedman test for pregnancy: An analysis of the results of a year's experience and a suggested modification." *Guy's Hospital Report* 83: 308–335.

Bishop, P. M. F. 1934. "Pregnancy diagnosis." *BMJ* 1: 1186.

Bishop, P. M. F. 1951. *Gynaecological Endocrinology for the Practitioner*. 2nd ed. Edinburgh: Livingstone.

Bivins, B. 2015. *Contagious Communities: Medicine, Migration, and the NHS in Post-War Britain*. Oxford: Oxford University Press.

Bivins, R., and H. Marland. 2016. "Weighting for health: Management, measurement and self-surveillance in the modern household." *Social History of Medicine* 29: 757–780.

Björk, R. 2001. "Inside the Nobel Committee on Medicine: Prize competition procedures 1901–1950 and the fate of Carl Neuberg." *Minerva* 39: 393–408.

Black, L. 2004. "*Which?*Craft in post-war Britain: The Consumers' Association and the politics of affluence." *Albion* 36: 52–82.

Black, L., H. Pemberton, and P. Thane, eds. 2013. *Reassessing 1970s Britain*. Manchester: Manchester University Press.

Blacker, C. P. 1939. "Maternity and child welfare work and the population problem." *Eugenics Review* 31: 91–95.

Blackwell, B., and K. P. Walker. 1969. "Possible pregnancy in a general practice." *Practitioner* 202: 691–695.

Blaikley, J. B., and G. Chamberlain. 1971. "Chelsea hospital for women, 1871–1971." *Journal of Obstetrics and Gynaecology of the British Commonwealth* 78: 1053–1057.

Blair-Bell, W. 1930. *Some Aspects of the Cancer Problem*. London: Baillière.

Blair-Bell, W. 1934. *Principles of Gynaecology: A Textbook for Students and Practitioners*. 4th ed. London: Baillière.

Blake, D. M. 1971. "Pregnancy tests." *Chemist & Druggist* 196: 841.

Blanarsch, M. 2009. "Die Arzt-Patienten-Beziehung zu Beginn des 18. Jahrhunderts, untersucht anhand Johann Storchs Kasuistik zu Molenschwangerschaften." *Medizin, Gesellschaft und Geschichte* 28: 121–152.

Bland, L. 2012. "'Hunnish scenes' and a 'virgin birth': A 1920s case of sexual and bodily ignorance." *History Workshop Journal* 73: 118–143.

Block, B. P. 1970. "Views on the pregnancy test." *Pharmaceutical Journal* 204: 4–5.

Blumenfeld, S. 1935. *Jew Boy*. London: Jonathan Cape.

Bolman, B. 2022. "Dogs for life: Beagles, drugs, and capital in the twentieth century." *Journal of the History of Biology* 55: 147–179.

Bonah, C., D. Cantor, and A. Laukötter, eds. 2018. *Health Education Films in the Twentieth Century*. Rochester, NY: University of Rochester Press.

Bonhams. 2017. *Voices of the 20th Century*. New York: Bonhams.

Borell, M. 1985. "Organotherapy and the emergence of reproductive endocrinology." *Journal of the History of Biology* 18: 1–30.

Borell, M. 1987. "Biologists and the promotion of birth control research, 1918–1938." *Journal of the History of Biology* 19: 51–87.

Borge, J. 2020. *Protective Practices: A History of the London Rubber Company and the Condom Business*. Montreal: McGill-Queen's University Press.

Bosman, L. P. 1937. "Early diagnosis of pregnancy." *BMJ* 2: 939.

Bourke, J. 2003. "Fear and anxiety: Writing about emotion in modern history." *History Workshop Journal* 55: 111–133.

Bourne, A. 1935. *Midwifery for Nurses*. London: Churchill.

Bowler, P. J. 2006. "Experts and publishers: Writing popular science in early twentieth-century Britain, writing popular history of science now." *British Journal for the History of Science* 39: 159–187.

Bowler, P. J. 2009. *Science For All: The Popularization of Science in Early Twentieth-Century Britain*. Chicago: University of Chicago Press.

Boyden, S. V. 1951. "The adsorption of proteins on erythrocytes treated with tannic acid and subsequent hemagglutination by antiprotein sera." *Journal of Experimental Medicine* 93: 107–120.

Bracke, M. A. 2022a. "Family planning, the pill, and reproductive agency in Italy, 1945–1971: From 'conscious procreation' to 'a new fundamental right'?" *European Review of History* 29: 88–108.

Bracke, M. A. 2022b. "Women's rights, family planning, and population control: The emergence of reproductive rights in the United Nations (1960s–70s)." *International History Review* 44: 751–777.

Brattebø, G. 1988. "Occult biochemical pregnancies." *BMJ* 297: 921.

Braybrooke, P. 1927. *Some Goddesses of the Pen*. London: Daniel.

Brendon, P. 1991. *Thomas Cook: 150 Years of Popular Tourism*. London: Secker & Warburg.

Brewer, H. F. 1934. "The Aschheim-Zondek reaction." *St. Bartholomew's Hospital Journal* 41: 65–69.

Brews, A. 1935. "A follow-up survey of the cases of hydatidiform mole and chorion-epithelioma treated at the London Hospital since 1912." *Proceedings of the Royal Society of Medicine* 28: 1213–1228.

Brews, A. 1939. "Blair-Bell Memorial Lecture January 27th, 1939. Hydatidiform mole and chorion-epithelioma." *Journal of Obstetrics and Gynaecology of the British Empire* 46: 815–835.

Bright, S., ed. 2013. *Home Truths: Photography and Motherhood*. London: Art/Books.

Brittain, V. 1957. *Testament of Experience: An Autobiographical Story of the Years 1925–1950*. London: Gollancz.

Britton, H. G. 1956a. "Pregnancy test." *BMJ* 2: 419.

Britton, H. G. 1956b. "Diagnosis of early pregnancy." *BMJ* 2: 1118.

Brockman, R. S. L. 1913. "The diagnostic value of Abderhalden's method in carcinoma." *Lancet* 182: 1385–1387.

Brodie, J., and H. Mellis. 1966. "A comparative study of pregnancy tests." *Practitioner* 196: 821–825.

Brodie, J., and I. A. Porter. 1971. "Pregnancy testing." *BMJ* 4: 489.

Bröer, R. 2004. "Motes, protective ferments and hormones: Pregnancy testing from antiquity to the present day." In *Senses, Sensors and Systems: A Journey through the History of Laboratory Diagnosis*, edited by M. Barthels, 129–143. Basel: Roche.

Brooke, S. 2001a. "'A new world for women'? Abortion law reform in Britain during the 1930s." *American Historical Review* 106: 431–459.

Brooke, S. 2001b. "Gender and working class identity in Britain during the 1950s." *Journal of Social History* 34: 773–795.

Brooke, S. 2011. *Sexual Politics: Sexuality, Family Planning, and the British Left from the 1880s to the Present Day*. Oxford: Oxford University Press.

Brooke, S. 2014. "Living in 'new times': Historicizing 1980s Britain." *History Compass* 12: 20–32.

Brookes, B. L. 1988. *Abortion in England, 1900–1967*. London: Croom Helm.

Brookes, B. L., and P. Roth. 1994. "*Rex v. Bourne* and the medicalization of abortion." In *Legal Medicine in History*, edited by M. Clark and C. Crawford, 314–343. Cambridge: Cambridge University Press.

Brown, A. C. 2003. "Centennial history of the Zoology Department, University of Cape Town, 1903–2003: A personal memoir." *Transactions of the Royal Society of South Africa* 58: 11–34.

Brown, E., and M. Grover. 2012. *Middlebrow Literary Cultures: The Battle of the Brows, 1920–1960*. Basingstoke: Palgrave Macmillan.

Brown, K. 2003. "The maternal physician: Teaching American mothers to put the baby in the bathwater." In *Right Living: An Anglo-American Tradition of Self-help Medicine and Hygiene*, edited by C. E. Rosenberg, 88–111. Baltimore: Johns Hopkins University Press.

Brown, P. S. 1977. "Female pills and the reputation of iron as an abortifacient." *Medical History* 21: 291–304.

Brown, P. S. 1991. "Medically qualified naturopaths and the General Medical Council." *Medical History* 35: 50–77.

Brown, S., L. R. Fraga, G. Cameron, L. Erskine, and N. Vargesson. 2018. "The Primodos components norethisterone acetate and ethinyl estradiol induce developmental abnormalities in zebrafish embryos." *Scientific Reports* 8: 2917.

Browne, S. 1939. "Pregnancy tests and notification." *Lancet* 233: 478.

Bruehl, F. S. 1952. "The development of pregnancy tests." *American Journal of Nursing* 52: 591–593.

Brumberg, J. J. 1993. "'Something happens to girls': Menarche and the emergence of the modern American hygiene imperative." *Journal of the History of Sexuality* 4: 99–127.

Brunton, D. 2004. "The rise of laboratory medicine." In *Medicine Transformed: Health, Disease and Society in Europe 1800–1930*, edited by D. Brunton, 92–117. Manchester: Manchester University Press.

Bryder, L. 1984. "Papworth Village Settlement—a unique experiment in the treatment and care of the tuberculous?" *Medical History* 28: 372–390.

Buklijas, T. 2014. "Food, growth and time: Elsie Widdowson's and Robert McCance's research into prenatal and early postnatal growth." *Studies in History and Philosophy of Biological and Biomedical Sciences* 47: 267–277.

Bull, T. 1877. *Hints to Mothers*. 14th ed. London: Longmans, Green.

Buller, R. E., ed. 2012. *Reconciling Art and Mothering*. Burlington, VT: Ashgate.

Bullock, W. E. 1915. "A critical study of the basis of Abderhalden's serum reaction." *Lancet* 185: 223–228.

Bullough, V. L. 1994. *Science in the Bedroom: A History of Sex Research*. New York: Basic.

Burlbaw, J. 2004. "Obstetric sonography—that's entertainment?" *Journal of Diagnostic Medical Sonography* 20: 444–448.

Burney, I., and N. Pemberton. 2011. "Bruised witness: Bernard Spilsbury and the performance of early twentieth-century English forensic pathology." *Medical History* 55: 41–60.

Burt, C. 1977. *The Subnormal Mind*. 3rd ed. Oxford: Oxford University Press.

Bury, V. 1953. "African clawed frogs (*Xenopus laevis*)." *Water Life and Aquaria World* 8: 145.

Butler, S. M. 2019. "More than mothers: Juries of matrons and pleas of the belly in medieval England." *Law and Historical Review* 37: 353–396.

Butt, W. R. 1963. "Immunological pregnancy tests." *Lancet* 282: 1331–1332.

Button, C. 2018. "James Cossar Ewart and the origins of the Animal Breeding Research Department in Edinburgh, 1895–1920." *Journal of the History of Biology* 51: 445–477.

Cahen, F. 2019. "Who was afraid of pregnancy tests? Gestational information and reproductive politics in France (1920–50)." *Medical History* 63: 134–152.

Calder, A. B. 1906. *Questions and Answers on Midwifery for Midwives*. 2nd ed. London: Ballière.

Cambrosio, A., and P. Keating. 1995. *Exquisite Specificity: The Monoclonal Antibody Revolution*. Oxford: Oxford University Press.

Cameron, A. T. 1936. *Recent Advances in Endocrinology*. 3rd ed. London: Churchill.

Cameron, G. D. W. 1945. "The organization of laboratory services: Observations from a recent visit to Britain." *Canadian Journal of Public Health* 36: 58–64.

Cameron, S. J., A. N. McLellan, R. A. Lennie, and J. Hewitt. 1924. *A Glasgow Manual of Obstetrics*. London: Arnold.

Campion, S. 1950. *National Baby*. London: Ernest Benn.

Canguilhem, G. 1991. *The Normal and the Pathological*. New York: Zone.

Cantor, D., ed. 2008. *Cancer in the Twentieth Century*. Baltimore: Johns Hopkins University Press.

Capel, L. 1958. "Morning sickness." *Family Doctor* 8: 381–383.

Carmen, A., and H. Moody. 1973. *Abortion Counseling and Social Change from Illegal Act to Medical Practice: The Story of the Clergy Consultation Service on Abortion*. Valley Forge, PA: Judson Press.

Carruthers, J. 2010. "Scientists in society: The Royal Society of South Africa." *Transactions of the Royal Society of South Africa* 63: 1–30.

Cartwright, L. 1995. *Screening the Body: Tracing Medicine's Visual Culture*. Minneapolis: University of Minnesota Press.

Casper, M. J. 1998. *The Making of the Unborn Patient: A Social Anatomy of Fetal Surgery*. New Brunswick, NJ: Rutgers University Press.

Casper, M. J. 2007. "Fetal surgery then and now." *Conscience* 28: 24–27.

Casper, M. J., and A. E. Clarke. 1998. "Making the Pap smear into the 'right tool' for the job." *Social Studies of Science* 28: 255–290.

Cesarani, D., and T. Kushner. 1993. *Internment of Aliens in Twentieth Century Britain*. London: Frank Cass.

Chan, J. 2019. "The brave new worlds of birth control: Women's travel in Soviet Russia and Naomi Mitchison's *We Have Been Warned*." *Journal of Modern Literature* 42: 38–56.

Chard, T. 1991. "Frequency of implantation and early pregnancy loss in natural cycles." *Baillière's Clinical Obstetrics and Gynaecology* 5: 179–189.

Chard, T. 1992. "Pregnancy tests: A review." *Human Reproduction* 7: 701–710.

Charlish, A. 1989. *A Woman's Guide to Birth-Tech: Tests and Technology in Pregnancy and Birth*. London: Christopher Helm.

Chen, W. 1992. "The laboratory as business: Sir Almroth Wright's vaccine programme and the construction of penicillin." In *The Laboratory Revolution in Medicine*, edited by A. Cunningham and P. Williams, 245–297. Cambridge: Cambridge University Press.

Childerhose, J. E., and M. E. MacDonald. 2013. "Health consumption as work: The home pregnancy test as a domesticated health tool." *Social Science & Medicine* 86: 1–8.

Chisolm, A. E. 1930. "Diagnosis in gynaecology." *Practitioner* 124: 561–569.

Church, R., and E. M. Tansey. 2007. *Burroughs Wellcome & Co. Knowledge, Trust, Profit and the Transformation of the British Pharmaceutical Industry, 1880–1940*. Lancaster: Crucible.

Claes, T. 2022. "How silence became 'outdated': Secrecy, anonymity and artificial insemination by donor in Belgium, 1950s–1990s." *Journal of Family History*, online ahead of print: September 21.

Clarke, A. 1987. "Moral reform and the anti-abortion movement." *Sociological Review* 35: 123–149.

Clarke, A. E. 1987. "Research materials and reproductive science in the United States, 1910–1940." In *Physiology in the American Context, 1850–1940*, edited by G. L. Geison, 323–350. Baltimore: American Physiological Society.

Clarke, A. E. 1998. *Disciplining Reproduction: Modernity, American Life Sciences, and "the Problems of Sex."* Berkeley: University of California Press.

Clarke, A. E. 2000. "Maverick reproductive scientists and the production of contraceptives, 1915–2000+." In *Bodies of Technology: Women's Involvement with Reproductive Medicine*, edited by A. Saetnan, N. Oudshoorn, and M. Kirejczyk, 37–89. Columbus: Ohio State University Press.

Clarke, A. E. 2007. "Reflections on the reproductive sciences in agriculture in the UK and US, ca. 1900–2000+." *Studies in History and Philosophy of Biological and Biomedical Sciences* 38: 316–339.

Clarke, A. E., and M. J. Casper. 1996. "From simple technology to complex arena: Classification of Pap smears, 1917–90." *Medical Anthropology Quarterly* 10: 601–623.

Clay, C., M. DiCenzo, B. Green, and F. Hackney, eds. 2017. *Women's Periodicals and Print Culture in Britain, 1918–1939: The Interwar Period*. Edinburgh: Edinburgh University Press.

Claye, A. M. 1936. "Haemorrhage in early pregnancy." *Clinical Journal* 65: 309–314.

Close-Koening, T. 2011. *Betwixt and Between: Production and Commodification of Knowledge in a Medical School Pathological Anatomy Laboratory in Strasbourg (Mid-19th Century to 1939).* PhD diss., University of Strasbourg.

Clow, B. 2003. "'An illness of nine months' duration': Pregnancy and thalidomide use in Canada and the United States." In *Women Health and Nation: Canada and the United States since 1945*, edited by G. Feldberg, M. Ladd-Taylor, A. Li, and M. McPherson, 45–66. Montreal: McGill-Queen's University Press.

Cockburn, C., and S. Ormrod. 1993. *Gender and Technology in the Making.* London: Sage.

Cofield, L., B. Mechen, and M. Worley. 2022. "History from the top shelf: The cultural politics of sex in post-war Britain." *Contemporary British History* 36: 165–173.

Collins, H. M. 1985. *Changing Order: Replication and Induction in Scientific Practice.* London: Sage.

Collins, K. E. 1988. *Go and Learn: The International Story of Jews and Medicine in Scotland.* Aberdeen: Aberdeen University Press.

Collins, K. 2009. "European refugee physicians in Scotland, 1933–1945." *Social History of Medicine* 22: 513–553.

Collins, M., ed. 2007. *The Permissive Society and Its Enemies: Sixties British Culture.* London: Rivers Oram.

Colpus, E. 2011. "*The Week's Good Cause*: Mass culture and cultures of philanthropy at the inter-war BBC." *Twentieth Century British History* 22: 305–329.

Comella, L. 2017. *Vibrator Nation: How Feminist Sex-Toy Stores Changed the Business of Pleasure.* Durham, NC: Duke University Press.

Comrie, J. D. 1931. *Woman's Own Home Doctor.* London: George Newnes.

Conboy, M. 2006. *Tabloid Britain: Constructing a Community through Language.* London: Routledge.

Consumers' Association. 1974. *Sex with Health: The Which? Guide to Contraception, Abortion and Sex-Related Diseases.* London: Consumers' Association.

Conway, L. 1989. "Frontier of medical diagnostics." *Unilever Magazine* 72: 35–39.

Cook, A. 2022. *The Politics of Decimalisation in the UK: Britain's Other D-Day.* Newcastle: Cambridge Scholars.

Cook, H. 2004. *The Long Sexual Revolution: English Women, Sex, and Contraception 1800–1975.* Oxford: Oxford University Press.

Cooper, J. E. 1989. "The guiding hand . . . foundations of lower vertebrate pathology. The first Edward Elkan Memorial Lecture." *Herpetopathologia* 1: 1–4.

Coopey, R., S. O'Connell, and D. Porter. 2005. *Mail Order Retailing in Britain: A Business and Social History*. Oxford: Oxford University Press.

Corea, G. 1985. *The Mother Machine: Reproductive Technologies from Artificial Insemination to Artificial Wombs*. New York: Harper & Row.

Cowan, R. S. 1983. *More Work for Mother: The Ironies of Household Technology from the Open Hearth to the Microwave*. New York: Basic.

Cowan, R. S. 2008. *Heredity and Hope: The Case for Genetic Screening*. Cambridge, MA: Harvard University Press.

Cowie, A. T. 1948. *Pregnancy Diagnosis Tests: A Review*. Edinburgh: Commonwealth Bureau of Animal Breeding and Genetics.

Cox, G. M. 1933. *The Woman's Book of Health: A Complete Guide to Physical Well-Being*. London: C. Arthur Pearson.

Cox-Maksimov, D. 1997. *The Making of the Clinical Trial in Britain, 1910–1945: Expertise, the State and the Public*. PhD diss., University of Cambridge.

Craft, I. 1975. *Specialized Futures: Essays in Honour of Sir George Godber*. London: Oxford University Press.

Crane, J., and J. Hand, eds. 2022. *Posters, Protests, and Prescriptions: Cultural Histories of the National Health Service in Britain*. Manchester: Manchester University Press.

Creager, A. N. H. 2008. "Molecular surveillance: A history of radioimmunoassays." In *Crafting Immunity: Working Histories of Clinical Immunology*, edited by K. Kroker, J. Keelan, and P. M. H. Mazumdar, 201–230. Burlington, VT: Ashgate.

Creager, A. N. H., and H. Landecker. 2009. "Technical matters: Method, knowledge and infrastructure in twentieth-century life science." *Nature Methods* 6: 701–705.

Crenner, C. 2006. "Private laboratories and medical expertise in Boston circa 1900." In *Devices and Designs: Medical Technologies in Historical Perspectives*, edited by C. Timmermann and J. Anderson, 61–73. Basingstoke: Palgrave Macmillan.

Crew, F. A. E. 1929a. "Biological diagnosis of pregnancy." *Lancet* 213: 688.

Crew, F. A. E. 1929b. "Diagnosis of early pregnancy." *BMJ* 1: 633.

Crew, F. A. E. 1930. "Pregnancy diagnosis station: Report on first year's working." *BMJ* 1: 662–664.

Crew, F. A. E. 1932. *An Introduction to the Study of Sex*. London: Gollancz.

Crew, F. A. E. 1933. *Sex Determination*. London: Methuen.

Crew, F. A. E. 1934. "Pregnancy diagnosis station." *BMJ* 2: 531.

Crew, F. A. E. 1936a. "Notes from a pregnancy diagnosis laboratory (1935)." *BMJ* 1: 993–996.

Crew, F. A. E. 1936b. "Laboratory tests for the early diagnosis of pregnancy." *BMJ* 2: 1092–1094.

Crew, F. A. E. 1937a. "Early diagnosis of pregnancy." *BMJ* 2: 1044.

Crew, F. A. E. 1937b. "Notes from a pregnancy diagnosis laboratory (1936)." *American Journal of Obstetrics and Gynecology* 33: 989–999.

Crew, F. A. E. 1939. "Biological pregnancy diagnosis tests: A comparison of the rabbit, the mouse, and the 'clawed toad' (*Xenopus laevis*) as the experimental animal." *BMJ* 1: 766–770.

Crew, F. A. E. 1971. "The genealogy of the poultry research centre, Edinburgh." *British Poultry Science* 12: 289–295.

Crosby, K. 2019. "Abolishing juries of matrons." *Oxford Journal of Legal Studies* 39: 259–284.

Crowe, N. 2021. *Forgotten Clones: The Birth of Cloning and the Biological Revolution*. Pittsburgh: University of Pittsburgh Press.

Cunningham, G. J. 1992. *History of British Pathology*. Bristol: White Tree.

D'Antonio, A. 2008. "Shopping for the Ballin Baby: Infant food and maternal authority in *Baby* magazine." *Nineteenth-Century Gender Studies* 41 (Spring).

Dakin, W. R. 1897. *A Handbook of Midwifery*. London: Longmans, Green.

Daniel, F. 1925. *The Expectant Mother*. London: C. W. Daniel.

Davidson, A. 1948. "Public health in Britain: The changing science." *American Journal of Public Health and the Nation's Health* 38: 319–324.

Davidson, H. A. 1949. "Male subfertility: Interim report of 3,182 cases." *BMJ* 2: 1328–1332.

Davidson, N., and J. Rakusen. 1982. *Out of Our Hands*. London: Pan.

Davies, A., B. Jackson, and F. Sutcliffe-Braithwaite, eds. 2021. *The Neoliberal Age? Britain since the 1970s*. London: UCL Press.

Davin, A. 1978. "Imperialism and motherhood." *History Workshop* 5: 9–65.

Davis, A. 1963. "A single-injection pregnancy test." *Practitioner* 191: 70–71.

Davis, A. 2012. *Modern Motherhood: Women and Family in England, 1945–2000*. Manchester: Manchester University Press.

Davis, G. 2008. *"The Cruel Madness of Love": Sex, Syphilis and Psychiatry in Scotland, 1880–1930*. Amsterdam: Rodopi.

Davis, G. 2013. "Test tubes and turpitude: Medical responses to the infertile patient in mid-twentieth-century Scotland." In *Western Maternity and Medicine, 1880–1990*, edited by J. Greenlees and L. Bryder, 113–127. London: Pickering & Chatto.

Davis, G. 2017. "'A tragedy as old as history': Medical responses to infertility and artificial insemination by donor in 1950s Britain." In *The Palgrave Handbook of Infertility in History: Approaches, Contexts and Perspectives*, edited by G. Davis and T. Loughran, 359–381. London: Palgrave Macmillan.

Davis, G., and R. Davidson. 2006. "'A fifth freedom' or 'hideous atheistic expediency'? The medical community and abortion law reform in Scotland, *c.*1960–1975." *Medical History* 50: 29–48.

de Chadarevian, S. 2002. *Designs for Life: Molecular Biology after World War II*. Cambridge: Cambridge University Press.

de Chadarevian, S. 2011. "The making of an entrepreneurial science: Biotechnology in Britain, 1975–1995." *Isis* 102: 601–633.

de Chadarevian, S. 2020. *Heredity Under the Microscope: Chromosomes and the Study of the Human Genome*. Chicago: University of Chicago Press.

Debenham, C. 2014. *Birth Control and the Rights of Women: Post-Suffrage Feminism in the Early Twentieth Century*. London: I. B. Tauris.

Dee, O. 2019. *The Anti-Abortion Campaign in England, 1966–1989*. London: Routledge.

Deeson, T. 1995. *Parke-Davis in Britain: The First Hundred Years*. Eastleigh: Parke-Davis & Co.

Deichmann, U. 2007. "'I detest his way of working': Leonor Michaelis (1875–1949), Emil Abderhalden (1877–1950) and Jewish and non-Jewish biochemists in Germany." In *Jews and Sciences in German Contexts: Case Studies from the 19th and 20th Centuries*, edited by U. Charpa and U. Deichmann, 101–126. Tübingen: Mohr Siebeck.

Delap, L. 2020. *Feminisms: A Global History*. Chicago: University of Chicago Press.

Dick-Read, G. 1942. *Revelation of Childbirth: The Principles and Practice of Natural Childbirth*. London: Heinemann.

Digby, A., and N. Bosanquet. 1988. "Doctors and patients in an era of national health insurance and private practice, 1913–1938." *Economic History Review* 41: 74–94.

Dodd, P. H. 1914. *Chavasse's Advice to a Wife: On the Management of Her Own Health and on the Treatment of Some of the Complaints Incidental to Pregnancy, Labour, and Suckling*. 16th ed. London: Churchill.

Dodds, G. H. 1930. "Value of the bromine test for diagnosis of pregnancy." *BMJ* 1: 948–949.

Dodds, G. H. 1931. "Rapid laboratory test for pregnancy." *BMJ* 2: 700–701.

Dodds, G. H. 1936. "The Visscher-Bowman test for pregnancy." *BMJ* 2: 224–225.

Donaldson, M. 1986. "Testing pregnancy tests." *Pharmaceutical Journal* 236: 529.

Donaldson, M., S. Cade, W. D. Harmer, R. O. Ward, and A. T. Edwards. 1936. *The Early Diagnosis of Malignant Disease for the Use of General Practitioners*. London: Oxford University Press.

Donnelly, M. 2005. *Sixties Britain: Culture, Society and Politics*. London: Routledge.

Dopson, L. 1971. *The Changing Scene in General Practice*. London: Johnson.

Dorland, W. A. N., and M. J. Hubeny. 1926. *The X-Ray in Embryology and Obstetrics*. London: Henry Kimpton.

Dose, R. 2003. "The World League for Sexual Reform: Some possible approaches." *Journal of the History of Sexuality* 12: 1–15.

Doublilet, P. M. 2005. "Entertainment ultrasound." *Journal of Ultrasound in Medicine* 24: 251–253.

Dougall, M. 2021. "'What does a socialist woman do?': Birth control and the body politic in Naomi Mitchison's *We Have Been Warned*." *Cambridge Quarterly* 50: 18–37.

Dow, K. 2019. "Looking into the test tube: The birth of IVF on British television." *Medical History* 63: 189–208.

Drucker, D. J. 2020. *Contraception: A Concise History*. Cambridge, MA: MIT Press.

Dry, S. 2006. "The population as patient: Alice Stewart and the controversy over low-level radiation in the 1950s." In *The Risks of Medical Innovation: Risk Perception and Assessment in Historical Context*, edited by T. Schlich and U. Tröhler, 108–122. New York: Routledge.

Dubow, S. 2006. *A Commonwealth of Knowledge: Science, Sensibility, and White South Africa, 1820–2000*. Oxford: Oxford University Press.

Dubow, S. 2011. *Ourselves Unborn: A History of the Fetus in Modern America*. Oxford: Oxford University Press.

Dubow, S. 2020. "Global science, national horizons: South Africa in deep time and space." *History Journal* 63: 1079–1106.

Dubowitz, V. 1962. "Virilisation and malformation of a female infant." *Lancet* 280: 405–406.

Dubowitz, V. 2005. *Ramblings of a Peripatetic Paediatrician: Victor Dubowitz*. Stanhope: Memoir Club.

Duden, B. 1991. *The Woman Beneath the Skin: A Doctor's Patients in Eighteenth-Century Germany*. Cambridge, MA: Harvard University Press.

Dukes, C. 1936. "Clinical examination of the urine." *Practitioner* 137: 900–905.

Duncan, M. N. S. 1952. "The general practitioner and the laboratory service." *Lancet* 260: 379–380.

Dunn, C. L. 1952–1953. *The Emergency Medical Services*. 2 vols. London: HMSO.

Dupree, M. W. 1997. "Other than healing: Medical practitioners and the business of life assurance during the nineteenth and early twentieth centuries." *Social History of Medicine* 10: 79–103.

Durham, M. 1991. *Sex and Politics: The Family and Morality in the Thatcher Years*. Basingstoke: Macmillan.

Eden, T. W. 1897. "The diagnosis of early pregnancy." *American Journal of the Medical Sciences* 114: 629–745.

Eden, T. W. 1906. *A Manual of Midwifery*. London: Churchill.

Eden, T. W. 1919. *A Manual of Midwifery*. 5th ed. London: Churchill.

Eder, S. 2022. *How the Clinic Made Gender: The Medical History of a Transformative Idea*. Chicago: University of Chicago Press.

Edgerton, D. 2006. *The Shock of the Old: Technology and Global History since 1900*. London: Profile.

Edgerton, D. 2010. "Innovation, technology, or history: What is the historiography of technology about?" *Technology and Culture* 51: 680–697.

Edgerton, D. 2012. "Time, money, and history." *Isis* 103: 316–327.

Edwards, A. 2022. *Are We Rich Yet? The Rise of Mass Investment Culture in Contemporary Britain*. Oakland: University of California Press.

Edwards, M. 2007. *Control and the Therapeutic Trial: Rhetoric and Experimentation in Britain, 1918–48*. Amsterdam: Rodopi.

Edwards, R. G. 1985. "Causes of early embryonic loss in human pregnancy." *Human Reproduction* 1: 185–198.

Eisenberg, Z. 2013. *The Whole Nine Months: Women, Men, and the Making of Modern Pregnancy in America*. PhD diss., Yale University.

Elkan, E. 1938a. "The *Xenopus* pregnancy test." *Practitioner* 140: 312–315.

Elkan, E. 1938b. "The *Xenopus* pregnancy test." *BMJ* 2: 1253–1256.

Elkan, E. 1939a. "L'épreuve de grossesse *Xenopus*." *La Presse Médicale* 47: 308–310.

Elkan, E. 1939b. "Biological pregnancy diagnosis tests." *BMJ* 1: 899–900.

Elkan, E. 1939c. "*Xenopus laevis*, Daudin." *Aquarist and Pond-Keeper* 9: 95–100.

Elkan, E. 1940. "Pregnancy tests in war-time." *Lancet* 236: 697.

Elkan, E. 1946a. "A new test for pregnancy." *Postgraduate Medical Journal* 22: 87–93.

Elkan, E. 1946b. "The *Xenopus laevis* pregnancy test." *Human Fertility* 11: 103–108.

Elkan, E. 1947a. "*Xenopus laevis* Daudin." In *The UFAW Handbook on the Care and Management of Laboratory Animals*, edited by A. N. Worden, 251–262. London: Baillière.

Elkan, E. 1947b. "A useful pregnancy test." *Marriage Hygiene* 1: 104–105.

Elkan, E. 1960. "Some interesting pathological cases in amphibians." *Proceedings of the Zoological Society of London* 134: 274–296.

Elliot, R. 2014. "Miscarriage, abortion or criminal feticide: Understandings of early pregnancy loss in Britain, 1900–1950." *Studies in History and Philosophy of Biological and Biomedical Sciences* 47: 248–256.

Elston, M. A. 1987. "Women and anti-vivisection in Victorian England, 1870–1900." In *Vivisection in Historical Perspective*, edited by N. A. Rupke, 259–293. London: Croom Helm.

Emin, T. 2005. *Strangeland*. London: Sceptre.

Ennis, J. E. 1963. "Immunological pregnancy tests." *Lancet* 282: 1379.

Eraso, Y. 2007. "Biotypology, endocrinology, and sterilization: The practice of eugenics in the treatment of Argentinian women during the 1930s." *Bulletin of the History of Medicine* 81: 793–822.

Erlingsson, S. J. 2016. "'Enfant terrible': Lancelot Hogben's life and work in the 1920s." *Journal of the History of Biology* 49: 495–526.

Erskine, C. G. 1929. *Why Be Childless?* London: Christophers.

Evans, B. 1984. *Freedom to Choose: The Life and Work of Dr. Helena Wright, Pioneer of Contraception*. London: Bodley Head.

Evens, E. 1983. "Pregnancy testing and other matters." *Pharmaceutical Journal* 231: 170–171.

Fairweather, D. V. I., and A. W. Cremer. 1972. "Do-it-yourself pregnancy test." *BMJ* 1: 747–748.

Ferguson, H. 1913. "Unusual case of hydatidiform mole dealt with by abdominal hysterectomy." *Transactions of the Edinburgh Obstetrical Society* 38: 260–267.

Ferguson, M. 1998. *Animal Advocacy and Englishwomen, 1780–1900: Patriots, Nation, and Empire*. Ann Arbor: University of Michigan Press.

Ferreira, H. P. 1954. "The relative merits of the various biological tests for pregnancy." *Postgraduate Medical Journal* 30: 355–369.

Finkelstein, M., and B. Zondek. 1966. "Professor Bernhard Zondek: An interview." *Journal of Reproduction and Fertility* 12: 3–19.

Fisher, K. 1998. "Women's experience of abortion before the 1967 act." In *Abortion, Law and Politics Today*, edited by E. Lee, 27–42. Basingstoke: Macmillan.

Fisher, K. 1999. "'Didn't stop to think, I just didn't want another one': The culture of abortion in interwar South Wales." In *Sexual Cultures in Europe: Themes in Sexuality*, edited by F. X. Eder, L. A. Hall, and G. Hekma, 213–232. Manchester: Manchester University Press.

Fisher, K. 2006. *Birth Control, Sex, and Marriage in Britain, 1918–1960*. Oxford: Oxford University Press.

Fisher, K. 2018. "Modern ignorance." In *Reproduction: Antiquity to the Present Day*, edited by N. Hopwood, R. Flemming, and L. Kassell, 471–484. Cambridge: Cambridge University Press.

Fisher, M. M., and A. Winick, eds. 2021. *Designing Motherhood: Things That Make and Break Our Births*. Cambridge, MA: MIT Press.

Fissell, M. E. 2003. "Making a masterpiece: The Aristotle texts in vernacular medical culture." In *Right Living: An Anglo-American Tradition of Self-Help Medicine and Hygiene*, edited by C. E. Rosenberg, 59–87. Baltimore: Johns Hopkins University Press.

Fleck, L. 1979. *Genesis and Development of a Scientific Fact*. Chicago: University of Chicago Press.

Fleming, J. B. 1966. "Montgomery and the follicles of the areola as a sign of pregnancy (1837)." *Irish Journal of Medical Science* 6: 169–182.

Florence, L. S. 1956. *Progress Report on Birth Control*. London: Heinemann.

Forbes, R. 1948. *Sixty Years of Medical Defence*. London: Medical Defence Union.

Forbes, T. R. 1957. "Early pregnancy and fertility tests." *Yale Journal of Biology and Medicine* 30: 16–29.

Ford, C. 1985. "Biotechnology products start to roll." *Unilever Magazine* 57: 10–11.

Ford, R. K. 1935. *A Short Ante-Natal and Post-Natal Handbook*. London: Oxford University Press.

Forster, D. H. 1959. "Early diagnosis of pregnancy." *Lancet* 274: 242.

Forster, L., and J. Hollows, eds. 2020. *Women's Periodicals and Print Culture in Britain, 1940s–2000s: The Postwar and Contemporary Period*. Edinburgh: Edinburgh University Press.

Foster, W. D. 1961. *A Short History of Clinical Pathology*. Edinburgh: Livingstone.

Foster, W. D. 1983. *Pathology as a Profession in Great Britain and the Early History of the Royal College of Pathologists*. London: Royal College of Pathologists.

Fothergill, W. E. 1900. *Manual of Midwifery*. 2nd ed. Edinburgh: William F. Clay.

Frank, R. T. 1929. *Female Sex Hormone*. London: Ballière.

Franklin, S. 1997. *Embodied Progress: A Cultural Account of Assisted Conception*. New York: Routledge.

Franklin, S. 2013. *Biological Relatives: IVF, Stem Cells, and the Future of Kinship*. Durham, NC: Duke University Press.

Franklin, S., and C. Roberts. 2006. *Born and Made: An Ethnography of Preimplantation Genetic Diagnosis*. Princeton, NJ: Princeton University Press.

Franklin, S., C. Lury, and J. Stacey, eds. 1991. *Off-Centre: Feminism and Cultural Studies*. London: Harper Collins.

Frazer, J. F. D. 1950. "The use of amphibia in pregnancy tests: A review." *British Journal of Herpetology* 1: 84.

Frazer, J. F. D., and F. X. Wohlzogen. 1949. "Male-toad pregnancy test." *Lancet* 2: 134.

Frazer, J. F. D., and F. X. Wohlzogen. 1950. "Use of the male British toad as a pregnancy test animal." *BMJ* 2: 330–331.

Freidenfelds, L. 2009. *The Modern Period: Menstruation in Twentieth-Century America*. Baltimore: Johns Hopkins University Press.

Freidenfelds, L. 2020. *The Myth of the Perfect Pregnancy: A History of Miscarriage in America*. New York: Oxford University Press.

Friedman, M. H., and M. E. Lapham. 1931. "A simple, rapid procedure for the laboratory diagnosis of early pregnancies." *American Journal of Obstetrics and Gynecology* 21: 405–410.

Fulthorpe, A. J., J. A. C. Park, J. E. Tovey, and J. C. Monckton. 1963. "Pregnancy diagnosis by a one-stage passive haemagglutination inhibition method." *BMJ* 1: 1049–1054.

Furedi, A. 1999. "The public health implications of the 1995 'pill scare.'" *Human Reproduction Update* 5: 621–626.

Gal, E., and I. Gal. 1975. *Human Congenital Malformations: The Design of a Computer-Aided Study*. London: Butterworths.

Gal, I. 1979. *A Clinical and Aetiological Survey of Spina-bifida and Hydrocephalus*. PhD diss., University of London.

Gal, I., B. Kirman, and J. Stern. 1967. "Hormonal pregnancy tests and congenital malformation." *Nature* 216: 83.

Galabin, A. L. 1900. *A Manual of Midwifery*. 5th ed. London: Churchill.

Galabin, A. L., and G. Blacker. 1910. *The Practice of Midwifery*. 7th ed. London: Churchill.

Galli Mainini, C. 1947a. "Reacción diagnóstica de embarazo en la que se usa el sapo macho como animal reactivo." *Semana Médica* 54: 337–340.

Galli Mainini, C. 1947b. "Pregnancy test using the male toad." *Journal of Clinical Endocrinology* 7: 653–658.

Galli Mainini, C. 1948a. *El diagnostico del embarazo con batracios machos*. Buenos-Aires: Editorial Impaglione.

Galli Mainini, C. 1948b. "Pregnancy test using the male batrachian." *JAMA* 138: 121–125.

Galli Mainini, C. 1948c. "Pregnancy test with the male batrachian." *Endocrinology* 43: 349–350.

Gallwey, A. 2011. *Lone Motherhood in England, 1945–1990: Economy, Agency and Identity*. PhD diss., University of Warwick.

Gallwey, A. 2013. "The rewards of using archived oral histories in research: The case of the Millennium Memory Bank." *Oral History Society Journal* 41: 37–50.

Garden, N. R., H. Miller, and J. A. Durant. 1963. "Serological pregnancy test using latex particles: An assessment of its reliability and convenience." *BMJ* 2: 480–483.

Gaudillière, J. P. 2001. "The pharmaceutical industry in the biotech century: Toward a history of science, technology and business?" *Studies in History and Philosophy of Biological and Biomedical Sciences* 32: 191–201.

Gaudillière, J. P. 2002. *Inventer la biomédecine: La France, l'Amérique et la production des savoirs du vivant (1945–1965)*. Paris: Découverte.

Gaudillière, J. P. 2005. "Better prepared than synthesized: Adolf Butenandt, Schering Ag and the transformation of sex steroids into drugs (1930–1946)." *Studies in History and Philosophy of Biological and Biomedical Sciences* 36: 612–644.

Gaudillière, J. P. 2010. "The visible industrialist: Standards and the manufacture of sex hormones." In *Evaluating and Standardizing Therapeutic Agents, 1890–1950*, edited by C. Gradmann and J. Simon, 174–201. Basingstoke: Palgrave Macmillan.

Gaudillière, J. P. 2018. "Sex hormones, pharmacy and the reproductive sciences." In *Reproduction: Antiquity to the Present Day*, edited by N. Hopwood, R. Flemming, and L. Kassell, 521–534. Cambridge: Cambridge University Press.

Gaudillière, J. P., and I. Löwy. 1998. *The Invisible Industrialist: Manufacture and the Construction of Scientific Knowledge*. Basingstoke: Macmillan.

Gavronsky, J. O. 1915. "Results of clinical investigations based on the methods of Abderhalden." *Lancet* 185: 119–123.

Gavronsky, J. O. 1918. "The Abderhalden reaction." *Lancet* 192: 830.

Geppert, A. C. T. 1998. "Divine sex, happy marriage, regenerated marriage: Marie Stopes's marital manual *Married Love* and the making of a best-seller, 1918–1955." *Journal of the History of Sexuality* 8: 389–433.

Gerritsma, K. W. 1980. "Emeritaat Professor Polderman." *Pharmaceutisch Weekblad* 2: 82–85.

Gevers, W. 1998. "Henri (Harry) Zwarenstein 1900–1905." *Transactions of the Royal Society of South Africa* 52: 433–434.

Giles, A. B. 1971. "Pregnancy testing." *BMJ* 4: 623.

Gill, A. M., and J. Howkins. 1937. "The Antuitrin S intradermal pregnancy test." *BMJ* 2: 1069.

Glasson, C. J. 1901. *Motherhood.* London: John Bale, Sons & Danielsson.

Glucksmann, M. 1990. *Women Assemble: Women Workers and the New Industries in Inter-War Britain.* London: Routledge.

Golden, J. L. 2005. *Message in a Bottle: The Making of Fetal Alcohol Syndrome.* Cambridge, MA: Harvard University Press.

Gooday, G. 2008. "Placing or replacing the laboratory in the history of science?" *Isis* 99: 783–795.

Gooday, G. 2015. *Domesticating Electricity: Technology, Uncertainty and Gender, 1880–1914.* London: Routledge.

Gordon, B. M. 2012. "Reinventions of a spa town: The unique case of Vichy." *Journal of Tourism History* 4: 35–55.

Gorsky, M. 2013. "'Searching for the people in charge': Appraising the 1983 Griffiths NHS management inquiry." *Medical History* 57: 87–107.

Gossel, P. P. 1999. "Packaging the pill." In *Manifesting Medicine: Bodies and Machines,* edited by R. Bud, 105–121. New York: Taylor & Francis.

Gowing, L. 2003. *Common Bodies: Women, Touch and Power in Seventeenth-Century England.* New Haven, CT: Yale University Press.

Gradmann, C., and J. Simon, eds. 2010. *Evaluating and Standardizing Therapeutic Agents, 1890–1950.* Basingstoke: Palgrave Macmillan.

Greaves, R. I. N. 1968. "Serum-plasma preservation: Historical review." *Cryobiology* 5: 76–86.

Green, K. 1991. "Shaping technologies and shaping markets: Creating demand for biotechnology." *Technology Analysis & Strategic Management* 3: 57–76.

Green, N., and L. D. Platt. 2005. "Nonmedical use of ultrasound: Greater harm than good?" *Journal of Ultrasound in Medicine* 24: 123–125.

Green-Armytage, V. B. 1934. "Obstetrical regrets." *Clinical Journal* 63: 53–59.

Greene, G. 1999. *The Woman Who Knew Too Much: Alice Stewart and the Secrets of Radiation.* Ann Arbor: University of Michigan Press.

Greene, J. A. 2022. *The Doctor Who Wasn't There: Technology, History, and the Limits of Telehealth.* Chicago: University of Chicago Press.

Greene, J. A., and E. S. Watkins, eds. 2012. *Prescribed: Writing, Filling, Using, and Abusing the Prescription in Modern.* Baltimore: Johns Hopkins University Press.

Greenfield, J., and C. Reid. 1998. "Women's magazines and the commercial orchestration of femininity in the 1930s: Evidence from *Woman's Own*." *Media History* 4: 161–174.

Greer, G. 1984. *Sex and Destiny: The Politics of Human Fertility*. London: Secker & Warburg.

Gregg, R. 1995. *Pregnancy in a High-Tech Age: Paradoxes of Choice*. New York: NYU Press.

Grier, D. A. 2005. *When Computers Were Human*. Princeton, NJ: Princeton University Press.

Grier, K. C. 2006. *Pets in America: A History*. Chapel Hill: University of North Carolina Press.

Griffith, E. F. 1939. *The Childless Family: Its Cause and Cure*. London: Kegan Paul.

Griffiths, J. 1997. "Creating, transmitting, and transforming a corporate culture in a public sector enterprise: The General Post Office, 1920–1990." *Business and Economic History* 26: 676–685.

Grigg, J. 1995. *The History of the Times: Vol. 6. The Thomson years, 1966–1981*. London: Harper Collins.

Grob, P. R., B. T. B. Manners, G. P. J. Beynon, and F. J. Gibbs. 1968. "Do-it-yourself pregnancy test." *BMJ* 1: 112–113.

Grossmann, A. 1995. *Reforming Sex: The German Movement for Birth Control and Abortion Reform, 1920–1950*. Oxford: Oxford University Press.

Gunn, J. W. C. 1939a. "*Xenopus* test for pregnancy." *BMJ* 1: 1258.

Gunn, J. W. C. 1939b. "*Xenopus* test for pregnancy." *BMJ* 2: 580.

Gurdon, J. B., and N. Hopwood. 2000. "The introduction of *Xenopus laevis* into developmental biology: Of empire, pregnancy testing and ribosomal genes." *International Journal of Developmental Biology* 44: 43–50.

Gurney, P. 2017. *The Making of Consumer Culture in Modern Britain*. London: Bloomsbury.

Ha, N. Q. 2011. "The riddle of sex: Biological theories of sexual difference in the early twentieth-century." *Journal of the History of Biology* 44: 505–546.

Haarburger, D., and T. S. Pillay. 2011. "Historical perspectives in diagnostic clinical pathology: Development of the pregnancy test." *Journal of Clinical Pathology* 64: 546–548.

Hackney, F. 2007. "'They opened up a whole new world': Narrative, text and image in British women's magazines in the 1930s." *Working Papers on Design* 2: 1–24.

Hackney, F. 2008. "'Women are news': British women's magazines 1919–1939." In *Transatlantic Print Culture, 1880–1940: Emerging Media, Emerging Modernisms*, edited by A. Ardis and P. Collier, 114–133. New York: Palgrave Macmillan.

Hackney, F. 2010. *"They Opened Up a Whole New World": Feminine Modernity and the Feminine Imagination in Women's Magazines, 1919–1939*. PhD diss., Goldsmiths, University of London.

Hadley, A., ed. 1999. *Tough Choices: Young Women Talk Pregnancy*. London: Livewire.

Haines, M. 1948. "The male-toad test for pregnancy." *Lancet* 252: 923–926.

Hajo, C. M. 2010. *Birth Control on Main Street: Organizing Clinics in the United States, 1916–1939*. Urbana: University of Illinois Press.

Hall, A. 2021. *Evolution on British Television and Radio: Transmissions and Transmutations*. Basingstoke: Palgrave Macmillan.

Hall, L. A. 2000. "'Not a domestic utensil but a woman and a citizen': Stella Browne on women, health and society." *Clio Medica* 60: 275–302.

Hall, L. A. 2011. *The Life and Times of Stella Browne: Feminist and Free Spirit*. London: I. B. Tauris.

Hall, L. A. 2013. *Sex, Gender, and Social Change in Britain since 1880*. 2nd ed. New York: Palgrave Macmillan.

Hall, R., ed. 1978. *Dear Dr. Stopes: Sex in the 1920s*. London: Deutsch.

Hamera, J. 2012. *Parlor Ponds: The Cultural Work of the American Home Aquarium, 1850–1970*. Ann Arbor: University of Michigan Press.

Hamilton, D. 1986. *The Monkey Gland Affair*. London: Chatto & Windus.

Han, S. 2013. *Pregnancy in Practice: Expectations and Experience in the Contemporary US*. New York: Berghahn.

Hanley, A. R. 2017. *Medicine, Knowledge and Venereal Diseases in England, 1886–1916*. Cham: Palgrave Macmillan.

Hannan, J. H. 1930. "The detection of the presence of the hormone of the anterior pituitary body in the urine as an aid to the diagnosis of pregnancy." *Proceedings of the Royal Society of Medicine* 23: 634–639.

Hannan, J. H. 1933. "Chorionic carcinoma." *BMJ* 2: 1046–1047.

Hansen, B. 2009. *Picturing Medical Progress from Pasteur to Polio: A History of Mass Media Images and Popular Attitudes in America*. New Brunswick, NJ: Rutgers University Press.

Hanson, C. 2004. *A Cultural History of Pregnancy: Pregnancy, Medicine, and Culture, 1750–2000*. Basingstoke: Palgrave Macmillan.

Harding, T. S. 1931. *The Degradation of Science*. New York: Farrar & Rinehart.

Hardyment, C. 1983. *Dream Babies: Child Care from Locke to Spock*. Oxford: Oxford University Press.

Haultain, W. F., and E. C. Fahmy. 1931. *Ante-Natal Care, Including the Abnormalities Associated with Pregnancy and a Section on Postnatal Care*. Edinburgh: Livingstone.

Healey, J. 2016. *Sooner or Later: Age, Pregnancy, and the Reproductive Revolution in Late Twentieth-Century America*. PhD diss., Yale University.

Heneghan, C., and J. Aronson. 2021. "Forgotten Primodos story and the roles of general practitioners." *BMJ Evidence-Based Medicine* 26: 79–81.

Henig, R. M. 2004. *Pandora's Baby: How the First Test Tube Babies Sparked the Reproductive Revolution*. Boston: Houghton Mifflin.

Henriksen, E. 1941. "Pregnancy tests of the past and present." *Western Journal of Surgical Obstetric Gynecology* 85: 610–618.

Henry, J. B., and W. A. Little. 1962. "Immunological test for pregnancy." *JAMA* 182: 230–233.

Herschkorn-Barnu, P. 2002. "Wie der Fötus einen klinischen Status erhielt: Bedingungen und Verfahren der Produktion eines medizinischen Fachwissens, Paris 1832–1848." In *Geschichte des Ungeborenen: Zur Erfahrungs- und Wissenschaftsgeschichte der Schwangerschaft, 17.–20. Jahrhundert*, edited by B. Duden, J. Schlumbohm, and P. Veit, 167–203. Göttingen: Vandenhoeck & Ruprecht.

Hess, V. 2005. "Standardizing body temperature: Quantification in hospitals and daily life." In *Body Counts: Medical Quantification in Historical and Sociological Perspectives*, edited by G. Jorland, A. Opinel, G. Weisz, 109–126. Montreal: McGill-Queens University Press.

Hey, D. 1949. "A report on the culture of the South African clawed frog *Xenopus laevis* (Daudin) at the Jonkershoek inland fish hatchery." *Transactions of the Royal Society of South Africa* 32: 45–54.

Hey, D. 1977. *Water: Source of Life*. Oxford: Oxford University Press.

Hicks, M. 2017. *Programmed Inequality: How Britain Discarded Women Technologists and Lost Its Edge in Computing*. Cambridge, MA: MIT Press.

Hiddinga, A. 1995. *Changing Normality: Pregnancy and Scientific Knowledge Claims 1920–1950, with Special Reference to the USA*. PhD diss., University of Amsterdam.

Higgins, G. L., and W. R. Sadler. 1960. "A two-tablet oral pregnancy test." *Practitioner* 185: 677–680.

Hilliard, C. 2006. *To Exercise Our Talents: The Democratization of Writing in Britain*. Cambridge, MA: Harvard University Press.

Hilton, M. 2002. "The female consumer and the politics of consumption in twentieth-century Britain." *Historical Journal* 45: 103–128.

Hilton, M. 2003. *Consumerism in Twentieth-Century Britain: The Search for a Historical Movement*. Cambridge: Cambridge University Press.

Hilton, M., C. Moores, and F. Sutcliffe-Braithwaite. 2017. "*New Times* revisited: Britain in the 1980s." *Contemporary British History* 31: 145–165.

Hinds, H. 2009. "Domestic disappointments: Feminine middlebrow fiction of the interwar years." *Home Cultures* 6: 199–211.

Hines, B. 1966. *The Blinder*. London: Joseph.

Hirst, J. C., and C. F. Long. 1926. "The early diagnosis of pregnancy by methods of precision: Further observations on sugar tolerance tests: Final report." *American Journal of the Medical Sciences* 171: 846–853.

Hitchens, R. A. N., and C. R. Lowe. 1966. "Laboratory services in general practice." *Medical Care* 4: 142–149.

Hobson, B. M. 1952. "Routine pregnancy diagnosis and quantitative estimation of chorionic gonadotrophin using female *Xenopus laevis*." *Journal of Obstetrics and Gynaecology of the British Empire* 59: 352–362.

Hobson, B. M. 1959. "Early diagnosis of pregnancy." *Lancet* 274: 409.

Hobson, B. M. 1963a. "A better test for pregnancy." *New Scientist* 325: 309.

Hobson, B. M. 1963b. "Immunological pregnancy tests." *BMJ* 1: 1606–1607.

Hobson, B. M. 1963c. "Latex tests for pregnancy." *BMJ* 2: 749.

Hobson, B. M. 1963d. "Immunological pregnancy tests." *BMJ* 2: 1203.

Hobson, B. M. 1966. "Review: Pregnancy diagnosis." *Journal of Reproduction and Fertility* 12: 33–48.

Hobson, B. M. 1968. "Pregnancy diagnosis using the Pregnosticon haemagglutination inhibition test." *Journal of Obstetrics and Gynaecology of the British Commonwealth* 75: 718–723.

Hobson, B. M. 1987. "Pregnancy test." *New Scientist* 113: 68.

Hodge, A. E. 1927. *Tropical Aquarium-Fishes: How to Breed and Rear Them. A Practical Guide for the Amateur*. London: Witherby.

Hodgekiss, A. 2005. "Pregnancy tests 'not to blame.'" *Chemist & Druggist* 264: 8.

Hogben, A., and A. Hogben. 1998. *Lancelot Hogben: Scientific Humanist. An Unauthorised Autobiography*. Rendlesham: Merlin Press.

Hogben, L. T. 1930. *Principles of Animal Biology*. London: Christophers.

Hogben, L. T. 1936. *Mathematics for the Million: A Popular Self-Educator*. London: Allen and Unwin.

Hogben, L. T. 1939. "*Xenopus* test for pregnancy." *BMJ* 2: 38–39.

Hogben, L. T. 1946a. "History of the Hogben test." *BMJ* 2: 554.

Hogben, L. T. 1946b. "The Hogben test." *BMJ* 2: 962–963.

Hogben, L. T. 1974. "Francis Albert Eley Crew, 1886–1973." *Biographical Memoirs of Fellows of the Royal Society* 20: 135–153.

Hogben, L. T., and D. Slome. 1931. "The pigmentary effector system. VI. The dual character of endocrine co-ordination in amphibian colour change." *Proceedings of the Royal Society B* 108: 10–53.

Hogg, D. 1959. "Withdrawal bleeding in diagnosis of pregnancy." *Practitioner* 182: 612–614.

Hoggart, R. 1957. *The Uses of Literacy: Aspects of Working-Class Life with Special Reference to Publications and Entertainments.* London: Penguin.

Holland, W. W., and T. P. Whitehead. 1973. "Value of new laboratory tests in diagnosis and treatment." *Lancet* 304: 391–394.

Holmes, H. B., ed. 1994. *Issues in Reproductive Technology.* New York: NYU Press.

Holt, D., and D. Cameron. 2012. *Cultural Strategy: Using Innovative Ideologies to Build Breakthrough Brands.* Oxford: Oxford University Press.

Hopwood, N. 2015. *Haeckel's Embryos: Images, Evolution, and Fraud.* Chicago: University of Chicago Press.

Hopwood, N. 2018. "Artificial fertilization." In *Reproduction: Antiquity to the Present Day,* edited by N. Hopwood, R. Flemming, and L. Kassell, 581–596. Cambridge: Cambridge University Press.

Hopwood, N., P. M. Jones, L. Kassell, and J. Secord. 2015. "Introduction: Communicating reproduction." *Bulletin of the History of Medicine* 89: 379–404.

Horowitz, R., and A. Mohun. 1998. *His and Hers: Gender, Consumption, and Technology.* Charlottesville: University Press of Virginia.

Howell, J. D. 1995. *Technology in the Hospital: Transforming Patient Care in the Early Twentieth Century.* Baltimore: Johns Hopkins University Press.

Hull, A. 2007. "Teamwork, clinical research, and the development of scientific medicine in interwar Britain: The 'Glasgow school' revisited." *Bulletin of the History of Medicine* 81: 569–593.

Humble, N. 2001. *The Feminine Middlebrow Novel, 1920s to 1950s: Class, Domesticity, and Bohemianism.* Oxford: Oxford University Press.

Hume, M. C. W. 1947. "Law and practice: The rights of laboratory animals." In *The UFAW Handbook on the Care and Management of Laboratory Animals,* edited by A. N. Worden, 1–20. London: Baillière.

Hume, M. C. W. 1962. *Experiments on Animals in Great Britain.* 2nd ed. London: UFAW.

Humphries, S. 1988. *A Secret World of Sex: Forbidden Fruit: The British Experience, 1900–1950.* London: Sidgwick & Jackson.

Hüntelmann, A. C. 2013. "A different mode of marketing? The importance of scientific articles in the marketing process of Salvarsan." *History and Technology* 29: 116–134.

Hunter, C. S. 1969. "Doctor defends new pregnancy testing scheme." *Pharmaceutical Journal* 203: 392.

Hutt, F. B. 1931. "The work of the Department of Animal Genetics of the University of Edinburgh." *Journal of Animal Science* 1931: 102–105.

Hyams, E. 1963. *The New Statesman: The History of the First Fifty Years, 1913–1963*. London: Longmans.

Igo, S. E. 2007. *The Averaged American: Surveys, Citizens, and the Making of a Mass Public*. Cambridge, MA: Harvard University Press.

Ingman, H. 1998. *Women's Fiction between the Wars: Mothers, Daughters and Writing*. Edinburgh: Edinburgh University Press.

Inman, W. H. W. 1999. *Don't Tell the Patient: Behind the Drug Safety Net*. Bishops Waltham: Highland Park Productions.

Inman, W. H. W. 2006. *Feeling Better Doctor?* Southampton: Highland Park Productions.

Jackson, M. 2013. *The Age of Stress: Science and the Search for Stability*. Oxford: Oxford University Press.

Jacyna, L. S. 2001. "'A host of experienced microscopists': The establishment of histology in nineteenth-century Edinburgh." *Bulletin of the History of Medicine* 75: 225–253.

Jaffé, G. V. 1960. "Early diagnosis of pregnancy." *Lancet* 275: 654.

Jalland, P. 1988. *Women, Marriage and Politics 1860–1914*. Oxford: Clarendon.

Jardine, R. 1905. *Clinical Obstetrics*. 2nd ed. London: Rebman.

Jeffree, G. M. 1953. "The comparative sensitivity of the male toad pregnancy test and the Friedman test." *Journal of Clinical Pathology* 6: 150–154.

Jeffreys, D. B. 2001. "The regulation of medical devices and the role of the Medical Devices Agency." *British Journal of Clinical Pharmacology* 52: 229–235.

Jeffries, L. M. 1935. "Treatment of repeated abortion." *BMJ* 1: 1265.

Jenkinson, J. 2012. "A 'crutch to assist in gaining an honest living': Dispensary shopkeeping by Scottish general practitioners and the responses of the British medical elite, *ca*. 1852–1911." *Bulletin of the History of Medicine* 86: 1–36.

Jennings, R. 2017. "Lesbian motherhood and the artificial insemination by donor scandal of 1978." *Twentieth Century British History* 28: 570–594.

Jesse, F. T. 1934. *A Pin to See the Peep Show*. London: Heinemann.

Johnson, A., E. M. Stim, and A. Yankauer. 1976. "Do-it-yourself pregnancy testing." *American Journal of Public Health* 66: 129–132.

Johnson, D. K. 2019. *Buying Gay: How Physique Entrepreneurs Sparked a Movement*. New York: Columbia University Press.

Johnstone, R. W. 1913. *A Text-Book of Midwifery for Students and Practitioners*. London: Black.

Johnstone, R. W. 1923. *A Text-Book of Midwifery for Students and Practitioners*. 4th ed. London: Black.

Johnstone, R. W. 1929a. "Biological diagnosis of pregnancy." *Lancet* 213: 306.

Johnstone, R. W. 1929b. "Diagnosis of early pregnancy." *BMJ* 1: 264.

Johnstone, R. W. 1930. "The new physiology of menstruation and its practical implications in obstetrics and gynecology. The Joseph Price Foundation Lecture." *American Journal of Obstetrics and Gynecology* 19: 167–180.

Johnstone, R. W. 1932. *A Text-Book of Midwifery for Students and Practitioners*. 6th ed. London: Black.

Johnstone, R. W. 1933. "Gynaecological aspects of endocrinology." *BMJ* 2: 557–560.

Johnstone, R. W. 1934. *A Text-Book of Midwifery for Students and Practitioners*. 7th ed. London: Black.

Johnstone, R. W. 1947. "Ballantyne's ghost." *Transactions of the Edinburgh Obstetrical Society* 54: 1–13.

Johnstone, R. W. 1954. "The diagnosis of pregnancy." In *Historical Review of British Obstetrics and Gynaecology, 1800–1950*, edited by J. M. Munro Kerr, R. W. Johnstone, and M. H. Phillips, 92–93. Edinburgh: Livingstone.

Johnstone, R. W., and R. J. Kellar. 1968. *A Text-Book of Midwifery for Students and Practitioners*. 21st ed. London: Black.

Johnstone, R. W., B. P. Wiesner, and P. G. Marshall. 1932. "The therapeutic application of gonadotropic hormones." *Lancet* 220: 509–511.

Jolly, M. 2019. *Sisterhood and After: An Oral History of the UK Women's Liberation Movement, 1968–Present*. Oxford: Oxford University Press.

Jones, C. 2020. *The Business of Birth Control: Contraception and Commerce in Britain before the Sexual Revolution*. Manchester: Manchester University Press.

Jones, E. L. 2007. *Abortion in England, 1861–1967*. PhD diss., Royal Holloway, University of London.

Jones, E. L. 2011. "Attitudes to abortion in the era of reform: Evidence from the Abortion Law Reform Association correspondence." *Women's History Review* 20: 283–298.

Jones, G. 2005. *Renewing Unilever: Transformation and Tradition*. Oxford: Oxford University Press.

Jones, G., and A. Kraft. 2004. "Corporate venturing: The origins of Unilever's pregnancy test." *Business History* 46: 100–122.

Jones, T. 1981. "Small business development and the Asian community in Britain." *Journal of Ethnic and Migration Studies* 9: 467–477.

Jordan, B. 1977. "The self-diagnosis of early pregnancy: An investigation of lay competence." *Medical Anthropology* 2: 1–38.

Jouannou, M., ed. 2013. *The History of British Women's Writing, 1920–1945*. Basingstoke: Palgrave Macmillan.

Jülich, S. 2011. "Fetal photography in the age of cool media." In *History of Participatory Media: Politics and Publics, 1750–2000*, edited by A. Ekström, S. Jülich, F. Lungren, and P. Wisselgren, 125–141. New York: Routledge.

Jülich, S. 2015. "The making of a best-selling book on reproduction: Lennart Nilsson's *A Child Is Born*." *Bulletin of the History of Medicine* 89: 491–525.

Jülich, S., and E. Björklund, eds. 2024. *Rethinking the Public Fetus: Historical Perspectives on the Visual Culture of Pregnancy*. Rochester: University of Rochester Press.

Jüttemann, A. 2020. *Auf dem Weg zum Schwangerschaftstest: Das Laboratorium der II. Gynäkologischen Universitätsklinik unter Selmar Aschheim und Bernhard Zondek*. Berlin: be.bra.

Kaasch, M. 2000. "Sensation, Irrtum, Betrug? Emil Abderhalden und die Geschichte der Abwehrfermente." *Acta Historica Leopoldina* 36: 145–210.

Kamnitzer, H., and S. Joseph. 1921a. "Zur biologischen Diagnostik der Schwangerschaft." *Therapie der Gegenwart* 62: 312–324.

Kamnitzer, H., and S. Joseph. 1921b. "Zur Phloridzindiagnostik der Frühgravität." *Therapie der Gegenwart* 62: 459–461.

Kanter, A. E., C. P. Bauer, and A. H. Klawans. 1934. "A new biologic test for hormones in pregnancy urine. Preliminary report." *JAMA* 103: 2026–2027.

Karpin, I., and S. Kristin. 2012. *Perfecting Pregnancy: Law, Disability, and the Future of Reproduction*. Cambridge: Cambridge University Press.

Kassell, L. 2018. "Fruitful bodies and astrological medicine." In *Reproduction: Antiquity to the Present Day*, edited by N. Hopwood, R. Flemming, and L. Kassell, 225–240. Cambridge: Cambridge University Press.

Kavanagh, J. 1999. "The BBC's written archives as a source for media history." *Media History* 5: 81–86.

Keating, J. 2009. *A Child for Keeps: The History of Adoption in England, 1918–1945*. Basingstoke: Palgrave MacMillan.

Keating, P., and A. Cambrosio. 2003. *Biomedical Platforms: Realigning the Normal and the Pathological in Late-Twentieth-Century Medicine*. Cambridge, MA: MIT Press.

Keenan, M. 1971. "Do-it-yourself pregnancy test." *Times*, September 15, 16.

Keevil, N. L. 1937. "Antuitrin S intradermal pregnancy test." *BMJ* 2: 1199.

Kemp, R. 1959. "The friendly *Bufo bufo* and his chums." *Family Doctor* 9: 280–281.

Kenton, R. J. 1959. "Early diagnosis of pregnancy." *Lancet* 274: 409–410.

Keown, J. 1988. *Abortion, Doctors and the Law: Some Aspects of the Legal Regulation of Abortion in England from 1803 to 1982*. Cambridge: Cambridge University Press.

Kevin, C. 2017. "Maternal responsibility and traceable loss: Medicine and miscarriage in twentieth-century Australia." *Women's History Review* 26: 840–856.

Kiernan, K., H. Land, and J. Lewis. 1998. *Lone Motherhood in Twentieth-Century Britain: From Footnote to Front Page*. Oxford: Clarendon Press.

King, L. 2017. "Hiding in the pub to cutting the cord? Men's presence at childbirth in Britain *c*.1940s–2000s." *Social History of Medicine* 30: 389–407.

King, H. 2007. *Midwifery, Obstetrics and the Rise of Gynaecology: The Uses of a Sixteenth-Century Compendium*. Aldershot: Ashgate.

King, L. 2015. *Family Men: Fatherhood and Masculinity in Britain, 1914–1960*. Oxford: Oxford University Press.

King, W. W. 1913. "The serum reaction in pregnancy and cancer by the coagulation method." *Journal of Obstetrics and Gynaecology of the British Empire* 24: 296–303.

Kingdom, J. C. P., T. Kelly, A. B. MacLean, and E. J. McAllister. 1991. "Rapid one step urine test for human chorionic gonadotrophin in evaluating suspected complications of early pregnancy." *BMJ* 302: 1308–1311.

Kirk, R. G. W. 2008. "'Wanted—standard guinea pigs': Standardisation and the experimental animal market in Britain ca. 1919–1947." *Studies in History and Philosophy of Biological and Biomedical Sciences* 39: 280–291.

Kirk, R. G. W. 2009. "Between the clinic and the laboratory: Ethology and pharmacology in the work of Michael Robin Alexander Chance, *c*.1946–1964." *Medical History* 53: 513–536.

Kirk, R. G. W. 2010. "A brave new animal for a brave new world: The British Laboratory Animals Bureau and the constitution of international standards of laboratory animal production and use, circa 1947–1968." *Isis* 101: 62–94.

Kirksey, E., ed. 2014. *The Multispecies Salon*. Durham, NC: Duke University Press.

Kirksey, E., D. Hannah, C. Lotterman, and L. J. Moore. 2016. "The *Xenopus* pregnancy test: A performative experiment." *Environmental Humanities* 8: 37–56.

Kitzinger, S. 1980. *Pregnancy and Childbirth*. London: Joseph.

Klee, A. J. 2003. *The Toy Fish: A History of the Aquarium Hobby in America. The First One Hundred Years*. Pascoag, RI: Finley Aquatic.

Klein, R. D. 1984. "Doing it ourselves: Self insemination." In *Test-Tube Women: What Future for Motherhood?*, edited by R. Arditti, R. D. Klein, and S. Minden, 382–390. London: Pandora.

Kline, W. 2019. *Coming Home: How Midwives Changed Birth.* Oxford: Oxford University Press.

Klopper, A., and H. Frank. 1949. "The use of English male toads in pregnancy tests." *Lancet* 254: 9–11.

Knight, P. 1977. "Women and abortion in Victorian and Edwardian England." *History Workshop* 4: 57–68.

Knott, S. 2019. *Mother: An Unconventional History.* London: Penguin.

Kohler, R. E. 1994. *Lords of the Fly: Drosophila Genetics and the Experimental Life.* Chicago: University of Chicago Press.

Kohler, R. E. 2008. "Lab history." *Isis* 99: 761–768.

Kracke, R. R. 1938. *Textbook of Clinical Pathology.* London: Baillière.

Kroker, K., J. Keelan, and P. M. H. Mazumdar, eds. 2008. *Crafting Immunity: Working Histories of Clinical Immunology.* Aldershot: Ashgate.

Kynaston, D. 2010. *Austerity Britain, 1945–51.* London: Bloomsbury.

L'Aimable, G. 1996. "Supreme survivor." *Chemist & Druggist* 246: 256–257.

LaFollette, M. C. 2008. *Science on the Air: Popularizers and Personalities on Radio and Early Television.* Chicago: University of Chicago Press.

Lambe, J. 2022. "The multiple 'epidemic': Debating responsibility in US medicine, media and the law." *Social History of Medicine* 35: 1285–1309.

Lambert, T. D. 1956. "Diagnosis of early pregnancy." *BMJ* 2: 1118.

Landgrebe, F. W. 1939. "The maintenance of reproductive activity in *Xenopus laevis* for pregnancy diagnosis." *Journal of Experimental Biology* 16: 89–95.

Landgrebe, F. W. 1946. "The Hogben test." *BMJ* 2: 963.

Lane-Petter, W. 1961. *Provision of Laboratory Animals for Research: A Practical Guide.* Amsterdam: Elsevier.

Lane-Roberts, C. S. L., A. Sharman, K. Walker, and B. P. Wiesner. 1939. *Sterility and Impaired Fertility: Pathogenesis, Diagnosis and Treatment.* London: Hamish Hamilton.

Lane-Roberts, C. S. L., A. Sharman, K. Walker, and B. P. Wiesner. 1948. *Sterility and Impaired Fertility: Pathogenesis, Diagnosis and Treatment.* 2nd ed. London: Hamish Hamilton.

Langer, J. 1969. "Pregnancy testing." *Pharmaceutical Journal* 203: 90, 138–139.

Langhamer, C. 2000. *Women's Leisure in England, 1920–60*. Manchester: Manchester University Press.

Langhamer, C. 2007. *Mass Observation and Histories of Women*. Mass Observation.

Lansbury, C. 1985. *The Old Brown Dog: Women, Workers, and Vivisection in Edwardian England*. Madison: University of Wisconsin Press.

Larbie, J. 1985. *Black Women and Maternity Services: A Survey of Thirty Young Afro-Caribbean Women's Experiences and Perceptions of Pregnancy and Childbirth*. London: Training in Health and Race.

Larkin, I. M. 1955. "The nursery requirements of *Xenopus laevis*." *Journal of Obstetrics and Gynaecology of the British Empire* 62: 67–69.

Latham, M. 2002. *Regulating Reproduction: A Century of Conflict in Britain and France*. Manchester: Manchester University Press.

Lawrence, C. 1987. "Cinema verité: The image of William Harvey's experiments in 1928." In *Vivisection in Historical Perspective*, edited by N. A. Rupke, 295–313. London: Croom Helm.

Lawrence, C. 1998. "Still incommunicable: Clinical holists and medical knowledge in inter-war Britain." In *Greater Than the Parts: Holism in Biomedicine, 1920–1950*, edited by C. Lawrence and G. Weisz, 94–111. Oxford: Oxford University Press.

Lawrence, C. 1999. "A tale of two sciences: Bedside and bench in twentieth-century Britain." *Medical History* 43: 421–449.

Lawrence, C. 2005. *Rockefeller Money, the Laboratory, and Medicine in Edinburgh, 1919–1930: New Science in an Old Country*. Rochester, NY: University of Rochester Press.

Layne, L. L. 2002. *Motherhood Lost: A Feminist Account of Pregnancy Loss in America*. New York: Routledge.

Layne, L. L. 2010. "Why the home pregnancy test isn't the feminist technology it's cracked up to be and how to make it better." In *Feminist Technology*, edited by L. L. Layne, S. L. Vostral, and K. Boyer, 89–118. Urbana: University of Illinois Press.

LB, Jade. 2021. *Keisha the Sket*. London: #Merky.

Leap, N., and B. Hunter. 1993. *The Midwife's Tale: An Oral History from Handywoman to Professional Midwife*. London: Scarlet Press.

Leathard, A. 1980. *The Fight for Family Planning: The Development of Family Planning Services in Britain 1921–74*. London: Macmillan.

Leavitt, J. W. 1988. *Brought to Bed: Childbearing in America, 1750–1950*. New York: Oxford University Press.

Leavitt, J. W. 2009. *Make Room for Daddy: The Journey from Waiting Room to Birthing Room*. Chapel Hill: University of North Carolina Press.

Leavitt, S. A. 2006. "'A private little revolution': The home pregnancy test in American culture." *Bulletin of the History of Medicine* 80: 317–345.

Lederer, S. E. 1992. "Political animals: The shaping of biomedical research literature in twentieth-century America." *Isis* 83: 61–79.

Lehmann, R. 1936. *The Weather in the Streets*. London: Collins.

Leitch, A. 1914. "The serum diagnosis of pregnancy and of cancer: A critical study of Abderhalden's method." *BMJ* 2: 330–335.

LeMahieu, D. L. 1988. *A Culture for Democracy: Mass Communication and the Cultivated Mind in Britain between the Wars*. Oxford: Clarendon.

Lennerhed, L. 2002. *Sex i folkhemmet: RFSUs tidiga historia*. Uppsala: Gidlunds.

Lennerhed, L. 2009. "Taking the middle way: Sex education debates in Sweden in the early twentieth century." In *Shaping Sexual Knowledge: A Cultural History of Sex Education in Twentieth Century Europe*, edited by L. D. H. Sauerteig and R. Davidson, 55–70. New York: Routledge.

Lerman, N. E., R. Oldenziel, and A. Mohun. 2003. *Gender and Technology: A Reader*. Baltimore: Johns Hopkins University Press.

Lewis, J. 1980. *The Politics of Motherhood: Child and Maternal Welfare in England, 1900–1939*. London: Croom Helm.

Li, A. 2003. *J. B. Collip and the Development of Medical Research in Canada: Extracts and Enterprise*. Montreal: McGill-Queen's University Press.

Lieberman, H. 2017. *Buzz: A Stimulating History of the Sex Toy*. New York: Pegasus.

Lind, T., and P. G. Whittaker. 1988. "How positive is a positive pregnancy test?" *BMJ* 296: 730–731.

Lindroth, S. 1976. *A History of Uppsala University, 1477–1977*. Stockholm: Almqvist & Wiksell.

Liss, A. 2009. *Feminist Art and the Maternal*. Minneapolis: University of Minnesota Press.

Lister, U. M. 1962. "Tests for the early diagnosis of pregnancy." In *Calling the Laboratory*, edited by W. A. R. Thomson, 83–86. Edinburgh: Livingstone.

Lodge, D. 1965. *The British Museum Is Falling Down*. London: MacGibbon & Kee.

Logan, K. 2022. "Customers who don't buy anything! The introduction of free dispensing at Boots the Chemists." In *Posters, Protests, and Prescriptions: Cultural Histories of the National Health Service in Britain*, edited by J. Crane and J. Hand, 150–177. Manchester: Manchester University Press.

Long, C. F., and J. C. Hirst. 1923. "Ingestion glycosuria, an aid to early diagnosis of pregnancy: Preliminary report." *New York Medical Journal and Medical Record* 118: 543–546.

Long, M. 2022. *Mass Media and Public Understandings of Nature in Interwar Britain.* PhD diss., University of Cambridge.

Longford, E. 1986. *The Pebbled Shore: The Memoirs of Elizabeth Longford.* London: Weidenfeld & Nicolson.

Loudon, I., and M. Drury. 1988. "Some aspects of clinical care in general practice." In *General Practice under the National Health Service, 1948–1997: The First Fifty Years,* edited by I. Loudon, J. Horder, and C. Webster, 92–127. Oxford: Oxford University Press.

Loughlin, K. 2005. "Networks of mass communication: Reporting science, health and medicine in the 1950s and 60s." In *Making Health Policy: Networks in Research and Policy after 1945,* edited by V. Berridge, 295–322. Amsterdam: Rodopi.

Loughran, T. 2020. "'The most helpful friends in the world': Letters pages, expertise and emotion in British women's magazines, *c.*1960–80." In *Women's Periodicals and Print Culture in Britain, 1940s–2000s: The Postwar and Contemporary Period,* edited by L. Forster and J. Hollows, 133–149. Edinburgh: Edinburgh University Press.

Lourie, H. 1962. *A Question of Abortion.* London: Bodley Head.

Lovat, A. 1921. *Marriage and Motherhood.* London: Oates & Washbourne.

Love, G. 2022. "Trans pregnancy in a repronormative world." In *Towards Gender Equality in Law: An Analysis of State Failures from a Global Perspective,* edited by G. Guney, D. Davies, and P. H. Lee, 35–38. Cham: Palgrave Macmillan.

Löwy, I. 1993. "Testing for a sexually transmissible disease, 1907–1970: The history of the Wassermann reaction." In *AIDS and Contemporary History,* edited by V. Berridge and P. Strong, 74–92. Cambridge: Cambridge University Press.

Löwy, I. 1996. *Between Bench and Bedside: Science, Healing, and Interleukin-2 in a Cancer Ward.* Cambridge, MA: Harvard University Press.

Löwy, I. 2004. "'A river that is cutting its own bed': The serology of syphilis between laboratory, society and the law." *Studies in History and Philosophy of Biological and Biomedical Sciences* 35: 509–524.

Löwy, I. 2009. "Producing pharmaceutical standards at the margins: Chemical contraceptives between the laboratory and the field." In *Harmonizing Drugs: Standards in 20th-Century Pharmaceutical History,* edited by C. Bonah, C. Masutti, A. Rasmussen, and J. Simon, 297–322. Paris: Glyphe.

Löwy, I. 2010. *Preventive Strikes: Women, Precancer and Prophylactic Surgery.* Baltimore: Johns Hopkins University Press.

Löwy, I. 2011. "'Sexual chemistry' before the pill: Science, industry and chemical contraceptives, 1920–1960." *British Journal for the History of Science* 44: 245–274.

Löwy, I. 2012. "Defusing the population bomb in the 1950s: Foam tablets in India." *Studies in History and Philosophy of Biological and Biomedical Sciences* 43: 583–593.

Löwy, I. 2014. "Prenatal diagnosis: The irresistible rise of the 'visible fetus.'" *Studies in History and Philosophy of Biological and Biomedical Sciences* 47: 290–299.

Löwy, I. 2017. *Imperfect Pregnancies: A History of Birth Defects and Prenatal Diagnosis*. Baltimore: Johns Hopkins University Press.

Löwy, I. 2018a. "Prenatal diagnosis, surveillance and risk." In *Reproduction: Antiquity to the Present Day*, edited by N. Hopwood, R. Flemming, and L. Kassell, 567–580. Cambridge: Cambridge University Press.

Löwy, I. 2018b. *Tangled Diagnoses: Prenatal Testing, Women, and Risk*. Chicago: University of Chicago Press.

Lublin, N. 1998. *Pandora's Box: Feminism Confronts Reproductive Technology*. Lanham, MD: Rowman & Littlefield.

Luckin, B. 1990. *Questions of Power: Electricity and Environment in Inter-War Britain*. Manchester: Manchester University Press.

Lunay, G. G. 1969. "Pregnancy tests over the counter." *BMJ* 4: 561.

Lynn, D. 2022. "World's oddest toads: *Xenopus* pregnancy tests and animal commodities in 'capitalist ruins.'" *Capitalism Nature Socialism* 33: 103–119.

Macaulay, R. 1926. *Crewe Train*. London: Collins.

Macdonald, G. 1980. *One Hundred Years: Wellcome: In Pursuit of Excellence*. London: Wellcome Foundation.

Macdonald, K., ed. 2018. *Rose Macaulay, Gender, and Modernity*. New York: Routledge.

Macdonald, R. R., and A. Sharman. 1959. "Early diagnosis of pregnancy." *Lancet* 274: 566–567.

Mackenzie Wallis, R. L. 1913. "The serum diagnosis of pregnancy." *Journal of Obstetrics and Gynaecology of the British Empire* 24: 249–256.

Mackenzie Wallis, R. L. 1914. "The value of Abderhalden's tests in the diagnosis of pregnancy." *Journal of Obstetrics and Gynaecology of the British Empire* 25: 53–73.

MacKenzie, D., and J. Wajcman, eds. 1999. *The Social Shaping of Technology*. 2nd ed. Buckingham: Open University Press.

Mackenzie Wallis, R. L. 1916. "Critical review: The protective ferments of the body." *QJM* 9: 138–163.

Macklon, N. S., J. P. M. Geraedts, and C. J. M. Fauser. 2002. "Conception to ongoing pregnancy: The 'black box' of early pregnancy loss." *Human Reproduction Update* 8: 333–343.

Maclean, U. 2004. "History of the Usher Institute." In *Public Health: Past, Present and Future. Celebrating Academic Public Health in Edinburgh, 1902–2002*, edited by R. Bhopal and J. Last, 49–69. London: Nuffield Trust.

Macleod, D. 1936. "Obstetrics and gynaecology." *Practitioner* 137: 554–568.

Maddox, B. 2002. *Rosalind Franklin: The Dark Lady of DNA*. London: HarperCollins.

Maguire, T. A. 1989. "Pregnancy and ovulation testing." *Pharmaceutical Journal* 242: 531–534.

Maisin, J. H. 1968. *History of the International Union against Cancer*. Geneva: UICC.

Malich, L. 2017. *Die Gefühle der Schwangeren: Eine Geschichte somatischer Emotionalität (1780–2010)*. Bielefeld: Transcript.

Malleson, J. 1939. "Criminal abortion: A suggestion for lessening its incidence." *Lancet* 233: 366–367.

Mamo, L. 2007. *Queering Reproduction: Achieving Pregnancy in an Age of Technoscience*. Durham, NC: Duke University Press.

Marcus, P. 2011. "The evolution of the urine pregnancy test." *Female Patient* 36: 40–43.

Marcuse, M. 1927. *Sexualforschung: Verhandlungen des 1. internationalen Kongresses Berlin, 10.–16. Oktober 1926*. Vol. 2. Berlin: A. Marcus & E. Weber's Verlag.

Marie, J. 2004. *The Importance of Place: A History of Genetics in 1930s Britain*. PhD diss., University College London.

Marks, L. V. 2010. *Sexual Chemistry: A History of the Contraceptive Pill*. New Haven, CT: Yale University Press.

Marks, L. V. 2015. *The Lock and Key of Medicine: Monoclonal Antibodies and the Transformation of Healthcare*. New Haven, CT: Yale University Press.

Marsh, W., and W. Ronner. 2008. *The Fertility Doctor: John Rock and the Reproductive Revolution*. Baltimore: Johns Hopkins University Press.

Martin, J. 2010. "The wild rabbit: Plague, polices and pestilence in England and Wales, 1931–1955." *Agricultural History Review* 58: 255–276.

Mathieu, A. 1929. "Pregnancy tests." *Medical Sentinel* 37: 623–632.

Matthew, G. D. 1956. "Simple clinical test for the diagnosis of early pregnancy." *BMJ* 2: 979.

Matthew, G. D., and B. M. Hobson. 1953. "Observations on progesterone-oestrogen withdrawal bleeding and the Hogben test in the diagnosis of pregnancy." *Journal of Obstetrics and Gynaecology of the British Empire* 60: 363–367.

Matthews, S., and L. Wexler. 2000. *Pregnant Pictures*. London: Routledge.

Mazumdar, P. M. H. 1989. *Immunology 1930–1980: Essays on the History of Immunology*. Toronto: Wall & Thompson.

Mazumdar, P. M. H. 2002. *Species and Specificity: An Interpretation of the History of Immunology*. Cambridge: Cambridge University Press.

McAleer, J. 1992. *Popular Reading and Publishing in Britain, 1914–50*. Oxford: Clarendon.

McCarthy, H. 2020. *Double Lives: A History of Working Motherhood in Modern Britain*. London: Bloomsbury.

McClive, C. 2002. "The hidden truths of the belly: The uncertainties of pregnancy in early modern Europe." *Social History of Medicine* 15: 209–227.

McClive, C. 2015. *Menstruation and Procreation in Early Modern France*. Farnham: Ashgate.

McClive, C., and H. King. 2007. "When is a foetus not a foetus? Diagnosing false conceptions in early modern France." In *L'embryon humain à travers l'histoire: Images, savoirs et rites*, edited by V. Dasen, 223–238. Paris: Infolio.

McCrae, A. M. 2003. *The National Health Service in Scotland: Origins and Ideals, 1900–1950*. East Linton: Tuckwell.

McEnroe, N., and S. Hurley. 2019. *The Medicine Cabinet: The Story of Health and Disease Told through Extraordinary Objects*. London: André Deutsch.

McIntosh, T. 2000. "'An abortionist city': Maternal mortality, abortion, and birth control in Sheffield, 1920–1940." *Medical History* 44: 75–96.

McIntosh, T. 2012. *A Social History of Maternity and Childbirth: Key Themes in Maternity Care*. New York: Routledge.

McLaren, A. 2012. *Sex, Robots, Trees, and Test-Tube Babies in Interwar Britain*. Chicago: University of Chicago Press.

McNish, H. 2016. *Nobody Told Me: Poetry and Parenthood*. London: Blackfriars.

McTavish, D. 2005. "Scottish and English health policy from 1948 to the 1973 reforms: Management through a UK prism." *Scottish Affairs* 51: 59–78.

Mechen, B. 2017. "'Closer together': Durex condoms and contraceptive consumerism in 1970s Britain." In *Perceptions of Pregnancy from the Seventeenth to the Twentieth Century*, edited by J. Evans and C. Meehan, 213–236. Cham: Palgrave Macmillan.

Mechling, J. 1975. "Advice to historians on advice to mothers." *Journal of Social History* 9: 44–63.

Medeiros, A. 2016. *Heightened Expectations: The Rise of the Human Growth Hormone Industry in America*. Tuscaloosa: University of Alabama Press.

Medina-Domenech, R., and C. Castañeda. 2007. "Redefining cancer during the interwar period: British Medical Officers of Health, state policy, managerialism, and public health." *American Journal of Public Health* 97: 1563–1571.

Melody, M. E., and L. M. Peterson. 1999. *Teaching America about Sex: Marriage Guides and Sex Manuals from the Late Victorians to Dr. Ruth*. New York: NYU Press.

Meredith, H. T. D. 1935. *The Modern Home Doctor*. London: Daily Express.

Meryman, H. T. 1976. "Historical recollections of freeze-drying." *Developments in Biological Standardization* 36: 29–32.

Méthot, P. O., with J. Sholl, eds. 2020. *Vital Norms: Canguilhem's "The Normal and the Pathological" in the Twenty-First Century*. Paris: Hermann.

Mèuller, G. B., ed. 2017. *Vivarium: Experimental, Quantitative, and Theoretical Biology at Vienna's Biologische Versuchsanstalt*. Cambridge, MA: MIT Press.

Michaels, P. A. 2014. *Lamaze: An International History*. Oxford: Oxford University Press.

Milanich, N. B. 2019. *Paternity: The Elusive Quest for the Father*. Cambridge, MA: Harvard University Press.

Millar, D. R., and G. J. Jarvis. 1980. "Pregnancy diagnosis." *Lancet* 315: 875.

Millington, G., and A. Smith. 1969. "Changes in conduct rules." *Pharmaceutical Journal* 203: 428.

Millward, G. 2022. *Sick Note: A History of the British Welfare State*. Oxford: Oxford University Press.

Milnor, G. C., and E. A. Fennel. 1924. "The diagnosis of pregnancy." *JAMA* 82: 538–540.

Milton, R. F. 1946. "The *Xenopus* pregnancy test." *BMJ* 2: 328.

Mitchison, N. 1935. *We Have Been Warned*. London: Constable.

Mitchison, N. 1979. *You May Well Ask: A Memoir, 1920–1940*. London: Gollancz.

Mold, A. 2015. *Making the Patient-Consumer: Patient Organisations and Health Consumerism in Britain*. Manchester: Manchester University Press.

Moore, M. 2022. "Waiting for the doctor: Managing time and emotion in the British National Health Service, 1948–80." *Twentieth Century British History* 33: 203–229.

Morgan, L. M. 2009. *Icons of Life: A Cultural History of Human Embryos*. Berkeley: University of California Press.

Morgan, L. M., and M. W. Michaels, eds. 1999. *Fetal Subjects, Feminist Positions*. Philadelphia: University of Pennsylvania Press.

Morrice, A. A. G. 1994. "'The medical pundits': Doctors and indirect advertising in the lay press, 1922–1927." *Medical History* 38: 255–280.

Mort, F. 2010. *Capital Affairs: London and the Making of the Permissive Society*. New Haven, CT: Yale University Press.

Mort, F. 2011. "The permissive society revisited." *Twentieth-Century British History* 22: 269–298.

Moscucci, O. 1990. *The Science of Woman: Gynaecology and Gender in England 1800–1929*. Cambridge: Cambridge University Press.

Moscucci, O. 2003. "Holistic obstetrics: The origins of 'natural childbirth' in Britain." *Postgraduate Medical Journal* 79: 168–173.

Moscucci, O. 2009. "The British fight against cancer: Publicity and education, 1900–1948." *Social History of Medicine* 23: 356–373.

Moulder, C. 1990. *Miscarriage: Women's Experiences and Needs.* London: Pandora.

Mulkay, M. 1997. *The Embryo Research Debate: Science and the Politics of Reproduction.* Cambridge: Cambridge University Press.

Munro Kerr, J. M., J. H. Ferguson, J. Young, and J. Hendry, eds. 1923. *A Combined Textbook of Obstetrics and Gynaecology.* Edinburgh: Livingstone.

Myelnikov, D. 2015. *Transforming Mice: Technique and Communication in the Making of Transgenic Animals, 1974–1988.* PhD diss., University of Cambridge.

Myelnikov, D. 2017. "Cuts and the cutting edge: British science funding and the making of animal biotechnology in 1980s Edinburgh." *British Journal for the History of Science* 50: 701–728.

Myers, M. 2013. *Alcines Clair Siddall, M.D.* Oberlin: Oberlin Heritage Center.

Naldrett, P. 2007. *A Bun in the Oven: The Pregnancy Diary of a Dad to Be.* Liskeard: Exposure.

Nathoo, A. 2009. *Hearts Exposed: Transplants and the Media in 1960s Britain.* Basingstoke: Palgrave Macmillan.

Neavill, G. B. 1971. "Victor Gollancz and the Left Book Club." *Library Quarterly* 41: 197–215.

Nemec, B. "Chosing the best apple: Counselling leaflets and technologies of communication in the history of reproduction." *Medical Humanities,* online ahead of print: July 4.

Nemec, B., and H. Dron. 2022. "The environments of reproductive and birth defects research in the U.S. and West Germany (c. 1955–1975)." *Studies in History and Philosophy of Science* 95: 50–63.

Nemec, B., and J. Olszynko-Gryn. 2022. "The Duogynon controversy and ignorance production in post-thalidomide West Germany." *Reproductive Biomedicine & Society Online* 14: 75–87.

Nevett, T. R. 1982. *Advertising in Britain: A History.* London: Heinemann.

Newman, A. 1966. *The Cage.* London: Blond.

Newman, L. 2020. "'Death germs through the post': Postal pathology and workplace experiences of disease in Britain, c.1895–1935." *Social History of Medicine* 33: 1211–1223.

Nicholas, S. 1995. "'Sly demagogues' and wartime radio: J. B. Priestley and the BBC." *Twentieth Century British History* 6: 247–266.

Nicholson, M. 1995. *What Did You Do in the War, Mummy?* London: Pimlico.

Nicolson, M. 2011. "James Young Simpson and the development of physical diagnosis." In *James Young Simpson: Lad o pairts*, edited by A. Nuttall and R. Mander, 57–75. Erskine: Scottish History Press.

Nicolson, M., and J. Fleming. 2013. *Imaging and Imagining the Fetus: The Development of Obstetric Ultrasound*. Baltimore: John Hopkins University Press.

Nixon, W. C. W. 1956. "Antenatal care." In *Refresher Course for General Practitioners: Specially Commissioned Articles from the British Medical Journal*, 3rd collection, 311–322. London: BMA.

Noll, R. 2006. "The blood of the insane." *History of Psychiatry* 17: 395–418.

Nordlund, C. 2011. *Hormones of Life: Endocrinology, the Pharmaceutical Industry, and the Dream of a Remedy for Sterility, 1930–1970*. Sagamore Beach: Science History Publications.

Nottingham, C. 2000. *The NHS in Scotland: The Legacy of the Past and the Prospect of the Future*. Aldershot: Ashgate.

Novick, T. 2023. *Milk and Honey: Technologies of Plenty in the Making of a Holy Land*. Cambridge, MA: MIT Press.

Nunn, J. F. 2002. *Ancient Egyptian Medicine*. Norman: University of Oklahoma Press.

O'Dowd, M. J., and E. E. Philipp. 2000. *The History of Obstetrics and Gynecology*. 2nd ed. Carnforth: Parthenon.

O'Hara, G. 2013. "The complexities of 'consumerism': Choice, collectivism and participation within Britain's National Health Service, *c.*1961–*c.*1979." *Social History of Medicine* 26: 288–304.

Oakley, A. 1979. *Becoming a Mother*. Oxford: Martin Robinson.

Oakley, A. 1984. *The Captured Womb: A History of the Medical Care of Pregnant Women*. New York: Blackwell.

Oakley, A., A. McPherson, and H. Roberts. 1990. *Miscarriage*. Rev. ed. London: Penguin.

Obermer, E. 1935a. *Health and a Changing Civilisation*. London: John Lane.

Obermer, E. 1935b. *Individual Health: A Technique for the Study of Individual Constitution and Its Application to Health*. London: Chapman & Hall.

Obermer, E. 1947. "Vitamin-D requirements in pregnancy." *BMJ* 2: 927.

Obermer, E. 1948. "Wartime rationing and nutrition in pregnancy. First communication." *Journal of Hygiene* 46: 325–330.

Oliver, J. 2006. *Minus Nine to One: The Diary of an Honest Mum*. London: Penguin.

Oliver, R. A. M. 1963. "Haemagglutination-inhibition tests in the diagnosis of pregnancy." *Lancet* 282: 840.

Olszynko-Gryn, J. 2013. "When pregnancy tests were toads: The *Xenopus* test in the early NHS." *Wellcome History* 51: 1–3.

Olszynko-Gryn, J. 2014a. "Laparoscopy as a technology of population control: A use-centred history of surgical sterilization." In *A World of Populations: Transnational Perspectives on Demography in the Twentieth Century*, edited by H. Hartmann and C. R. Unger, 147–177. New York: Berghahn.

Olszynko-Gryn, J. 2014b. "The demand for pregnancy testing: The Aschheim-Zondek reaction, diagnostic versatility, and laboratory services in 1930s Britain." *Studies in History and Philosophy of Biological and Biomedical Sciences* 47: 233–247.

Olszynko-Gryn, J. 2017. "Thin blue lines: Clearblue and the drama of pregnancy testing in British cinema and television." *British Journal for the History of Science* 50: 495–520.

Olszynko-Gryn, J. 2018. "Technologies of contraception and abortion." In *Reproduction: Antiquity to the Present Day*, edited by N. Hopwood, R. Flemming, and L. Kassell, 535–551. Cambridge: Cambridge University Press.

Olszynko-Gryn, J. 2019. "The feminist appropriation of pregnancy testing in 1970s Britain." *Women's History Review* 28: 869–894.

Olszynko-Gryn, J. 2020a. "Predictor: The first home pregnancy test." *Journal of British Studies* 59: 638–642.

Olszynko-Gryn, J. 2020b. "Predictor, Montreal, and Meg Crane: A very brief history of pregnancy testing in Canada." *Osler Library Newsletter* 132: 7–8.

Olszynko-Gryn, J. 2021. "Filming fly eggs: Time-lapse cinematography as an intermedial practice." *Isis* 112: 307–314.

Olszynko-Gryn, J., and P. Ellis. 2017. "'A machine for recreating life': An introduction to reproduction on film." *British Journal for the History of Science* 50: 383–409.

Olszynko-Gryn, J., E. Bjørvik, M. Weßel, S. Jülich, and C. Jean. 2018. "A historical argument for regulatory failure in the case of Primodos and other 'hormone pregnancy tests.'" *Reproductive BioMedicine & Society Online* 6: 34–44.

Ostherr, K. 2013. *Medical Visions: Producing the Patient through Film, Television and Imaging Technologies*. Oxford: Oxford University Press.

Oudshoorn, N. 1994. *Beyond the Natural Body: An Archaeology of Sex Hormones*. London: Routledge.

Oudshoorn, N., and T. Pinch, eds. 2003. *How Users Matter: The Co-Construction of Users and Technology*. Cambridge, MA: MIT Press.

Overy, R. J. 2009. *The Morbid Age: Britain Between the Wars*. London: Allen Lane.

Owen, S. 2005. "*The Abuse of Literacy* and the feeling heart: The trials of Richard Hoggart." *Cambridge Quarterly* 34: 147–176.

Panton, P. N. 1951. *Leaves from a Doctor's Life.* London: Heinemann.

Park, H. 2019. *An Epigenetic Century: The History and Future of a New Science of Life.* PhD diss., Yale University.

Parkes, A. S. 1947. "*Xenopus laevis.*" *School Science Review* 28: 219–223.

Parkes, A. S. 1966. "The rise of reproductive endocrinology, 1926–1940." *Proceedings of the Society of Endocrinology* 34: 20–32.

Passty, J. N. 1988. *Eros and Androgyny: The Legacy of Rose Macaulay.* London: Associated University Press.

Patterson, J. 1937. "The chemical diagnosis of early pregnancy: A method based upon the detection of oestriol in the urine." *BMJ* 2: 522–524.

Payling, D. 2022. "Selling shame: Feminine hygiene advertising and the boundaries of permissiveness in 1970s Britain." *Gender & History,* online ahead of print: June 22.

Pedersen, S. 2020. *The Politicization of Mumsnet.* Bingley: Emerald.

Peel, J. 1976. *The Lives of the Fellows of the Royal College of Obstetricians and Gynaecologists, 1929–1969.* London: Heinemann.

Peel, J. 1986. *William Blair-Bell: Father and Founder.* London: Royal College of Obstetricians and Gynaecologists.

Pembleton, J. T. 1969. "Pregnancy tests." *BMJ* 4: 497.

Perriton, L. 2007. "Forgotten feminists: The Federation of British Professional and Business Women, 1933–1969." *Women's History Review* 16: 79–97.

Petchesky, R. P. 1987. "Fetal images: The power of visual culture in the politics of reproduction." *Feminist Studies* 13: 263–292.

Peterson, E. L. 2016. *The Life Organic: The Theoretical Biology Club and the Roots of Epigenetics.* Pittsburgh: University of Pittsburgh Press.

Pfeffer, N. 1987. "Artificial insemination, in-vitro fertilization and the stigma of infertility." In *Reproductive Technologies: Gender, Motherhood and Medicine,* edited by M. Stanworth, 81–97. Cambridge: Polity.

Pfeffer, N. 1993. *The Stork and the Syringe: A Political History of Reproductive Medicine.* Cambridge: Polity Press.

Pfeffer, N. 2000. "Regulating reproduction." In *Bodies of Technology: Women's Involvement with Reproductive Medicine,* edited by A. R. Saetnan, N. Oudshoorn, and M. S. M. Kirejczyk, 254–277. Columbus: Ohio State University Press.

Philips, A., and J. Rakusen. 1989. *The New Our Bodies, Ourselves.* London: Penguin.

Phillips, H. 1993. *The University of Cape Town, 1918–1948: The Formative Years.* Cape Town: University of Cape Town Press.

Pinch, T. 1993. "Testing, one, two, three-testing: Towards a sociology of testing." *Science, Technology & Human Values* 18: 25–41.

Pink, C. V. 1930. *The Ideal Management of Pregnancy: Talks with Expectant Mothers.* London: Cassell.

Pink, C. V. 1941. *The Foundations of Motherhood.* London: Cassell.

Pitman, J. 1993. "John Dixon Comrie: His lectures and journal." *Proceedings of the Royal College of Physicians of Edinburgh* 23: 63–67.

Platt, S. 1995. "Testing times on TV." *Community Pharmacy*, November, 12–14.

Playfair, W. S. 1893. *A Treatise on the Science and Practice of Midwifery.* 8th ed. London: Smith, Elder.

Porter, D. 1997. "The decline of social medicine in Britain in the 1960s." In *Social Medicine and Medical Sociology in the Twentieth Century*, edited by D. Porter, 97–119. Amsterdam: Rodopi.

Porter, R., and L. Hall. 1995. *The Facts of Life: Creation of Sexual Knowledge in Britain, 1650–1950.* New Haven, CT: Yale University Press.

Preece, J. F. 1974. "Helping the pharmacist: An idea from general practice." *Journal of the Royal College of General Practitioners* 24: 209–212.

Press, B. 2000. *The Murder of the Jews in Latvia, 1941–1945.* Evanston, IL: Northwestern University Press.

Proctor, R. 1988. *Racial Hygiene: Medicine under the Nazis.* Cambridge, MA: Harvard University Press.

Pross, C., and R. Winau. 1984. *Nicht mißhandeln: Das Krankenhaus Moabit. 1920–1933: Ein Zentrum jüdischer Ärzte in Berlin; 1933–1945: Verfolgung, Widerstand, Zerstörung.* Berlin: Edition Hentrich.

Prüll, C. R. 2003. *Medizin am Toten oder am Lebenden? Pathologie in Berlin und in London, 1900–1945.* Basel: Schwabe.

Prüll, C. R., ed. 1998. *Pathology in the 19th and 20th Centuries: The Relationship between Theory and Practice.* Sheffield: European Association for the History of Medicine and Health Publications.

Purcell, S. 1997. "What's in a name?" *Community Pharmacy*, November, 21–22.

Pye, M., and H. Thomas. 1990. "An assessment of a hospital-based pregnancy testing service." *Journal of Public Health* 12: 186–189.

Quirke, V. 2008. *Collaboration in the Pharmaceutical Industry: Changing Relationships in Britain and France, 1935–1965.* New York: Routledge.

Quirke, V. 2009. "The material culture of British pharmaceutical laboratories in the golden age of drug discovery (*c.*1935–75)." *International Journal for the History of Engineering & Technology* 79: 280–299.

Quirke, V., and J. P. Gaudillière. 2008. "The era of biomedicine: Science, medicine, and public health in Britain and France after the Second World War." *Medical History* 52: 441–452.

Rader, K. 2004. *Making Mice: Standardizing Animals for American Biomedical Research, 1900–1955*. Princeton, NJ: Princeton University Press.

Radin, J. 2017. *Life on Ice: A History of New Uses for Cold Blood*. Chicago: University of Chicago Press.

Radin, J., and E. Kowal, eds. 2017. *Cryopolitics: Frozen Life in a Melting World*. Cambridge, MA: MIT Press.

Ralph, R. S. 1934. "Hormone tests for pregnancy." *Journal of Clinical Research* 20: 54–59.

Ramaswamy, C. 2016. *Expecting: The Inner Life of Pregnancy*. Glasgow: Saraband.

Ramsden, E. 2011. "From rodent utopia to urban hell: Population, pathology, and the crowded rats of NIMH." *Isis* 102: 659–688.

Ramsey, M. 2021. *The Swedish Abortion Pill: Co-Producing Medical Abortion and Values, ca. 1965–1992*. Uppsala: Acta Universitatis Upsaliensis.

Raphael, A. J. 2010. *Natural Childbirth in Twentieth Century England: A History of Alternative Approaches to Birth from the 1940s to the 1990s*. PhD diss., Queen Mary University of London.

Rapp, R. 1999. *Testing Women, Testing the Fetus: The Social Impact of Amniocentesis in America*. London: Routledge.

Rasmussen, N. 1997. *Picture Control: The Electron Microscope and the Transformation of Biology in America, 1940–1960*. Stanford, CA: Stanford University Press.

Ratmoko, C. 2010. *Damit die Chemie stimmt: die Anfänge der industriellen Herstellung von weiblichen und männlichen Sexualhormonen 1914–1938*. Zürich: Chronos.

Rautelin, M. 2015. "Cryptic pregnancies and their legal consequences in pre-modern Finland." *Social History of Medicine* 28: 663–685.

Rawlins, M. D. 1995. "Pharmacovigilance: Paradise lost, regained or postponed?" *Journal of the Royal College of Physicians of London* 29: 41–49.

Raymond, J. G. 1995. *Women as Wombs: Reproductive Technologies and the Battle over Women's Freedom*. North Melbourne: Spinifex Press.

Rayner, C. 2003. *How Did I Get Here from There?* London: Virago.

Read, C. H., and D. B. Stone. 1958. "An immunologic assay for minute amounts of human pituitary growth hormone." *Journal of Diseases of Children* 96: 538.

Read, D. 1960. "Early diagnosis of pregnancy." *Lancet* 276: 158.

Reagan, L. 2010. *Dangerous Pregnancies: Mothers, Disabilities, and Abortion in Modern America*. Berkeley: University of California Press.

Redhead, G. 2021. "'A British problem affecting British people': Sickle cell anaemia, medical activism and race in the National Health Service, 1975–1993." *Twentieth Century British History* 32: 189–211.

Rentetzi, M. 2011. "Packaging radium, selling science: Boxes, bottles and other mundane things in the world of science." *Annals of Science* 68: 375–399.

Renwick, C. 2013. "Completing the circle of social sciences? William Beveridge and social biology at London School of Economics during the 1930s." *Philosophy of the Social Sciences* 20: 1–19.

Rheinberger, H. J. 2010. *An Epistemology of the Concrete: Twentieth-Century Histories of Life*. Durham, NC: Duke University Press.

Richards, M. 2008. "Artificial insemination and eugenics: Celibate motherhood, eutelegenesis and germinal choice." *Studies in History and Philosophy of Biological and Biomedical Sciences* 39: 211–221.

Richardson, S. S. 2013. *Sex Itself: The Search for Male and Female in the Human Genome*. Chicago: University of Chicago Press.

Richardson, S. S. 2021. *The Maternal Imprint: The Contested Science of Maternal-Fetal Effects*. Chicago: University of Chicago Press.

Richmond, M. L. 2007. "The cell as the basis for heredity, development, and evolution: Richard Goldschmidt's program of physiological genetics." In *From Embryology to Evo-Devo: A History of Developmental Evolution*, edited by J. Maienschein and M. D. Laubichler, 169–211. Cambridge, MA: MIT Press.

Riddle, O. 1927. "The accomplishments of the First International Congress for Sex Research." *Journal of Social Hygiene* 13: 138–144.

Riley, D. 1981. "'The Free mothers': Pronatalism and working women in industry at the end of the last war in Britain." *History Workshop Journal* 11: 59–118.

Ritchie, J. 1953. *History of the Laboratory of Royal College of Physicians of Edinburgh*. Edinburgh: Royal College of Physicians.

Rivett, G. 1986. *The Development of the London Hospital System 1823–1982*. London: King Edward's Hospital Fund.

Roberts, J. 2012. *The Visualised Foetus: A Cultural and Political Analysis of Ultrasound Imagery*. Farnham: Ashgate.

Roberts, R. E. 1938. "Diagnosis of pregnancy and maternal pelvic deformities." In *A Text-Book of X-Ray Diagnosis by British Authors*, Vol. 2, edited by S. C. Shanks, 386–391. London: Lewis.

Robertson, E. M. 1930. "The practicability of the Aschheim-Zondek test for pregnancy." *Edinburgh Medical Journal* 37: 124–132.

Robertson, W. G. A. 1921. *Manual of Medical Jurisprudence and Toxicology*. 4th ed. London: Black.

Robinson, J. H. 2016. "Bringing the pregnancy test home from the hospital." *Social Studies of Science* 46: 649–674.

Robinson, J. H. 2020. "What the pregnancy test is testing." *British Journal of Sociology* 71: 460–473.

Robson, J. M. 1934a. "Pregnancy diagnosis in theory and practice." *BMJ* 1: 1063–1065.

Robson, J. M. 1934b. *Recent Advances in Sex and Reproductive Physiology*. London: Churchill.

Rock, J. 1932. "Progress in obstetrics." *New England Journal of Medicine* 206: 77–87.

Rogers, L. 1937. *The Truth about Vivisection*. London: Churchill.

Rogers, L. 2014. "Victims start 'new thalidomide' fight." *Sunday Times*, February 23, 9.

Rose, J. 2010. *The Intellectual Life of the British Working Classes*. 2nd ed. New Haven, CT: Yale University Press.

Rosenberg, C. E., ed. 2003. *Right Living: An Anglo-American Tradition of Self-Help Medicine and Hygiene*. Baltimore: Johns Hopkins University.

Rosenfeld, S., and V. W. Rosenfeld. 1944. "The reaction of *Xenopus laevis* Daudin (the South African clawed frog) to human pregnancy serum." *Journal of Laboratory and Clinical Medicine* 29: 527–529.

Ross, E. 1993. *Love and Toil: Motherhood in Outcast London*. New York: Oxford University Press.

Ross, E. 2015. *Exploring Tentativeness: Risk, Uncertainty and Ambiguity in First Time Pregnancy*. PhD diss., University of Edinburgh.

Ross, E. 2018. "Provisionally pregnant: Uncertainty and interpretive work in accounts of home pregnancy testing." *Health* 22: 87–105.

Røstvik, C. M. 2022. *Cash Flow: The Businesses of Menstruation*. London: UCL Press.

Rothman, B. K. 1986. *The Tentative Pregnancy: Prenatal Diagnosis and the Future of Motherhood*. New York: Viking.

Rowland, R. 1993. *Living Laboratories: Women and Reproductive Technology*. London: Cedar.

Russell, R. 2012. *Nausea and Vomiting: A History of Signs, Symptoms and Sickness in Nineteenth-Century Britain*. PhD diss., University of Manchester.

Russell, W. M. S., and R. L. Burch. 1959. *The Principles of Humane Experimental Technique*. London: Methuen.

Rusterholz, C. 2020. *Women's Medicine: Sex, Family Planning and British Female Doctors in Transnational Perspective, 1920–70*. Manchester: Manchester University Press.

Rusterholz, C. 2021. "Teenagers, sex and the Brook Advisory Centres, 1964–85." In *Children's Experiences of Welfare in Modern Britain*, edited by S. Pooley and J. Taylor, 247–271. London: University of London Press.

Rusterholz, C. 2022. "Youth sexuality, responsibility, and the opening of the Brook Advisory Centres in London and Birmingham in the 1960s." *Journal of British Studies* 61: 315–342.

Rustoyburu, C., and Y. Eraso. 2018. *Cuerpos Hormonales: Intersecciones entre laboratorio, clínica y Sociedad*. Mar del Plata: EUDEM.

Saetnan, R., N. Oudshoorn, and M. S. M. Kirejczyk, eds. 2000. *Bodies of Technology: Women's Involvement with Reproductive Medicine*. Columbus: Ohio State University Press.

Saffron, L. 1994. *Challenging Conceptions: Pregnancy and Parenting beyond the Traditional Family*. London: Cassell.

Sage, L. 2000. *Bad Blood*. London: Fourth Estate.

Samuels, S. 1966. "The Left Book Club." *Journal of Contemporary History* 1: 65–86.

Sánchez, M. S. 2015. "'I would not feel the pain if I were with you': Catalina Micaela and the cycle of pregnancy at the Court of Turin, 1585–1597." *Social History of Medicine* 28: 445–464.

Sauerteig, L. D. 2009. "Representations of pregnancy and childbirth in (West) German sex education books, 1900s–1970s." In *Shaping Sexual Knowledge: A Cultural History of Sex Education in Twentieth Century Europe*, edited by L. D. H. Sauerteig and R. Davidson, 129–160. New York: Routledge.

Schaefer, E., ed. 2014. *Sex Scene: Media and the Sexual Revolution*. Durham, NC: Duke University Press.

Schlich, T. 2010. *The Origins of Organ Transplantation: Surgery and Laboratory Science, 1880s–1930s*. Rochester, NY: University of Rochester Press.

Schmidt, S. 2020. *Midlife Crisis: The Feminist Origins of a Chauvinist Cliché*. Chicago: University of Chicago Press.

Schneck, P. 1997. "Selmar Aschheim (1878–1965) und Bernhard Zondek (1891–1966): Zum Schicksal zweier jüdischer Ärzte und Forscher an der Berliner Charité." *Zeitschrift für ärztliche Fortbildung und Qualitätssicherung* 91: 187–194.

Schneider, W. 2003. "Blood transfusion between the wars." *Journal of the History of Medicine and Allied Sciences* 58: 187–224.

Schonland, B. F. J. 1934. "Minutes of proceedings." *Transactions of the Royal Society of South Africa* 22: i–lxxv.

Schuster, N. H. 1983. *A Clinical Pathologist Day by Day, 1916–1960*. London.

Schwabacher, H. 1950. "Review of one year's work in Watford Pregnancy Diagnostic Centre." *BMJ* 2: 328–330.

Schwabacher, H. 1951. "The laboratory diagnosis of pregnancy." *Medical Press* 226: 82–88.

Schwabacher, H. 1956. "The laboratory diagnosis of pregnancy." *Postgraduate Medical Journal* 32: 392–397.

Scott, D. H. 1936. *The New People's Physician: A Compendium of Practical Information on Personal Health and Domestic Hygiene*. 5 vols. London: Waverley.

Seaton, A. 2023. *Our NHS: A History of Britain's Best Loved Institution*. New Haven, CT: Yale University Press.

Seigel, M. 2014. *The Rhetoric of Pregnancy*. Chicago: University of Chicago Press.

Sengoopta, C. 2006. *The Most Secret Quintessence of Life: Sex, Glands, and Hormones, 1850–1950*. Chicago: University of Chicago Press.

Serlin, D., ed. 2010. *Imagining Illness: Public Health and Visual Culture*. Minneapolis: University of Minnesota Press.

Sethna, C., and G. Davis, eds. 2019. *Abortion Across Borders: Transnational Travel and Access to Abortion Services*. Baltimore: Johns Hopkins University Press, 2019.

Shah, C. Q. 2004. "A thin blue line: The history of the pregnancy test kit." *Public Historian* 26: 131–133.

Shapin, S. 1989. "The invisible technician." *American Scientist* 77: 554–563.

Shapiro, H. A., and H. Zwarenstein. 1933. "Metabolic changes associated with endocrine activity and the reproductive cycle in *Xenopus laevis*. I. The effects of gonadectomy and hypophysectomy on the calcium content of the serum." *Journal of Experimental Biology* 10: 186–195.

Shapiro, H. A., and H. Zwarenstein. 1934. "A rapid test for pregnancy on *Xenopus laevis*." *Nature* 133: 762.

Shapiro, H. A., and H. Zwarenstein. 1935. "A test for the early diagnosis of pregnancy on the South African clawed toad (*Xenopus laevis*)." *South African Medical Journal* 9: 202–205.

Shapiro, H. A., and H. Zwarenstein. 1946. "The *Xenopus* test for pregnancy." *BMJ* 2: 752.

Sharman, A. 1937. "Antuitrin S intradermal pregnancy test." *BMJ* 2: 1199.

Sharman, A. 1938. "The Antuitrin S test for the diagnosis of pregnancy." *Journal of Obstetrics and Gynaecology of the British Empire* 45: 82–83.

Sharman, A. 1963. "Immunological pregnancy tests." *Lancet* 282: 1224–1225.

Sharman, A. 1964. "How do you know you are pregnant?" *Family Doctor* 4: 290–292.

Sharman, A. 1965. "Immunological tests for pregnancy." *Triangle* 7: 66.

Sharman, A., and T. Pearston. 1964. "Immunological tests for pregnancy." *Journal of Medical Laboratory Technology* 21: 271.

Sharpey-Schafer, E. 1930. "Research in animal genetics." *BMJ* 2: 30–32.

Shea, J., and A. J. N. Warrack. 1963. "The haemagglutination inhibition and the Hogben test in pregnancy." *Journal of Clinical Pathology* 16: 581–584.

Sheldon, S., G. Davis, J. O'Neill, and C. Parker, eds. 2022. *The Abortion Act 1967: A Biography of a UK Law*. Cambridge: Cambridge University Press.

Shepherd, R. 1994. *Iain Macleod: A Biography*. London: Hutchison.

Sher, N. 1946. "Causes of delayed menstruation and its treatment: An investigation in the Women's Auxiliary Air Force." *BMJ* 1: 347–349.

Sheridan, D. 1994. "Using the Mass Observation Archive as a source for women's studies." *Women's History Review* 3: 101–113.

Shetterly, M. L. 2016. *Hidden Figures: The Untold Story of the African American Women Who Helped Win the Space Race*. London: William Collins.

Shuttleworth, E. M. 1963. "Evaluation of the Prepuerin pregnancy test." *Lancet* 282: 699–700.

Siddall, A. C. 1928a. "A suggested test for pregnancy based on the action of gravid female blood serum on mouse uterus: Preliminary report." *JAMA* 90: 380–381.

Siddall, A. C. 1928b. "The hormone test for pregnancy: Report II." *JAMA* 91: 779–782.

Silberman, M. 2009. "The perfect storm: Late nineteenth-century Chicago sex radicals: Moses Harman, Ida Craddock, Alice Stockman and Comstock obscenity laws." *Journal of the Illinois State Historical Society* 102: 324–367.

Silverstein, A. 2009. *A History of Immunology*. 2nd ed. London: Academic Press.

Singer, J. M., and C. M. Plotz. 1956. "The latex fixation test: I. Application to the serologic diagnosis of rheumatoid arthritis." *American Journal of Medicine* 21: 888–892.

Singleton, V., and M. Michael. 1993. "Actor-networks and ambivalence: General practitioners in the UK cervical screening programme." *Social Studies of Science* 23: 227–264.

Skloot, R. 2010. *The Immortal Life of Henrietta Lacks*. New York: Crown.

Sleigh, C. 2012. *Frog*. London: Reaktion.

Slinn, J. 1995. "Research and development in the UK pharmaceutical industry from the nineteenth century to the 1960s." In *Drugs and Narcotics and History*, edited by R. Porter and M. Teich, 168–186. Cambridge: Cambridge University Press.

Slinn, J. 2005. "Drug-guzzlers or medicine junkies? Regulating the cost and consumption of prescription pharmaceuticals in the UK 1948–1967." *Business History* 47: 352–366.

Sloan, A. W. 1979. "Highlights of physiological research at the University of Cape Town, 1912–1978." *South African Medical Journal* 56: 974–976.

Smith, A. 1996. *The New Statesman: Portrait of a Political Weekly, 1913–1931*. London: Cass.

Smith, A. 2009. "Thomas Bassett Macaulay and the Bahamas: Racism, business and Canadian sub-imperialism." *Journal of Imperial and Commonwealth History* 37: 29–50.

Smith, M. 1980. *Woman's Own Birth Control: A Guide to Every Method of Contraception*. Feltham: Hamlyn.

Smith, M. C., T. G. Booth, and I. Jones. 1973. "Paying the pharmacist under the British National Health Service." *Inquiry* 10: 57–64.

Smith, R. G. 1993. "The development of ethical guidance for medical practitioners by the General Medical Council." *Medical History* 37: 56–67.

Smith, R. G. 1994. *Medical Discipline: The Professional Conduct Jurisdiction of the General Medical Council, 1858–1990*. Oxford: Clarendon.

Smith, S. 1925. *Forensic Medicine: A Text-Book for Students and Practitioners*. London: Churchill.

Smith, Z. 2000. *White Teeth*. London: Hamish Hamilton.

Smithells, R. W. 1965. "The problem of teratogenicity." *Practitioner* 194: 104–110.

Soloway, R. 1990. *Demography and Degeneration: Eugenics and the Declining Birthrate in Twentieth-Century Britain*. Chapel Hill: University of North Carolina Press.

Soloway, R. 1995. "'The perfect contraceptive': Eugenics and birth control research in Britain and America in the interwar years." *Journal of Contemporary History* 30: 637–665.

Soskin, S., H. Wacntel, and D. Hechter. 1940. "The treatment of delayed menstruation with prostigin: A therapeutic test for early pregnancy." *JAMA* 14: 2090.

Soto Laveaga, G. 2009. *Jungle Laboratories: Mexican Peasants, National Projects, and the Making of the Pill*. Durham, NC: Duke University Press.

Spensky, M. 1992. "Producers of illegitimacy: Homes for unmarried mothers in the 1950s." In *Regulating Womanhood: Historical Essays on Marriage, Motherhood and Sexuality*, edited by C. Smart, 100–118. London: Routledge.

Squier, S. M. 1994. *Babies in Bottles: Twentieth-Century Visions of Reproductive Technology*. New Brunswick, NJ: Rutgers University Press.

Squier, S. M. 2017. *Epigenetic Landscapes: Drawings as Metaphor*. Durham, NC: Duke University Press.

St. John, J. 1990. *William Heinemann: A Century of Publishing, 1890–1990*. London: Heinemann.

Stacey, M. 1992. *Regulating British Medicine: The General Medical Council.* Chichester: Wiley.

Stallworthy, J. 1951. "Treatment of threatened abortion." *Practitioner* 166: 118–123.

Stanworth, M. 1987. *Reproductive Technologies: Gender, Motherhood and Medicine.* Cambridge: Polity Press.

Stark, J. F. 2020. *The Cult of Youth: Anti-Ageing in Modern Britain.* Cambridge: Cambridge University Press.

Stephens, M. 1910. *Woman and Marriage: A Handbook.* London: T. F. Unwin.

Stern, A. M. 2012. *Telling Genes: The Story of Genetic Counseling in America.* Baltimore: Johns Hopkins University Press.

Stevenson, J. 1984. *British Society, 1914–45.* London: Allen Lane.

Stevenson, J., and C. Cook. 2010. *The Slump: Britain in the Great Depression.* 3rd ed. Harlow: Longman.

Stewart, J. 2003. "The National Health Service in Scotland, 1947–74: Scottish or British?" *Historical Research* 76: 389–410.

Stockham, A. B. 1918. *Tokology: A Book for Everywoman.* Rev. ed. London: Fowler.

Stoff, H. 2012. *Wirkstoffe: Eine Wissenschaftsgeschichte der Hormone, Vitamine und Enzyme, 1920–1970.* Stuttgart: Franz Steiner.

Stolberg, M. 2007. "The decline of uroscopy in early modern learned medicine (1500–1650)." *Science and Medicine* 12: 313–336.

Stolberg, M. 2015. *Uroscopy in Early Modern Europe.* Farnham: Ashgate.

Stolberg, M. 2020. "Enthüllungen: Die uroskopische Schwangerschaftsdiagnose und ihre Darstellung in der frühneuzeitlichen Kunst." In *Visualisierung des Ungeborenen: Interdisziplinäre Perspektiven,* edited by H. Fangerau and D. Hornuff, 51–67. Paderborn: Fink.

Stoller, S. E. 2023. *Inventing the Working Parent: Work, Gender, and Feminism in Neoliberal Britain.* Cambridge, MA: MIT Press.

Stone, H. M., and A. Stone. 1936. *A Marriage Manual: A Practical Guide-Book to Sex and Marriage.* London: Gollancz.

Stone, J. E. 1932. *Hospital Organization and Management: Including Planning and Construction.* London: Faber & Faber.

Stopes, M. C. 1920. *Radiant Motherhood: A Book for Those Who Are Creating the Future.* London: Putnam's Sons.

Stopes, M. C., ed. 1929. *Mother England: A Contemporary History Self-Written by Those Who Have Had No Historian.* London: Bale.

Sturdy, S. 2011. "Looking for trouble: Medical science and clinical practice in the historiography of modern medicine." *Social History of Medicine* 24: 739–757.

Sturdy, S., and R. Cooter. 1998. "Science, scientific management, and the transformation of medicine in Britain, *c.*1870–1950." *History of Science* 36: 421–466.

Stürzbecher, M. 1997. *125 Jahre Krankenhaus Moabit 1872–1997.* Berlin: Weidler.

Summerfield, P. 1998. *Reconstructing Women's Wartime Lives: Discourse and Subjectivity in Oral Histories of the Second World War.* Manchester: Manchester University Press.

Summerfield, P., and N. Crockett. 1992. "'You weren't taught that with welding': Lessons in sexuality in the Second World War." *Women's History Review* 1: 435–454.

Sutcliffe-Braithwaite, F. 2014. "Observing the 80s." *Twentieth Century British History* 25: 484–495.

Sutton, M. 1992. *"We Didn't Know Aught": A Study of Sexuality, Superstition and Death in Women's Lives in Lincolnshire during the 1930s, '40s and '50s.* Stamford: Paul Watkins.

Szreter, S. 1996. *Fertility, Class and Gender in Britain, 1860–1940.* Cambridge: Cambridge University Press.

Szreter, S., and K. Fisher. 2010. *Sex Before the Sexual Revolution: Intimate Life in England 1918–1963.* Cambridge: Cambridge University Press.

Szuhan, N. 2018. "Sex in the laboratory: The Family Planning Association and contraceptive science in Britain, 1929–1959." *British Journal for the History of Science* 51: 487–510.

Szuhan, N. 2022. *The Family Planning Association and Contraceptive Science and Technology in Mid-Twentieth-Century Britain.* Cham: Palgrave Macmillan.

Tansey, E. M. 1989. "The Wellcome Physiological Research Laboratories, 1894–1904: The Home Office, pharmaceutical firms and animal experiments." *Medical History* 33: 1–41.

Tansey, E. M. 1990. "The early history of the Wellcome Research Laboratories, 1894–1914." In *Pill Peddlers: Essays on the History of the Pharmaceutical Industry*, edited by J. Libenau, G. J. Higby, and E. C. Stroud, 91–106. Madison, WI: American Institute of the History of Pharmacy.

Tansey, E. M. 1994. "Protection against dog distemper and dogs protection bills: The Medical Research Council and anti-vivisectionist protest, 1911–1933." *Medical History* 38: 1–26.

Tansey, E. M. 2008. "Keeping the culture alive: The laboratory technician in mid-twentieth-century British medical research." *Notes and Records of the Royal Society* 62: 77–95.

Tansey, E. M., and L. A. Reynolds. 1997. *Wellcome Witnesses to Twentieth Century Medicine.* Vol. 1. London: Wellcome Trust.

Tausk, M. 1984. *Organon: The Story of an Unusual Pharmaceutical Enterprise*. Oss: Akzo Pharma.

Taylor, J. S. 2008. *The Public Life of the Fetal Sonogram: Technology, Consumption, and the Politics of Reproduction*. New Brunswick, NJ: Rutgers University Press.

te Heesen, A. 2014. *The Newspaper Clipping: A Modern Paper Object*. Manchester: Manchester University Press.

Thane, P., and T. Evans. 2012. *Sinners? Scroungers? Saints? Unmarried Motherhood in Twentieth-Century England*. Oxford: Oxford University Press.

Thesing, C. 1935. *School of Biology*. London: Routledge.

Thistlethwaite, T. 1997. *Independent and Bloody Minded: The Story of the Medical Journalists' Association, 1967–97*. London: MJA.

Thom, M. 1981. *Having a Baby: A Month-by-Month Guide through Your Pregnancy*. London: Octopus.

Thomas, M. 1997. *Post-War Mothers: Childbirth Letters to Grantly Dick-Read, 1946–1956*. Rochester, NY: University of Rochester Press.

Thomas, N. 2002. "Challenging myths of the 1960s: The case of student protest in Britain." *Twentieth Century British History* 13: 277–297.

Thomas, N. 2008. "Will the real 1950s please stand up?" *Cultural and Social History* 5: 227–235.

Thomson, A. L. 1975. *Half a Century of Medical Research*. 2 vols. London: HMSO.

Thursz, A. D. 1971. "Pregnancy testing." *BMJ* 3: 769.

Thurtle, D. 1940a. "Abortion: Right or wrong." *Eugenics Review* 32: 10–16.

Thurtle, D. 1940b. *Abortion: Right or Wrong?* London: T. W. Laurie.

Tinkler, P. 1995. *Constructing Girlhood: Popular Magazines for Girls Growing Up in England, 1920–50*. London: Taylor & Francis.

Tinkler, P., S. Spencer, and C. Langhamer. 2017. "Revisioning the history of girls and women in Britain in the long 1950s." *Women's History Review* 26: 1–8.

Todd, S. 2019. *Tastes of Honey: The Making of Shelagh Delaney and a Culture of Revolution*. London: Chatto & Windus.

Tomes, N. 2016. *Remaking the American Patient: How Madison Avenue and Modern Medicine Turned Patients into Consumers*. Chapel Hill: University of North Carolina Press.

Tone, A. 1996. "Contraceptive consumers: Gender and the political economy of birth control in the 1930s." *Journal of Social History* 29: 485–506.

Tone, A. 2000. "Black market birth control: Contraceptive entrepreneurship and criminality in the Gilded Age." *Journal of American History* 87: 435–459.

Tone, A. 2001. *Devices and Desires: A History of Contraceptives in America*. New York: Hill & Wang.

Tone, A. 2002. "Making room for rubbers: Gender, technology, and birth control before the pill." *History and Technology* 18: 51–76.

Tone, A. 2012. "Medicalizing reproduction: The pill and home pregnancy tests." *Journal of Sex Research* 49: 319–327.

Topley, W. W. C. 1917. "The examination of the blood." In *The New System of Gynaecology*, Vol. 1, edited by T. W. Eden and C. Lockyer, 191–214. London: Macmillan.

Usborne, C. 2007. *Cultures of Abortion in Weimar Germany*. New York: Berghahn.

Valier, H. K. 2002. *The Politics of Scientific Medicine in Manchester, c.1900–1960*. PhD diss., University of Manchester.

Valier, H. K., and C. Timmermann. 2008. "Clinical trials and the reorganization of medical research in post-Second World War Britain." *Medical History* 52: 493–510.

van Abbé, N. J. 1949. "Pregnancy diagnosis." *Pharmaceutical Journal* 163: 78.

van de Velde, T. H. 1931. *Fertility and Sterility in Marriage: Their Voluntary Promotion and Limitation*. London: Heinemann.

van de Wiel, L. 2020. *Freezing Fertility: Oocyte Cryopreservation and the Gender Politics of Aging*. New York: NYU Press.

van den Belt, H. 2011. "The collective construction of a scientific fact: A re-examination of the early period of the Wassermann reaction (1906–1912)." *Social Epistemology* 25: 311–339.

van Sittert, L., and G. J. Measey. 2016. "Historical perspectives on global exports and research of African clawed frogs (*Xenopus laevis*)." *Transactions of the Royal Society of South Africa* 71: 157–166.

Vargha, D. 2018. *Polio across the Iron Curtain: Hungary's Cold War with an Epidemic*. Cambridge: Cambridge University Press.

Versluis, A. 2008. "Sexual mysticisms in nineteenth century America: John Humphrey Noyes, Thomas Lake Harris, and Alice Bunker Stockham." In *Hidden Intercourse: Eros and Sexuality in the History of Western Esotericism*, edited by W. J. Hanegraaff and J. J. Kripal, 333–355. Leiden: Brill.

Vincent, D. 1998. *The Culture of Secrecy: Britain, 1832–1998*. Oxford: Oxford University Press.

Voge, C. I. B. 1929. "A simple chemical test for pregnancy." *BMJ* 2: 829–830.

von Schwerin, A., H. Stoff, and B. Wahrig, eds. 2013. *Biologics: A History of Agents Made from Living Organisms in the Twentieth Century*. London: Pickering & Chatto.

von Villiez, A. 2009. *Mit aller Kraft verdrängt: Entrechtung und Verfolgung "nicht arischer" Ärzte in Hamburg 1933 bis 1945*. Hamburg: Dölling und Galitz.

Voss, S. 1992. "The usage of pregnancy tests in one health district." *British Journal of Obstetrics and Gynaecology* 99: 1000–1003.

Vostral, S. L. 2008. *Under Wraps: A History of Menstrual Hygiene Technology*. New York: Lexington.

Waggoner, M. R. 2017. *The Zero Trimester: Pre-pregnancy Care and the Politics of Reproductive Risk*. Oakland: University of California Press.

Wajcman, J. 1991. *Feminism Confronts Technology*. Cambridge: Polity.

Walker, E. M., M. Lewis, W. Cooper, M. Marnie, and P. W. Howie. 1988. "Occult biochemical pregnancy: Fact or fiction?" *British Journal of Obstetrics and Gynaecology* 95: 659–663.

Warrack, A. J. N. 1963. "Latex test for pregnancy." *BMJ* 2: 869.

Warrack, A. J. N. 1965. "The laboratory diagnosis of early pregnancy." *Practitioner* 194: 731–737.

Watkins, E. S. 1998. *On the Pill: A Social History of Contraceptives, 1950–1970*. Baltimore: Johns Hopkins University Press.

Watkins, E. S. 2007. *The Estrogen Elixir: A History of Hormone Replacement Therapy in America*. Baltimore: Johns Hopkins University Press.

Watkins, E. S. 2012a. "Deciphering the prescription: Pharmacists and the patient package insert." In *Prescribed: Writing, Filling, Using, and Abusing the Prescription in Modern America*, edited by J. A. Greene and E. S. Watkins, 91–116. Baltimore: Johns Hopkins University Press.

Watkins, E. S. 2012b. "How the pill became a lifestyle drug: The pharmaceutical industry and birth control in the United States since 1960." *American Journal of Public Health* 102: 1462–1472.

Way, S. 1937. "Antuitrin S intradermal pregnancy test." *BMJ* 2: 1142–1143.

Webster, C. 2002. *The National Health Service: A Political History*. Oxford: Oxford University Press.

Weeks, J. 2018. *Sex, Politics and Society: The Regulation of Sexuality Since 1800*. 4th ed. Harlow: Pearson.

Weightman, G. 2011. *Children of Light: How Electricity Changed Britain Forever*. London: Atlantic.

Weindling, P. 2009a. "Medical refugees in Britain and the wider world, 1930–1960: Introduction." *Social History of Medicine* 22: 451–459.

Weindling, P. 2009b. "Medical refugees and the modernisation of British medicine, 1930–1960." *Social History of Medicine* 22: 489–511.

Weingarten, K. 2023. *Pregnancy Test.* London: Bloomsbury.

Weisman, A. I. 1940. "Evaluation of the prostigmine test for early pregnancy." *JAMA* 115: 551–552.

Weisman, A. I., and C. W. Coates. 1944. *The South African Frog (Xenopus laevis) in Pregnancy Diagnosis: A Research Bulletin.* New York: New York Biologic Research Foundation.

Weldon, C., L. H. du Preez, A. D. Hyatt, R. Muller, and R. Speare. 2004. "Origin of the amphibian chytrid fungus." *Emerging Infectious Diseases* 10: 2100–2105.

Wells, A. L. 1937. *Tropical Aquariums, Plants and Fishes.* London: Frederick Warne.

Wells, G. P. 1978. "Lancelot Thomas Hogben, 9 December 1895–22 August 1975." *Biographical Memoirs of Fellows of the Royal Society* 24: 183–221.

Wells, H. G., J. Huxley, and G. P. Wells. 1935. *Reproduction, Heredity and the Development of Sex.* London: Cassell.

Welshman, J. 1996. "In search of the 'problem family': Public health and social work in England and Wales 1940–70." *Social History of Medicine* 9: 447–465.

White, C. L. 1970. *Women's Magazines 1693–1968.* London: Michael Joseph.

Whitfield, N. 2013. "Who is my donor? The local propaganda techniques of London's Emergency Blood Transfusion Service, 1939–45." *Twentieth Century British History* 24: 542–572.

Wide, L. 1962. "An immunological method for the assay of human chorionic gonadotrophin." *Acta Endocrinologica* 41, Suppl. 70: 1–111.

Wide, L. 2005. "Inventions leading to the development of the diagnostic test kit industry: From the modern pregnancy test to the sandwich assays." *Upsala Journal of Medical Science* 110: 193–216.

Wide, L., and C. A. Gemzell. 1960. "An immunological pregnancy test." *Acta Endocrinologica* 35: 261–267.

Wide, L., and C. A. Gemzell. 1962. "Determination of gonadotropins in urine by a haemagglutination inhibition reaction." In *Ciba Foundation Symposium—Immunoassay of Hormones,* edited by G. E. W. Wolstenholme and M. P. Cameron, 296–309. London: Churchill.

Wiesner, B. P. 1931. "Pregnancy diagnosis station: Report on second year's working." *BMJ* 1: 860–861.

Wiesner, B. P. 1932. "Pregnancy diagnosis station: Report on third year's working." *BMJ* 2: 759.

Wiesner, B. P. 1933. "Pregnancy diagnosis: Report of fourth year's working." *BMJ* 2: 296.

Wiesner, B. P. 1936. *Sex*. London: Thornton Butterworth.

Wiesner, B. P., and N. Sheard. 1933. *Maternal Behaviour in the Rat*. Edinburgh: Oliver & Boyd.

Wikler, D., and N. J. Wikler. 1991. "Turkey-baster babies: The demedicalization of artificial insemination." *Milbank Quarterly* 69: 5–40.

Wilcox, A. P., and B. P. Block. 1969. "Pregnancy testing." *Pharmaceutical Journal* 203: 389.

Williams, D. L. 1994. "Dr. Edward Elkan: Pioneer in lower vertebrate pathobiology." *Journal of Zoo and Wildlife Medicine* 25: 117–118.

Williams, M. 2023. *A Taste of Honey*. London: Bloomsbury.

Williamson, H. 1913. "The value of Abderhalden's test for pregnancy." *Journal of Obstetrics and Gynaecology of the British Empire* 24: 211–217.

Wilmot, S. 2007. "From 'public service' to artificial insemination: Animal breeding science and reproductive research in early twentieth-century Britain." *Studies in History and Philosophy of Biological and Biomedical Sciences* 38: 411–441.

Wilson, A. 1995. *The Making of Man-Midwifery: Childbirth in England, 1660–1770*. London: UCL Press.

Wilson, D. 2014. *The Making of British Bioethics*. Manchester: Manchester University Press.

Wilson, G. S. 1948. "The Public Health Laboratory Service: Origins and development of public health laboratories." *BMJ* 1: 677–682.

Wilson, G. S. 1951. "The Public Health Laboratory Service." *BMJ* 7: 146–152.

Wilson, J. 1996. "Dr Isaacson, Twiggy and me: A memoir of North-West London in the 1960s." *Jewish Quarterly* 43: 14–17.

Winkelstein, A. 1942. "Prostigmine methylsulfate and delayed menstruation." *American Journal of Obstetrics and Gynecology* 44: 231.

Winston, R. 1989. *Getting Pregnant: The Complete Guide to Fertility and Infertility*. London: Anaya.

Winton, F. W. 1971. "Pregnancy testing." *BMJ* 4: 296.

Withycombe, S. 2019. *Lost: Miscarriage in Nineteenth-Century America*. New Brunswick, NJ: Rutgers University Press.

Wivel, A. 1998. "Abortion policy and politics on the Lane Committee of Enquiry, 1971–1974." *Social History of Medicine* 11: 109–135.

Wolf, J. 2018. *Cesarean Section: An American History of Risk, Technology, and Consequence*. Baltimore: Johns Hopkins University Press.

Worboys, M. 2018. *Private Clinical Laboratories in Britain: The Clinical Research Association, 1894–1914*. University of Manchester.

Wreszinski, W. 1909. *Der grosse medizinische Papyrus des Berliner Museums (Pap. Berl. 3038)*. Leipzig: J. C. Hinrichs.

Wright, H. B. 1949a. "Pregnancy diagnosis laboratory." *Pharmaceutical Journal* 163: 59.

Wright, H. B. 1949b. "Pregnancy diagnosis." *Pharmaceutical Journal* 163: 115.

Wright, W. M., and C. G. L. Wolf. 1930. "The serological diagnosis of cancer. Part I." *Journal of Cancer Research* 14: 370–393.

Yeniyurt, K. 2014. "When it hurts to look: Interpreting the interior of the Victorian woman." *Social History of Medicine* 27: 22–40.

Young, D. S., M. C. Berwick, and L. Jarett. 1997. "Evolution of the William Pepper Laboratory." *Clinical Chemistry* 43: 174–179.

Young, R. H. 2009. "The history of British gynaecological pathology." *Histopathology* 54: 144–155.

Zajicek, G. T. 1986. "Pregnancy testing." *Pharmaceutical Journal* 236: 642.

Zondek, B. 1931. *Die Hormone des Ovariums und des Hypophysenvorderlappens*. Berlin: Springer.

Zondek, B. 1942. "Simplified hormonal treatment of amenorrhea." *JAMA* 118: 705.

Zondek, B., and S. Rozin. 1948. "Induction of uterine bleeding in amenorrhea with a single injection of precipitates of estrone and progesterone." *Journal of Clinical Endocrinology* 8: 406–412.

Zwarenstein, H. 1985. "The frog pregnancy test: The first of its kind in the world." *Adler Museum Bulletin* 11: 9–10.

Zwarenstein, H. 1986. *Memoirs of an Academic*. Cape Town: University of Cape Town.

Zwarenstein, H., and H. A. Shapiro. 1933. "Metabolic changes associated with endocrine activity and the reproductive cycle in *Xenopus laevis*. III. Changes in the calcium content of the serum associated with captivity and the normal reproductive cycle." *Journal of Experimental Biology* 10: 372–378.

Zweiniger-Bargielowska, I. 2000. *Austerity in Britain: Rationing, Controls, and Consumption, 1939–1955*. Oxford: Oxford University Press.

Zweiniger-Bargielowska, I. 2005. "The culture of the abdomen: Obesity and reducing in Britain, circa 1900–1939." *Journal of British Studies* 44: 239–273.

Zweiniger-Bargielowska, I. 2007. "Raising a nation of 'good animals': The New Health Society and health education campaigns in interwar Britain." *Social History of Medicine* 20: 73–89.

Zweiniger-Bargielowska, I. 2011. "Fair shares? The limits of food policy in Britain during the Second World War." In *Food and War in Twentieth Century Europe*, edited by I. Zweiniger-Bargielowska, R. Duffett, and A. Drouard, 125–138. Farnham: Ashgate.

INDEX

Note: Page numbers followed by *t* indicate tables and *f* indicate figures.

Inside Technology Series

Edited by Wiebe E. Bijker and Rebecca Slayton

Jesse Olszynko-Gryn, *A Woman's Right to Know: Pregnancy Testing in Twentieth-Century Britain*

Dagmar Schäfer, Annapurna Mamidipudi, and Marius Buning, editors, *Ownership of Knowledge: Beyond Intellectual Property*

Tamar Novick, *Milk and Honey: Technologies of Plenty in the Making of a Holy Land*

Kristin Asdal and Tone Huse, *Nature-made Economy: Cod, Capital, and the Great Transformation of the Ocean*

Cyrus C. M. Mody, *The Squares: US Physical and Engineering Scientists in the Long 1970s*

Brice Laurent, *European Objects: The Troubled Dreams of Harmonization*

Florian Jaton, *The Constitution of Algorithms: Ground-Truthing, Programming, Formulating*

Kean Birch and Fabian Muniesa, *Turning Things into Assets*

David Demortain, *The Science of Bureaucracy: Risk Decision-Making and the US Environmental Protection Agency*

Nancy Campbell, *OD: Naloxone and the Politics of Overdose*

Lukas Engelmann and Christos Lynteris, *Sulphuric Utopias: The History of Maritime Fumigation*

Zara Mirmalek, *Making Time on Mars*

Joeri Bruynincx, *Listening in the Field: Recording and the Science of Birdsong*

Edward Jones-Imhotep, *The Unreliable Nation: Hostile Nature and Technological Failure in the Cold War*

Jennifer L. Lieberman, *Power Lines: Electricity in American Life and Letters, 1882–1952*

Jess Bier, *Mapping Israel, Mapping Palestine: Occupied Landscapes of International Technoscience*

Benoît Godin, *Models of Innovation: The History of an Idea*

Stephen Hilgartner, *Reordering Life: Knowledge and Control in the Genomics Revolution*

Brice Laurent, *Democratic Experiments: Problematizing Nanotechnology and Democracy in Europe and the United States*

Cyrus C. M. Mody, *The Long Arm of Moore's Law: Microelectronics and American Science*

Tiago Saraiva, *Fascist Pigs: Technoscientific Organisms and the History of Fascism*

Teun Zuiderent-Jerak, *Situated Interventions: Sociological Experiments in Healthcare*

Basile Zimmermann, *Technology and Cultural Difference: Electronic Music Devices, Social Networking Sites, and Computer Encodings in Contemporary China*

Andrew J. Nelson, *The Sound of Innovation: Stanford and the Computer Music Revolution*

Sonja D. Schmid, *Producing Power: The Pre-Chernobyl History of the Soviet Nuclear Industry*

Casey O'Donnell, *Developer's Dilemma: The Secret World of Videogame Creators*

Christina Dunbar-Hester, *Low Power to the People: Pirates, Protest, and Politics in FM Radio Activism*

Eden Medina, Ivan da Costa Marques, and Christina Holmes, editors, *Beyond Imported Magic: Essays on Science, Technology, and Society in Latin America*

Anique Hommels, Jessica Mesman, and Wiebe E. Bijker, editors, *Vulnerability in Technological Cultures: New Directions in Research and Governance*

Amit Prasad, *Imperial Technoscience: Transnational Histories of MRI in the United States, Britain, and India*

Charis Thompson, *Good Science: The Ethical Choreography of Stem Cell Research*

Tarleton Gillespie, Pablo J. Boczkowski, and Kirsten A. Foot, editors, *Media Technologies: Essays on Communication, Materiality, and Society*

Catelijne Coopmans, Janet Vertesi, Michael Lynch, and Steve Woolgar, editors, *Representation in Scientific Practice Revisited*

Rebecca Slayton, *Arguments that Count: Physics, Computing, and Missile Defense, 1949–2012*

Stathis Arapostathis and Graeme Gooday, *Patently Contestable: Electrical Technologies and Inventor Identities on Trial in Britain*

Jens Lachmund, *Greening Berlin: The Co-Production of Science, Politics, and Urban Nature*

Chikako Takeshita, *The Global Biopolitics of the IUD: How Science Constructs Contraceptive Users and Women's Bodies*

Cyrus C. M. Mody, *Instrumental Community: Probe Microscopy and the Path to Nanotechnology*

Morana Alač, *Handling Digital Brains: A Laboratory Study of Multimodal Semiotic Interaction in the Age of Computers*

Gabrielle Hecht, editor, *Entangled Geographies: Empire and Technopolitics in the Global Cold War*

Michael E. Gorman, editor, *Trading Zones and Interactional Expertise: Creating New Kinds of Collaboration*

Matthias Gross, *Ignorance and Surprise: Science, Society, and Ecological Design*

Andrew Feenberg, *Between Reason and Experience: Essays in Technology and Modernity*

Wiebe E. Bijker, Roland Bal, and Ruud Hendricks, *The Paradox of Scientific Authority: The Role of Scientific Advice in Democracies*

Park Doing, *Velvet Revolution at the Synchrotron: Biology, Physics, and Change in Science*

Gabrielle Hecht, *The Radiance of France: Nuclear Power and National Identity after World War II*

Richard Rottenburg, *Far-Fetched Facts: A Parable of Development Aid*

Michel Callon, Pierre Lascoumes, and Yannick Barthe, *Acting in an Uncertain World: An Essay on Technical Democracy*

Ruth Oldenziel and Karin Zachmann, editors, *Cold War Kitchen: Americanization, Technology, and European Users*

Deborah G. Johnson and Jameson W. Wetmore, editors, *Technology and Society: Building Our Sociotechnical Future*

Trevor Pinch and Richard Swedberg, editors, *Living in a Material World: Economic Sociology Meets Science and Technology Studies*

Christopher R. Henke, *Cultivating Science, Harvesting Power: Science and Industrial Agriculture in California*

Helga Nowotny, *Insatiable Curiosity: Innovation in a Fragile Future*

Karin Bijsterveld, *Mechanical Sound: Technology, Culture, and Public Problems of Noise in the Twentieth Century*

Peter D. Norton, *Fighting Traffic: The Dawn of the Motor Age in the American City*

Joshua M. Greenberg, *From Betamax to Blockbuster: Video Stores and the Invention of Movies on Video*

Mikael Hård and Thomas J. Misa, editors, *Urban Machinery: Inside Modern European Cities*

Christine Hine, *Systematics as Cyberscience: Computers, Change, and Continuity in Science*

Wesley Shrum, Joel Genuth, and Ivan Chompalov, *Structures of Scientific Collaboration*

Shobita Parthasarathy, *Building Genetic Medicine: Breast Cancer, Technology, and the Comparative Politics of Health Care*

Kristen Haring, *Ham Radio's Technical Culture*

Atsushi Akera, *Calculating a Natural World: Scientists, Engineers and Computers during the Rise of U.S. Cold War Research*

Donald MacKenzie, *An Engine, Not a Camera: How Financial Models Shape Markets*

Geoffrey C. Bowker, *Memory Practices in the Sciences*

Christophe Lécuyer, *Making Silicon Valley: Innovation and the Growth of High Tech, 1930–1970*

Anique Hommels, *Unbuilding Cities: Obduracy in Urban Sociotechnical Change*

David Kaiser, editor, *Pedagogy and the Practice of Science: Historical and Contemporary Perspectives*

Charis Thompson, *Making Parents: The Ontological Choreography of Reproductive Technology*

Pablo J. Boczkowski, *Digitizing the News: Innovation in Online Newspapers*

Dominique Vinck, editor, *Everyday Engineering: An Ethnography of Design and Innovation*

Nelly Oudshoorn and Trevor Pinch, editors, *How Users Matter: The Co-Construction of Users and Technology*

Peter Keating and Alberto Cambrosio, *Biomedical Platforms: Realigning the Normal and the Pathological in Late-Twentieth-Century Medicine*

Paul Rosen, *Framing Production: Technology, Culture, and Change in the British Bicycle Industry*

Maggie Mort, *Building the Trident Network: A Study of the Enrollment of People, Knowledge, and Machines*

Donald MacKenzie, *Mechanizing Proof: Computing, Risk, and Trust*

Geoffrey C. Bowker and Susan Leigh Star, *Sorting Things Out: Classification and Its Consequences*

Charles Bazerman, *The Languages of Edison's Light*

Janet Abbate, *Inventing the Internet*

Herbert Gottweis, *Governing Molecules: The Discursive Politics of Genetic Engineering in Europe and the United States*

Kathryn Henderson, *On Line and On Paper: Visual Representation, Visual Culture, and Computer Graphics in Design Engineering*

Susanne K. Schmidt and Raymund Werle, *Coordinating Technology: Studies in the International Standardization of Telecommunications*

Marc Berg, *Rationalizing Medical Work: Decision Support Techniques and Medical Practices*

Eda Kranakis, *Constructing a Bridge: An Exploration of Engineering Culture, Design, and Research in Nineteenth-Century France and America*

Paul N. Edwards, *The Closed World: Computers and the Politics of Discourse in Cold War America*

Donald MacKenzie, *Knowing Machines: Essays on Technical Change*

Wiebe E. Bijker, *Of Bicycles, Bakelites, and Bulbs: Toward a Theory of Sociotechnical Change*

Louis L. Bucciarelli, *Designing Engineers*

Geoffrey C. Bowker, *Science on the Run: Information Management and Industrial Geophysics at Schlumberger, 1920–1940*

Wiebe E. Bijker and John Law, editors, *Shaping Technology/Building Society: Studies in Sociotechnical Change*

Stuart Blume, *Insight and Industry: On the Dynamics of Technological Change in Medicine*

Donald MacKenzie, *Inventing Accuracy: A Historical Sociology of Nuclear Missile Guidance*

Pamela E. Mack, *Viewing the Earth: The Social Construction of the Landsat Satellite System*

H. M. Collins, *Artificial Experts: Social Knowledge and Intelligent Machines*

http://mitpress.mit.edu/books/series/inside-technology